Anatomy & Physiology Lab

D1498504

LOUISIANA TECH UNIVERSITY

Biology

create.mheducation.com

ISBN-13: 9781307715743

ISBN-10: 1307715745

Contents

CONTENTS

*These laboratory exercises are available for instructors via the Instructor Resources in Connect. They are available for students via the eBook in Connect.

GUIDED TOUR THROUGH A LAB EXERCISE

This laboratory manual was prepared to be used with the textbook *Hole's Human Anatomy & Physiology,* Sixteenth Edition, by Charles J. Welsh and Cynthia Prentice-Craver. As with the textbook, the laboratory manual is designed for students with minimal backgrounds in the physical and biological sciences who are pursuing careers in professional health fields. This lab manual is designed to be used in a traditional class format as well as a hybrid or online class format.

The laboratory manual contains sixty-nine laboratory exercises and reports, which are closely integrated with the chapters of the textbook and other resources from Connect. (Four of these exercises appear only online.) The exercises are planned to illustrate and review the anatomical and physiological facts and principles presented in the textbook and to help students investigate some of these ideas in greater detail.

Many of the laboratory exercises are short or are divided into several separate procedures. This allows an instructor to select those exercises or parts of exercises that will best meet the needs of a particular program. Also, exercises requiring a minimal amount of laboratory equipment have been included.

The laboratory exercises have a variety of special features that are designed to stimulate interest in the subject matter, to involve students in the learning process, and to guide them through the planned activities. These special features include the following.

Materials Needed

This section lists the laboratory materials that are required to complete the exercise and to perform the demonstrations and learning extensions.

⚠ Safety

A list of Laboratory Safety Guidelines appears in Appendix 1 of your laboratory manual. Each laboratory session that requires special safety guidelines has a safety section following "Materials Needed." The instructor might require some modifications of these guidelines.

Purpose of the Exercise

The purpose states the intent of the exercise—that is, what will be accomplished.

Learning Outcomes

The learning outcomes ① list what a student should be able to do after completing the exercise. Each learning outcome has a matching assessment indicated by a corresponding icon ① in the laboratory exercise or the laboratory report.

Introduction

The introduction briefly describes the subject of the exercise or the ideas that will be investigated.

Procedure

The procedure provides a set of detailed instructions for accomplishing the planned laboratory activities. Each procedure heading includes a practice icon indicating ways to accomplish the learning outcomes. Usually, these

instructions are presented in outline form, so that a student can proceed efficiently through the exercise in stepwise fashion. Often the student is referred to particular sections of a textbook for necessary background information or for review of subject matter presented previously.

The procedures include a wide variety of laboratory activities and, from time to time, direct the student to complete various tasks in the laboratory reports.

LABORATORY EXERCISE

12

Bone Structure and Classification

MATERIALS NEEDED

Textbook
Prepared microscope slide of ground compact bone
Human bone specimens, including long, short, flat, irregular, and sesamoid types
Human long bone, sectioned longitudinally
Fresh animal bones, sectioned longitudinally and transversely
Compound light microscope
Dissecting microscope
Fresh chicken bones (radius and ulna from wings)
Vinegar or dilute hydrochloric acid

⚠ SAFETY

- Wear disposable gloves for handling fresh bones and for the demonstration of a bone soaked in vinegar or dilute hydrochloric acid.
- Wash your hands before leaving the laboratory.

PURPOSE OF THE EXERCISE

To review the way bones are classified and to examine the structure of a long bone.

LEARNING OUTCOMES APR

After completing this exercise, you should be able to

1. Locate the major structures of a long bone.
2. Distinguish between compact and spongy bone.
3. Differentiate the special characteristics of compact bone tissue.
4. Arrange five groups of bones based on their shapes and identify an example for each group.
5. Describe the functions of various structures of a bone.

A bone represents an organ of the skeletal system. As such, it is composed of a variety of tissues, including bone tissue, cartilage, dense connective tissue, blood, and nervous tissue. Bones are not only alive but also multifunctional. They support and protect softer tissues, provide points of attachment for muscles, house blood-producing cells, and store inorganic salts.

Bones are classified according to their shapes as long, short, flat, or irregular. Although the bones of the skeleton vary greatly in size and shape, they have much in common structurally and functionally.

PRACTICE

PROCEDURE—Bone Structure and Classification

1. Review section 5.3 titled "Connective Tissues" in chapter 5 of the textbook and section 7.2 titled "Bone Shape and Structure" in chapter 7 of the textbook.
2. As a review activity, label figures 12.1 and 12.2.
3. Examine the microscopic structure of bone tissue by observing a prepared microscope slide of ground compact bone. Use figure 12.3, a micrograph of bone tissue, to locate the following features:

   osteon (Haversian system)—cylinder-shaped unit
   central canal (Haversian canal)—contains blood vessels and nerves
   lamella—concentric ring of matrix around central canal
   lacuna—small chamber for an osteocyte
   bone extracellular matrix—collagen and calcium phosphate
   canaliculus—minute tube containing cellular process

LEARN: ACTIVITY

Examine a fresh chicken bone and a chicken bone that has been soaked for several days in vinegar or overnight in dilute hydrochloric acid. Wear disposable gloves for handling these bones. This acid treatment removes the inorganic salts from the bone extracellular matrix. Rinse the bones in water and note the texture and flexibility of each (fig. 12.4a). The bone becomes soft and flexible without the support of the inorganic salts with calcium.

Examine the specimen of chicken bone that has been exposed to high temperature (baked at 121°C [250°F] for 2 hours). This treatment removes the protein and other organic substances from the bone extracellular matrix (fig. 12.4b). The bone becomes brittle and fragile without the benefit of the collagen fibers. A living bone with a combination of the qualities of inorganic and organic substances possesses tensile strength.

Learn: Activity

Learn activities appear in separate boxes. They describe specimens, specialized laboratory equipment, or other materials to enrich the student's laboratory experience.

LEARN: LAB IN MOTION

A long rubber band can be used to simulate muscle locations, origins, insertions, and actions on the human torso model, the skeleton, or a laboratory partner. Hold one end of the rubber band firmly on the origin location of a muscle; then slightly stretch the rubber band and hold the other end on the insertion site. Allow the insertion end to slowly move toward the origin end to simulate the contraction and action of the muscle.

Learn: Lab in Motion

Some learning activities involve a greater degree of student group or class participation. These learning activities are often more active with physiological concepts.

ASSESS

CRITICAL THINKING

Assessments involving critical thinking skills are incorporated within many of the laboratory exercises and laboratory reports to enhance valuable critical thinking skills that students will need throughout their lives.

ASSESS

CRITICAL THINKING

Explain the advantage for melanin granules being located in the deep layer of the epidermis.

_____

_____

_____

_____

_____

Laboratory Reports

A laboratory report to be completed by the student immediately follows each exercise. These reports include various types of review activities, spaces for sketches of microscopic objects, tables for recording observations and experimental results, and questions dealing with the analysis of such data. There are also several labeling exercises included in many of the lab reports that utilize images of commonly used anatomical models and cadaver images from the Practice Atlas for Anatomy and Physiology and the Anatomy & Physiology Revealed resources.

As a result of these activities, students will develop a better understanding of the structural and functional characteristics of their bodies and will increase their skills in gathering information by observation and experimentation. By completing all of the assessments in the laboratory report, students are able to determine if they have accomplished all of the learning outcomes.

GUIDED TOUR THROUGH A LAB EXERCISE

Illustrations

Diagrams from the textbook and diagrams similar to those in the textbook often are used as aids for reviewing subject matter. Other illustrations provide visual instructions for performing steps in procedures or are used to identify parts of instruments or specimens.

Frequent variations exist in anatomical structures among humans. The illustrations in the textbook and the laboratory manual represent normal (normal means the most common variation) anatomy. Variations from normal anatomy do not represent abnormal anatomy unless some function is impaired.

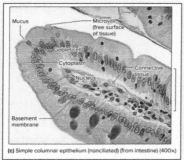

(c) Simple columnar epithelium (nonciliated) (from intestine) (400x)

©Victor P.Eroschenko

Micrographs are included to help students identify microscopic structures or to evaluate student understanding of tissues.

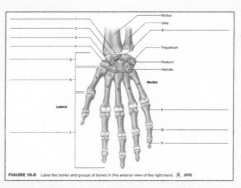

FIGURE 16.6 Label the bones and groups of bones in this anterior view of the right hand. APR

In some exercises, the figures include line drawings suitable for students to color with colored pencils. This activity may motivate students to observe the illustrations more carefully and help them to locate the special features represented in the figures. Students can check their work by referring to the corresponding full-color illustrations in the textbook.

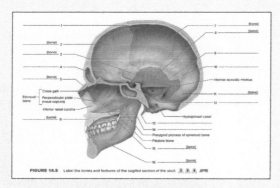

FIGURE 14.5 Label the bones and features of the sagittal section of the skull. APR

This sagittal section of the skull illustrates the more vivid colors used in the all new art program for the entire lab manual including the dissection labs.

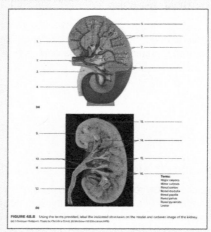

FIGURE 45.8 Using the terms provided, label the indicated structures on the model and cadaver image of the kidney.

Many lab exercises have images of anatomical models and cadaver specimens to help students learn the gross anatomy of certain organs. These images are taken from the Practice Atlas for Anatomy & Physiology. Not only does this help the student learn anatomy but also gives them a unique perspective of anatomy in general.

PREFACE

Changes to this Edition

Many of the changes in this new edition were made to enhance the students perspective of human anatomy by incorporating new labeling imagery utilizing anatomical models and cadaver images. In order to make more room for this imagery, and to make learning about human anatomy a more visual experience, the find-a-words from the previous edition have been relocated to a digital online resource; available on the online instructor resource site.

Specific Changes to Individual Laboratory Exercises

Lab 2: New figure 2.9
Lab 18: New figure 18.6
Lab 20: New figure 20.5
Lab 21: New figure 21.5b
Lab 23: New figures 23.8a and b
Lab 24: New figures 24.2 and 24.3b
Lab 26: New figure 26.6
Lab 28: New figure 28.7
Lab 32: New figure 32.9
Lab 34: New figure 34.11
Lab 42: New figure 42.3b
Lab 43: New figure 43.16
Lab 45: New figure 45.11
Lab 48: New figure 48.8
Lab 50: New figure 50.8
Lab 51: New figure 51.10

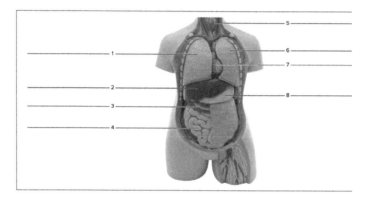

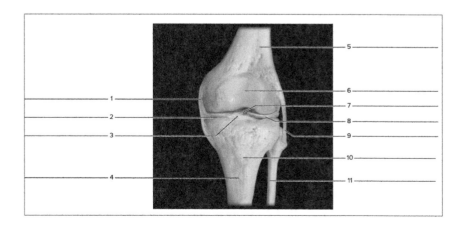

Acknowledgments

I would like to thank the wonderful people at McGraw-Hill Education, including Matthew Garcia, Krystal Faust, Jim Connely, Valerie Kramer, Fran Simon, Dr. Charles Welsh, Cynthia Prentice-Craver, Mary Powers, Ann Courtney, and Beth Cray, for your help and support and for allowing me the opportunity to help revise this lab manual.

I would like to say a special thank you to Terry Martin. Terry and I met for the first time in 2008 at a Human Anatomy and Physiology Society (HAPS) conference. It was an instant friendship. Not only has Terry been a friend to me but I also consider him my mentor. I am eternally grateful for our friendship.

I would also like to thank all the other instructors I work with for their dedication to their profession. A special thank you to my department chair person, Ms. Shirley Colvin for her support and encouragement.

Gratitude is extended to John W. Hole, Jr., for his years of dedicated effort in early editions of this classic work.

About the Authors

Courtesy of Shanae Snider

Phillip Snider

This Sixteenth edition is the first revision by Phillip Snider. Phillip has taught Human Anatomy and Physiology at the college level since 2001. He has also taught Human Gross Anatomy/ Pathophysiology since 2008. His student-centered classes are designed to encourage the student's fascination and knowledge of the human body to further their professional careers.

Phillip taught the very first human-cadaver-based gross anatomy course at the community college level in the state of Alabama. He was instrumental in constructing the curriculum for the course and continues his own education in gross anatomy whenever the opportunity arises. Among Phillip's awards are Who's Who Among America's Teachers, the Brenda Crowe Exceptional Achievement in Teaching Award, and the state of Alabama Chancellor's Award. Phillip's professional memberships include the Human Anatomy and Physiology Society (HAPS), the Alabama Conference of Educators (ACOE), the Textbook & Academic Authors Association (TAA), and the National Institute for Staff and Organizational Development (NISOD). He was also a contributor for the Terry Martin *Laboratory Manual for Human Anatomy & Physiology,* cat, main, and fetal pig versions, third editions. Phillip was also elected to be a faculty fellow at his college to help provide professional development and training to other faculty members.

Phillip has been married to his high school sweetheart, Shae, since 1991. He has two successful children, Collin and Camden, whom he is extremely proud of. Phillip enjoys spending time with his family, camping, and fishing.

Terry R. Martin

©J & J Photography

This Sixteenth edition is the ninth revision by Terry R. Martin of Kishwaukee College. Terry's teaching experience of over forty years, his interest in students and love for college instruction, and his innovative attitude and use of technology-based learning enhance the solid tradition of John Hole's laboratory manual. Among Terry's awards are the Kishwaukee College Outstanding Educator, Phi Theta Kappa Outstanding Instructor Award, Kishwaukee College ICCTA Outstanding Educator Award, Kishwaukee College Faculty Board of Trustees Award of Excellence, and John C. Roberts Community Service Award. Terry's professional memberships include the National Association of Biology Teachers, Human Anatomy and Physiology Society, and Nature Conservancy. In addition to writing many publications, he co-produced with Hassan Rastegar a videotape entitled *Introduction to the Human Cadaver and Prosection,* published by Wm. C. Brown Publishers. Terry revised the *Laboratory Manual to Accompany Hole's Human Anatomy and Physiology,* Fifteenth Edition, and authored *Laboratory Manual for Human Anatomy & Physiology,* cat, main, and fetal pig versions, third editions. Cadaver dissection experiences have been provided for his students for over thirty-five years. Terry also was a faculty exchange member in Ireland.

TO THE STUDENT

The exercises in this laboratory manual will provide you with opportunities to observe various anatomical parts and to investigate certain physiological phenomena. Such experiences should help you to relate specimens, models, microscope slides, cadaver images, and your body to what you have learned in the lecture and read about in the textbook.

The following list of suggestions and study skills may help make your laboratory activities more effective and profitable.

1. Prepare yourself before attending the laboratory session by reading the assigned exercise and reviewing the related sections of the textbook. It is important to have some understanding of what will be done in the laboratory before you come to class.
2. Bring your laboratory manual and textbook to each laboratory session. These books are closely integrated and will help you complete most of the exercises.
3. Be on time. During the first few minutes of the laboratory meeting, the instructor often will provide verbal instructions. Make special note of any changes in materials to be used or procedures to be followed. Also listen carefully for information about special techniques to be used and precautions to be taken.
4. Keep your work area clean and your materials neatly arranged so that you can locate needed items. This will enable you to proceed efficiently and will reduce the chances of making mistakes.
5. Pay particular attention to the purpose of the exercise, which states what you are to accomplish in general terms, and to the learning outcomes, which list what you should be able to do as a result of the laboratory experience. Then, before you leave the class, review the outcomes and make sure that you can meet all of the assessments.
6. Precisely follow the directions in the procedure and proceed only when you understand them clearly. Do not improvise procedures unless you have the approval of the laboratory instructor. Ask questions if you do not understand exactly what you are supposed to do and why you are doing it.
7. Handle all laboratory materials with care. Some of these materials are fragile and expensive to replace. Whenever you have questions about the proper treatment of equipment, ask the instructor.
8. Treat all living specimens humanely and try to minimize any discomfort they might experience.
9. Although at times you might work with a laboratory partner or a small group, try to remain independent when you are making observations, drawing conclusions, and completing the activities in the laboratory reports.
10. Read the Instructions for each section of the laboratory report before you begin to complete it. Think about the questions before you answer them. Your responses should be based on logical reasoning and phrased in clear and concise language.
11. Use the time allowed in the lab wisely. In some of your lab periods you will have extra time to study and observe models, review concepts, finish lab reports, etc. Take advantage of this time.
12. At the end of each laboratory period, clean your work area and the instruments you have used. Return all materials to their proper places and dispose of wastes, including glassware or microscope slides that have become contaminated with human blood or body fluids, as directed by the laboratory instructor. Wash your hands thoroughly before leaving the laboratory.

Study Skills for Anatomy and Physiology

Everyone has his or her own learning styles. There are techniques that work well for most students enrolled in Human Anatomy and Physiology. Using some of the skills listed here can make your course more enjoyable and rewarding.

1. **Note taking:** Look for the main ideas and briefly express them in your own words. Organize, edit, and review your notes soon after the lecture. Add textbook information to your notes as you reorganize them. Underline or highlight with different colors the important points, major headings, and key terms. Study your notes daily, as they provide sequential building blocks of the course content.
2. **Note cards/flash cards:** Make your own. Add labels and colors to enhance the material. Keep them with you; study them often and for short periods. Concentrate on a small number of cards at one time. Shuffle your cards and have someone quiz you on their content. As you become familiar with the material, you can set aside cards that don't require additional mastery.
3. **Chunking:** Organize information into logical groups or categories. Study and master one chunk of information at a time. For example, study the bones of the upper limb, lower limb, trunk, and head as separate study tasks.

4. **Mnemonic devices:** An *acrostic* is a combination of association and imagery to aid your memory. It is often in the form of a poem, rhyme, or jingle in which the first letter of each word corresponds to the first letters of the words you need to remember. **S**o **L**ong **T**op **P**art, **H**ere **C**omes **T**he **T**humb is an example of such a mnemonic device for remembering the eight carpals in a correct sequence. *Acronyms* are words formed by the first letters of the items to remember. *IPMAT* is an example of this type of mnemonic device to help you remember the phases of the cell cycle in the correct sequence. Try to create some of your own.

5. **Time management:** Prepare monthly, weekly, and daily schedules. Include dates of quizzes, exams, and projects on the calendar. On your daily schedule, budget several short study periods. Daily repetition alleviates cramming. Prioritize your tasks, so that you still have time for work and leisure activities. Find an appropriate study atmosphere with minimum distractions.

6. **Recording and recitation:** An auditory learner can benefit by recording lectures and review sessions with a digital recorder. Many students listen to the recorded sessions as they drive or just before they go to bed. Reading your notes aloud can help also. Explain the material to anyone (even if there are no listeners). Talk about anatomy and physiology in everyday conversations.

7. **Study groups:** Small study groups that meet periodically to review course material and compare notes have helped and encouraged many students. However, keep the group on the task at hand. This group often becomes a support group.

8. **Use all available resources.** Utilize the resources associated with your lab manual and textbook (i.e. Connect, Learnsmart, Anatomy & Physiology Revealed, Practice Atlas for Anatomy and Physiology) to help master the concepts covered in your class.

Practice sound study skills during your anatomy and physiology endeavor.

Correlation of Textbook Chapters and Laboratory Exercises

Textbook Chapters	Related Laboratory Exercises
Chapter 1 Introduction to Human Anatomy and Physiology	**Exercise 1** Scientific Method and Measurements
	Exercise 2 Body Organization and Terminology
Chapter 2 Chemical Basis of Life	**Exercise 3** Chemistry of Life
Chapter 3 Cells	**Exercise 4** Care and Use of the Microscope
	Exercise 5 Cell Structure and Function
	Exercise 6 Movements Through Membranes
	Exercise 7 Cell Cycle
Chapter 4 Cellular Metabolism	
Chapter 5 Tissues	**Exercise 8** Epithelial Tissues
	Exercise 9 Connective Tissues
	Exercise 10 Muscle and Nervous Tissues
Chapter 6 Integumentary System	**Exercise 11** Integumentary System
Chapter 7 Skeletal System	**Exercise 12** Bone Structure and Classification
	Exercise 13 Organization of the Skeleton
	Exercise 14 Skull
	Exercise 15 Vertebral Column and Thoracic Cage
	Exercise 16 Pectoral Girdle and Upper Limb
	Exercise 17 Pelvic Girdle and Lower Limb
Chapter 8 Joints of the Skeletal System	**Exercise 18** Joint Structure and Movements
Chapter 9 Muscular System	**Exercise 19** Skeletal Muscle Structure and Function
	Exercise 20 Muscles of the Head and Neck
	Exercise 21 Muscles of the Chest, Shoulder, and Upper Limb
	Exercise 22 Muscles of the Deep Back, Abdominal Wall, and Pelvic Floor
	Exercise 23 Muscles of the Hip and Lower Limb
	Exercise 24 Surface Anatomy
	Exercise 54 Cat Dissection: Musculature
	Exercise 60 Pig Dissection: Musculature
Chapter 10 Nervous System I: Basic Structure and Function	**Exercise 25** Nervous Tissue and Nerves
Chapter 11 Nervous System II: Divisions of the Nervous System	**Exercise 26** Brain and Cranial Nerves
	Exercise 27 Dissection of the Sheep Brain
	Exercise 28 Spinal Cord and Meninges
	Exercise 29 Reflex Arc and Reflexes
Chapter 12 Nervous System III: Senses	**Exercise 30** Receptors and General Senses
	Exercise 31 Smell and Taste
	Exercise 32 Ear and Hearing
	Exercise 33 Ear and Equilibrium
	Exercise 34 Eye Structure
	Exercise 35 Visual Tests and Demonstrations
Chapter 13 Endocrine System	**Exercise 36** Endocrine Histology and Diabetic Physiology
Chapter 14 Blood	**Exercise 37** Blood Cells and Blood Typing
Chapter 15 Cardiovascular System	**Exercise 38** Heart Structure
	Exercise 39 Cardiac Cycle
	Exercise 40 Blood Vessel Structure, Arteries, and Veins
	Exercise 41 Pulse Rate and Blood Pressure
	Exercise 55 Cat Dissection: Cardiovascular System
	Exercise 61 Pig Dissection: Cardiovascular System
Chapter 16 Lymphatic System and Immunity	**Exercise 42** Lymphatic System
Chapter 17 Digestive System	**Exercise 43** Digestive Organs
	Exercise 44 Action of a Digestive Enzyme
	Exercise 56 Cat Dissection: Digestive System
	Exercise 62 Pig Dissection: Digestive System
Chapter 18 Nutrition and Metabolism	

Textbook Chapters	Related Laboratory Exercises
Chapter 19 Respiratory System	**Exercise 45** Respiratory Organs
	Exercise 46 Breathing and Respiratory Volumes
	Exercise 47 Control of Breathing
	Exercise 57 Cat Dissection: Respiratory System
	Exercise 63 Pig Dissection: Respiratory System
Chapter 20 Urinary System	**Exercise 48** Urinary Organs
	Exercise 49 Urinalysis
	Exercise 58 Cat Dissection: Urinary System
	Exercise 64 Pig Dissection: Urinary System
Chapter 21 Water, Electrolyte, and Acid-Base Balance	
Chapter 22 Reproductive Systems	**Exercise 50** Male Reproductive System
	Exercise 51 Female Reproductive System
	Exercise 59 Cat Dissection: Reproductive Systems
	Exercise 65 Pig Dissection: Reproductive System
Chapter 23 Pregnancy, Growth, and Development	**Exercise 52** Fertilization and Early Development
Chapter 24 Genetics and Genomics	**Exercise 53** Genetics

Correlation of Textbook Chapters and Online Laboratory Exercises

Textbook Chapters	Related Laboratory Exercises
Chapter 9 Muscular System	*__Exercise 66__ Skeletal Muscle Contraction
Chapter 10 Nervous System I: Basic Structure and Function	*__Exercise 67__ Nerve Impulse Stimulation
Chapter 14 Blood	*__Exercise 68__ Blood Testing
Chapter 15 Cardiovascular System	*__Exercise 69__ Factors Affecting the Cardiac Cycle

*These laboratory exercises are available online for instructors via the Instructor Resources in Connect. They are available for students via the eBook in Connect.

The Use of Animals in Biology Education*

The National Association of Biology Teachers (NABT) believes that the study of organisms, including nonhuman animals, is essential to the understanding of life on Earth. NABT recommends the prudent and responsible use of animals in the life science classroom. NABT believes that biology teachers should foster a respect for life. Biology teachers also should teach about the interrelationship and interdependency of all things.

Classroom experiences that involve nonhuman animals range from observation to dissection. NABT supports these experiences as long as they are conducted within the long-established guidelines of proper care and use of animals, as developed by the scientific and educational community.

As with any instructional activity, the use of nonhuman animals in the biology classroom must have sound educational objectives. Any use of animals, whether for observation or for dissection, must convey substantive knowledge of biology. NABT believes that biology teachers are in the best position to make this determination for their students.

NABT acknowledges that no alternative can substitute for the actual experience of dissection or other use of animals and urges teachers to be aware of the limitations of alternatives. When the teacher determines that the most effective means to meet the objectives of the class do not require dissection, NABT accepts the use of alternatives to dissection, including models and the various forms of multimedia. The association encourages teachers to be sensitive to substantive student objections to dissection and to consider providing appropriate lessons for those students where necessary.

To implement this policy, NABT endorses and adopts the "Principles and Guidelines for the Use of Animals in Precollege Education" of the Institute of Laboratory Animals Resources (National Research Council). Copies of the "Principles and Guidelines" may be obtained from the ILAR (2101 Constitution Avenue, NW, Washington, DC, 20418; 202-334-2590).

---

*Adopted by the Board of Directors in October 1995. This policy supersedes and replaces all previous NABT statements regarding animals in biology education.

Fundamentals of Human Anatomy and Physiology

LABORATORY EXERCISE

1

Scientific Method and Measurements

MATERIALS NEEDED

Meterstick
Calculator
Human skeleton

PURPOSE OF THE EXERCISE

To become familiar with the scientific method of investigation, learn how to formulate sound conclusions, and provide opportunities to use the metric system of measurements.

LEARNING OUTCOMES

After completing this exercise, you should be able to

1. Convert English measurements to the metric system, and vice versa.

2. Calculate expected upper limb length and actual percentage of height from recorded upper limb lengths and heights.

3. Apply the scientific method to test the validity of a hypothesis concerning the direct, linear relationship between human upper limb length and height.

4. Design an experiment, formulate a hypothesis, and test it using the scientific method.

Scientific investigation involves a series of logical steps to arrive at explanations for various biological phenomena. It reflects a long history of asking questions and searching for knowledge. This technique, called the *scientific method,* is used in all disciplines of science. It allows scientists to draw logical and reliable conclusions about phenomena.

The scientific method begins with making *observations* related to the topic under investigation. This step commonly involves the accumulation of previously acquired information and/or your observations of the phenomenon. These observations are used to formulate a tentative explanation known as the *hypothesis.* An important attribute of a hypothesis is that it must be testable. The testing of the proposed hypothesis involves designing and performing a carefully controlled *experiment* to obtain data that can be used to support, reject, or modify the hypothesis.

During the experiment to test the proposed hypothesis, it is important to be able to examine only a single changeable factor known as a *variable.* An *independent variable* is one that can be changed, but is determined before the experiment occurs; a *dependent variable* is determined from the results of the experiment.

An *analysis of data* is conducted using sufficient information collected during the experiment. Data analysis may include organization and presentation of data as tables, graphs, and drawings. From the interpretation of the data analysis, *conclusions* are drawn. (If the data do not support the hypothesis, you must reexamine the experimental design and the data, and if needed, develop a new hypothesis.) The final presentation of the information is made from the conclusions. Results and conclusions are presented to the scientific community for evaluation through peer reviews, presentations at professional meetings, and published articles. If many investigators working independently can validate the hypothesis by arriving at the same conclusions, the explanation can become a *theory.* A theory serves as the explanation from a summary of known experiments and supporting evidence unless it is disproved by new information. The five components of the scientific method are summarized as

Observations
↓
Hypothesis
↓
Experiment
↓
Analysis of data
↓
Conclusions

Metric measurements are characteristic tools of scientific investigations. The English system of measurements is often used in the United States, so the investigator must make conversions from the English system to the metric system. Table 1.1 provides the conversion factors necessary to change from English units to metric units.

TABLE 1.1 Metric Measurement System and Conversions

Measurement	Unit & Abbreviation	Metric Equivalent	Conversion Factor Metric to English (approximate)	Conversion Factor English to Metric (approximate)
LENGTH	1 kilometer (km)	1,000 (10^3) m	1 km = 0.62 mile	1 mile = 1.61 km

	1 meter (m)	100 (10^2) cm
		1,000 (10^3) mm
			1 m = 1.1 yards
			    = 3.3 feet
			    = 39.4 inches
				1 yard = 0.9 m
				1 foot = 0.3 m

	1 decimeter (dm)	0.1 (10^{-1}) m	1 dm = 3.94 inches	1 inch = 0.25 dm
	1 centimeter (cm)	0.01 (10^{-2}) m	1 cm = 0.4 inches	1 foot = 30.5 cm 1 inch = 2.54 cm
	1 millimeter (mm)	0.001 (10^{-3}) m 0.1 (10^{-1}) cm	1 mm = 0.04 inches	
	1 micrometer (μm)	0.000001 (10^{-6}) m 0.001 (10^{-3}) mm		
MASS	1 metric ton (t)	1,000 (10^3) kg	1 t = 1.1 ton	1 ton = 0.91 t
	1 kilogram (kg)	1,000 (10^3) g	1 kg = 2.2 pounds	1 pound = 0.45 kg
	1 gram (g)	1,000 (10^3) mg	1 g = 0.04 ounce	1 pound = 454 g 1 ounce = 28.35 g
	1 milligram (mg)	0.001 (10^{-3}) g		
VOLUME (LIQUIDS AND GASES)	1 liter (L)	1,000 (10^3) mL	1 L = 1.06 quarts	1 gallon = 3.78 L 1 quart = 0.95 L
	1 milliliter (mL)	0.001 (10^{-3}) L 1 cubic centimeter (cc or cm^3)	1 mL = 0.03 fluid ounce 1 mL = 1/5 teaspoon 1 mL = 15–16 drops	1 quart = 946 mL 1 fluid ounce = 29.6 mL 1 teaspoon = 5 mL
TIME	1 second (s)	1/60 minute	Same	Same
	1 millisecond (ms)	0.001 (10^{-3}) s	Same	Same
TEMPERATURE	Degrees Celsius (°C)		°F = 9/5°C + 32	°C = 5/9 (°F − 32)

PRACTICE

PROCEDURE A—Using the Steps of the Scientific Method

1. Refer to Appendix A of the textbook for additional information on the scientific method.
2. Many people have observed a correlation between the length of the upper and lower limbs and the height (stature) of an individual. For example, a person who has long upper limbs (the arm, forearm, and hand combined) tends to be tall. Make some visual observations of other people in your class to observe a possible correlation.
3. From such observations, the following hypothesis can be formulated: The length of a person's upper limb is equal to 0.4 (40%) of the height of the person. To test this hypothesis, perform the following experiment.
4. Use a meterstick (fig. 1.1) to measure an upper limb length of ten subjects. Place the meterstick in the axilla (armpit) and record the length in centimeters to the end of the longest finger (fig. 1.2). Obtain the height of each person in centimeters by measuring them without shoes against a wall (fig. 1.3). The height of each person can also be calculated by multiplying each individual's height in inches by 2.54 to obtain his or her height in centimeters. Record all your measurements in Part A of Laboratory Report 1.

5. The data collected from all of the measurements can now be analyzed. The expected (predicted) correlation between upper limb length and height is determined using the following equation:

$$\text{Height} \times 0.4 = \text{expected upper limb length}$$

The observed (actual) correlation to be used to test the hypothesis is determined by

$$\text{Length of upper limb/height}$$
$$= \text{actual \% of height}$$

6. A graph is an excellent way to display a visual representation of the data. Plot the subjects' data in Part A of the laboratory report. Plot the upper limb length of each subject on the *x*-axis (independent variable)

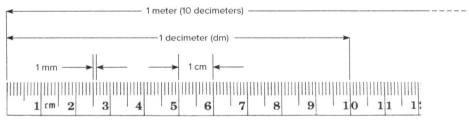

Metric ruler

FIGURE 1.1 Metric ruler with metric lengths indicated. A meterstick length would be 100 centimeters (10 decimeters). (The image size is approximately to scale.)

FIGURE 1.2 Measurement of upper limb length. (J and J Photography)

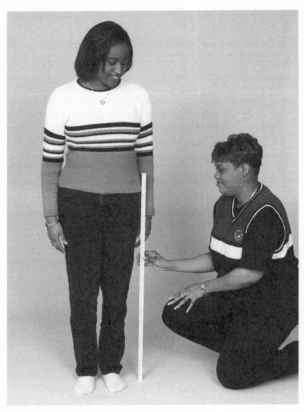

FIGURE 1.3   Measurement of height.   (J and J Photography)

and the height of each person on the *y*-axis (dependent variable). A line is already located on the graph that represents a hypothetical relationship of 0.4 (40%) upper limb length compared to height. This is a graphic representation of the original hypothesis.

7. Compare the distribution of all of the points (actual height and upper limb length) that you placed on the graph with the distribution of the expected correlation represented by the hypothesis.

8. Complete Part A of the laboratory report.

PRACTICE

PROCEDURE B—Design an Experiment

ASSESS

CRITICAL THINKING

You have probably concluded that there is some correlation of the length of body parts to height. Often when a skeleton is found, it is not complete. It is occasionally feasible to use the length of a single bone to estimate the height of an individual. Observe human skeletons and locate the humerus bone in an upper limb or the femur bone in a lower limb. Use your observations to identify a mathematical relationship between the length of the humerus or the femur and height. Formulate a hypothesis that can be tested. Make measurements, analyze data, and develop a conclusion from your experiment. Complete Part B of the laboratory report.

Name _____

Date _____

Section _____

The corresponds to the indicated Learning Outcome(s) found at the beginning of the laboratory exercise.

Scientific Method and Measurements

PART A ASSESSMENTS

1. Record measurements for the upper limb length and height of ten subjects. Use a calculator to determine the expected upper limb length and the actual percentage (as a decimal or a percentage) of the height for the ten subjects. Record your results in the following table.

Subject	Measured Upper Limb Length (cm)	Height* (cm)	Height × 0.4 = Expected Upper Limb Length (cm)	Actual % of Height = Measured Upper Limb Length (cm)/Height (cm)
1.				
2.				
3.				
4.				
5.				
6.				
7.				
8.				
9.				
10.				

*The height of each person can be calculated by multiplying each individual's height in inches by 2.54 to obtain his or her height in centimeters.

2. Plot the distribution of data (upper limb length and height) collected for the ten subjects on the following graph. The line located on the graph represents the *expected* 0.4 (40%) ratio of upper limb length to measured height (the original hypothesis). (The x-axis represents upper limb length and the y-axis represents height.) Draw a line of *best fit* through the distribution of points of the plotted data of the ten subjects. Compare the two distributions (expected line and the distribution line drawn for the ten subjects). △3

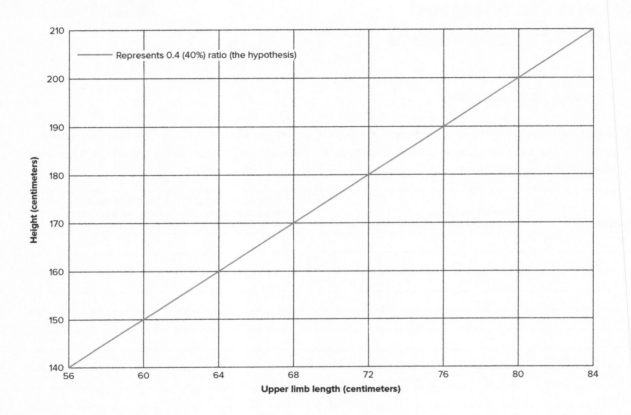

3. Does the distribution of the ten subjects' measured upper limb lengths support or reject the original hypothesis? _____ Explain your answer. △3

PART B ASSESSMENTS

1. Describe your observations of a possible correlation between the humerus or femur length and height. 🄰

2. Write a hypothesis based on your observations. 🄰

3. Describe the design of the experiment that you devised to test your hypothesis. 🄰

4. Place your analysis of the data in this space in the form of a table and a graph. 🄰
   a. Table:

   b. Graph:

5. Based on an analysis of your data, what can you conclude? Do these conclusions confirm or refute your original hypothesis? ⚠

6. Discuss your results and conclusions with classmates. What common conclusion can the class formulate about the correlation between the humerus or femur length and height? ⚠

LABORATORY EXERCISE

2

Body Organization and Terminology

MATERIALS NEEDED

Textbook
Dissectible human torso model (manikin)
Variety of specimens or models sectioned along
    various planes
Colored pencils

PURPOSE OF THE EXERCISE

To review the organizational pattern of the human body, to
review its organ systems and the organs included in each
system, and to become acquainted with the terms used to
describe the relative position of body parts, body sections,
and body regions.

LEARNING OUTCOMES APR

After completing this exercise, you should be able to

1. Locate and name the major body cavities and identify
   the membranes associated with each cavity.

2. Differentiate the general functions of the organ systems
   of the human body.

3. Associate the organs included within each system and
   locate the organs in a dissectible human torso model.

4. Select the terms used to describe the relative positions
   of body parts.

5. Match the terms used to identify body sections and
   identify the plane along which a particular specimen
   is cut.

6. Label the body regions and associate the terms used to
   identify body regions.

7. Recognize the anatomical terms pertaining to body
   organization.

The major features of the human body include certain cavities, a set of membranes associated with these cavities,
and a group of organ systems composed of related organs. In
order to communicate effectively with each other about the
body, scientists have devised names to describe these body
features. They also have developed terms to represent the relative positions of body parts, imaginary planes passing through
these parts, and body regions.

PRACTICE

PROCEDURE A—Body Cavities and Membranes

1. Review the concept headings titled "Body Cavities"
   and "Thoracic and Abdominopelvic Membranes" of
   section 1.6 in chapter 1 of the textbook.

2. As a review activity, label figures 2.1 and 2.2.

3. Locate the following features on the reference plates
   near the end of chapter 1 of the textbook and on the
   dissectible human torso model (fig. 2.3):

body cavities
    cranial cavity
    vertebral canal (spinal cavity)
    thoracic cavity
        mediastinum (region between the lungs;
            includes pericardial cavity)
        pleural cavities
    abdominopelvic cavity
        abdominal cavity
        pelvic cavity
diaphragm
smaller cavities within the head
    oral cavity
    nasal cavity with connected sinuses
    orbital cavity
    middle ear cavity
membranes and cavities
    pleural cavity
        parietal pleura
        visceral pleura

pericardial cavity
   parietal pericardium (covered by fibrous
     pericardium)
   visceral pericardium (epicardium)

peritoneal cavity
   parietal peritoneum
   visceral peritoneum

4. Complete Part A of Laboratory Report 2.

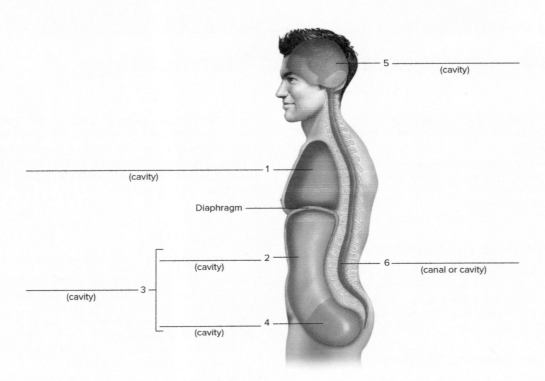

FIGURE 2.1   Label these major body cavities. APR

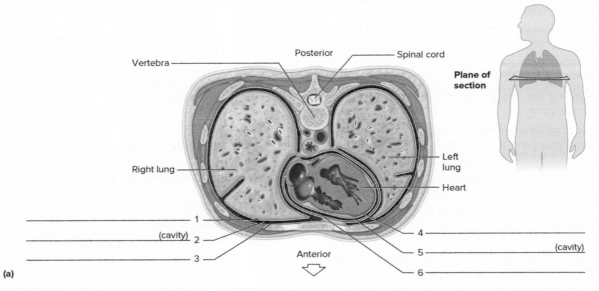

(a)

FIGURE 2.2   Label the thoracic membranes and cavities in (*a*) and the abdominopelvic membranes and cavity in (*b*), as shown in these superior views of transverse sections. APR

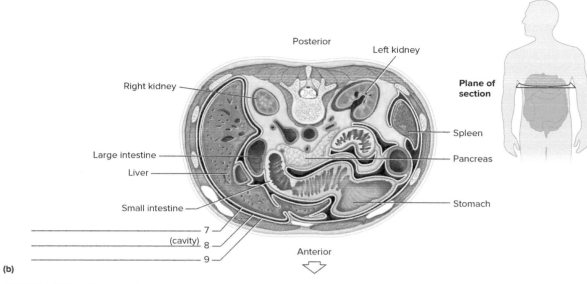

Posterior

Left kidney

Right kidney

Plane of section

Spleen

Large intestine

Pancreas

Liver

Small intestine

Stomach

7

(cavity) 8

9

Anterior

(b)

FIGURE 2.2 *Continued.* APR

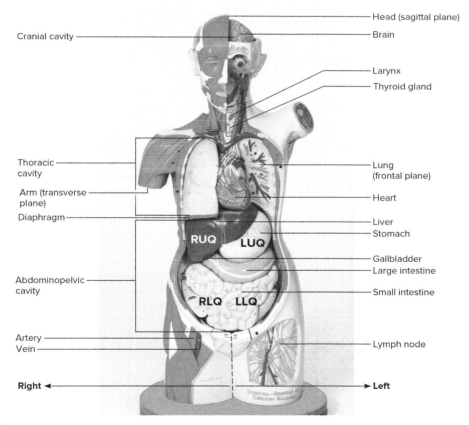

Head (sagittal plane)

Cranial cavity

Brain

Larynx

Thyroid gland

Thoracic cavity

Lung (frontal plane)

Arm (transverse plane)

Heart

Diaphragm

Liver

Stomach

RUQ   LUQ

Gallbladder

Large intestine

Abdominopelvic cavity

Small intestine

RLQ   LLQ

Artery

Vein

Lymph node

Right ◄   ► **Left**

FIGURE 2.3 Dissectible human torso model with body cavities, abdominopelvic quadrants, body planes, and major organs indicated. The abdominopelvic quadrants include right upper quadrant (RUQ), left upper quadrant (LUQ), right lower quadrant (RLQ), and left lower quadrant (LLQ). APR (J and J Photography)

PRACTICE

PROCEDURE B—Organ Systems

1. Review the concept heading titled "Organ Systems" of section 1.6 in chapter 1 of the textbook.
2. Use the reference plates near the end of chapter 1 of the textbook and the dissectible human torso model (fig. 2.3) to locate the following systems and their major organs:

integumentary system
  skin
  accessory organs such as hair and nails

skeletal system
  bones
  ligaments

muscular system
  skeletal muscles
  tendons

nervous system
  brain
  spinal cord
  nerves

endocrine system
  pituitary gland
  thyroid gland
  parathyroid glands
  adrenal glands
  pancreas
  ovaries
  testes
  pineal gland
  thymus

cardiovascular system
  heart
  arteries
  veins

lymphatic system
  lymphatic vessels
  lymph nodes
  thymus
  spleen

digestive system
  mouth
  tongue
  teeth
  salivary glands
  pharynx
  esophagus
  stomach
  liver
  gallbladder
  pancreas
  small intestine
  large intestine

respiratory system
  nasal cavity
  pharynx
  larynx
  trachea
  bronchi
  lungs

urinary system
  kidneys
  ureters
  urinary bladder
  urethra

male reproductive system
  scrotum
  testes
  penis
  urethra

female reproductive system
  ovaries
  uterine tubes (oviducts; fallopian tubes)
  uterus
  vagina

3. Complete Part B of the laboratory report.

 PRACTICE

PROCEDURE C—Relative Positions, Planes, Sections, and Regions

1. Observe the person standing in anatomical position (fig. 2.4). Anatomical terminology assumes the body is in anatomical position even though a person is often observed differently.
2. Review section 1.8 titled "Anatomical Terminology" in chapter 1 of the textbook.
3. As a review activity, label figures 2.5, 2.6, and 2.7.
4. Examine the sectioned specimens on the demonstration table and identify the plane along which each is cut. Cylindrical structures, such as a long bone or a blood vessel, may be cut in cross section, or longitudinal section. The same three sections can be demonstrated by three cuts of a banana (fig. 2.8).
5. Complete Parts C, D, E, F, and G of the laboratory report.

 LEARN: ACTIVITY

Use different colored pencils to distinguish body regions in figure 2.7.

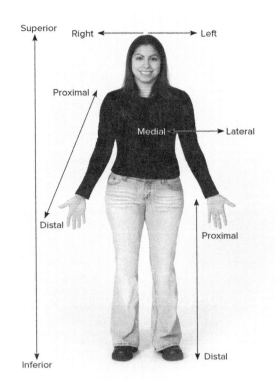

FIGURE 2.4 Anatomical position with directional terms indicated. The body is standing erect, face forward, with upper limbs at the sides and palms forward. When the palms are forward (supinated), the radius and ulna in the forearm are nearly parallel. This results in an anterior view of the body. Terms of relative position are used to describe the location of one body part with respect to another. **APR** (J and J Photography)

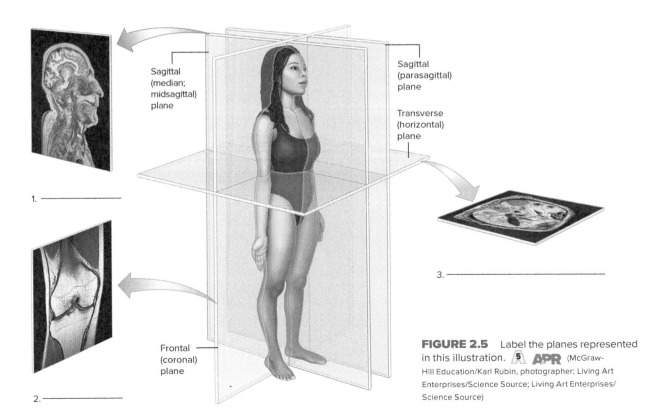

FIGURE 2.5 Label the planes represented in this illustration. **A** **APR** (McGraw-Hill Education/Karl Rubin, photographer; Living Art Enterprises/Science Source; Living Art Enterprises/Science Source)

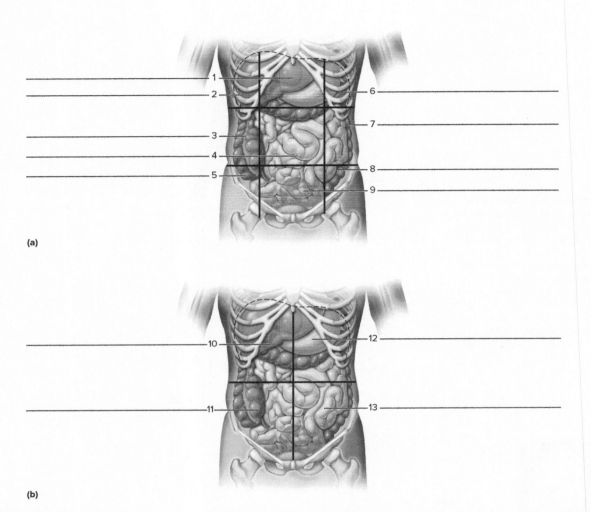

(a)

(b)

FIGURE 2.6 The abdominopelvic cavity is subdivided into either nine regions or four quadrants for purposes of locating organs, injuries, pain, or performed medical procedures. Label (*a*) the 9 regions and (*b*) the 4 quadrants of the abdominopelvic cavity. Ⓐ **APR**

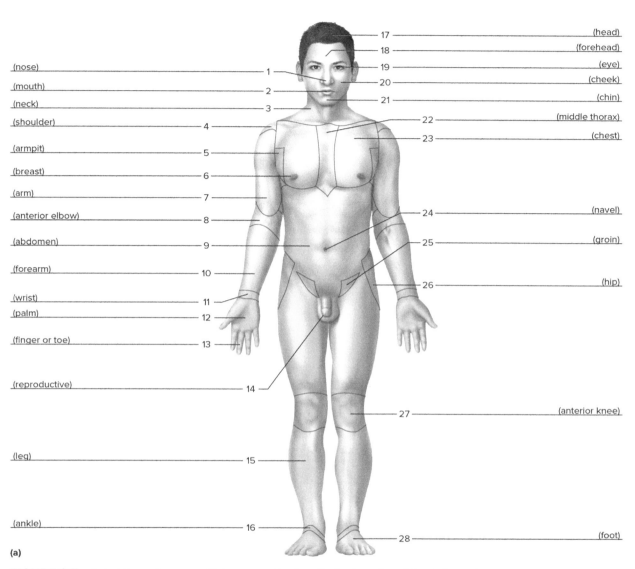

(nose) ———————————————— 1
(mouth) ——————————————— 2
(neck) ————————————————— 3
(shoulder) —————————————— 4
(armpit) ————————————————— 5
(breast) ————————————————— 6
(arm) ——————————————————— 7
(anterior elbow) —————————— 8
(abdomen) ————————————— 9
(forearm) ————————————— 10
(wrist) ——————————————— 11
(palm) ——————————————— 12
(finger or toe) ————————— 13
(reproductive) —————————— 14
(leg) ——————————————— 15
(ankle) ————————————— 16

17 ———————————————— (head)
18 ———————————————— (forehead)
19 ———————————————— (eye)
20 ———————————————— (cheek)
21 ———————————————— (chin)
22 ———————————————— (middle thorax)
23 ———————————————— (chest)
24 ———————————————— (navel)
25 ———————————————— (groin)
26 ———————————————— (hip)
27 ———————————————— (anterior knee)
28 ———————————————— (foot)

(a)

FIGURE 2.7   Label these diagrams with terms used to describe body regions: (*a*) anterior regions; (*b*) posterior regions. **6** **APR**

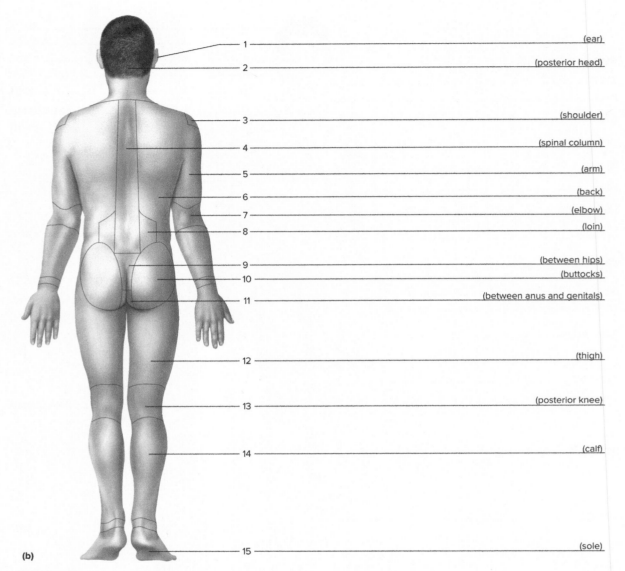

1 _____ (ear)

2 _____ (posterior head)

3 _____ (shoulder)

4 _____ (spinal column)

5 _____ (arm)

6 _____ (back)

7 _____ (elbow)

8 _____ (loin)

9 _____ (between hips)

10 _____ (buttocks)

11 _____ (between anus and genitals)

12 _____ (thigh)

13 _____ (posterior knee)

14 _____ (calf)

15 _____ (sole)

(b)

FIGURE 2.7   *Continued.* **APR**

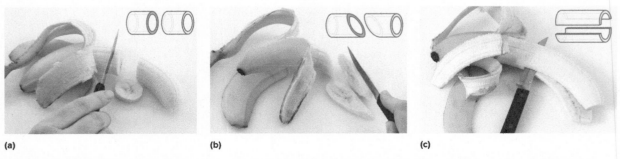

(a)   (b)   (c)

FIGURE 2.8   Three possible cuts of a banana: (*a*) cross section; (*b*) oblique section; (*c*) longitudinal section. Sections through an organ, as a body tube, frequently produce views similar to the cut banana. ((*a,b,c*) ©J & J Photography)

Name _____

Date _____

Section _____

The 🅰 corresponds to the indicated Learning Outcome(s) found at the beginning of the laboratory exercise.

LABORATORY
REPORT

2

Body Organization and Terminology

♻ PART A ASSESSMENTS

Match the body cavities in column A with the organs contained in the cavities in column B. Place the letter of your choice in the space provided. 🅰

Column A	Column B
a. Abdominal cavity	_____ **1.** Liver
b. Cranial cavity	_____ **2.** Lungs
c. Pelvic cavity	_____ **3.** Spleen
d. Thoracic cavity	_____ **4.** Stomach
e. Vertebral canal (spinal cavity)	_____ **5.** Brain
	_____ **6.** Internal reproductive organs
	_____ **7.** Gallbladder
	_____ **8.** Urinary bladder
	_____ **9.** Mediastinum
	_____ **10.** Spinal cord
	_____ **11.** Heart
	_____ **12.** Small intestine

♻ PART B ASSESSMENTS

Match the organ systems in column A with the general functions in column B. Place the letter of your choice in the space provided. 🄰

Column A

a. Cardiovascular system
b. Digestive system
c. Endocrine system
d. Integumentary system
e. Lymphatic system
f. Muscular system
g. Nervous system
h. Reproductive system
i. Respiratory system
j. Skeletal system
k. Urinary system

Column B

_____ **1.** Main system that secretes hormones that function in the integration and coordination of the body

_____ **2.** Provides an outer covering of the body for protection

_____ **3.** Produces gametes (eggs and sperm) and new organisms

_____ **4.** Stimulates muscles to contract, interprets information from sensory organs, and functions in the integration and coordination of the body

_____ **5.** Provides a framework and support for soft tissues and produces blood cells in red marrow

_____ **6.** Exchanges gases between air and blood

_____ **7.** Transports excess fluid from tissues to blood and helps defend the body against infections

_____ **8.** Involves contractions and movement of the joints of the body and creates most body heat

_____ **9.** Removes liquid and wastes from blood and transports them to the outside of the body

_____ **10.** Breaks down and converts food molecules into absorbable forms

_____ **11.** Transports nutrients, wastes, and gases throughout the body

PART C ASSESSMENTS

Write the correct term of relative position in the blank on the right that completes each sentence. (assume that the body is in the anatomical position as observed in fig. 2.4). 🔺

1. The mouth is _____ to the nose.   _____

2. The stomach is _____ to the diaphragm.   _____

3. The trachea is _____ to the spinal cord.   _____

4. The larynx is _____ to the esophagus.   _____

5. The heart is _____ to the lungs.   _____

6. The kidneys are _____ to the lungs.   _____

7. The hand is _____ to the elbow.   _____

8. The knee is _____ to the ankle.   _____

9. The thumb is the _____ digit of the hand.   _____

10. A _____ nerve passes from the spinal cord into the limbs.   _____

11. The spleen and gallbladder are _____.   _____

12. The popliteal surface region is _____ to the patellar region.   _____

 PART D ASSESSMENTS

Match the body regions in column A with the locations in column B. Place the letter of your choice in the space provided. 🔺

Column A	Column B
a. Antecubital	_____ 1. Wrist
b. Axillary	_____ 2. Reproductive organs
c. Brachial	_____ 3. Armpit
d. Buccal	_____ 4. Posterior region of elbow
e. Carpal	_____ 5. Buttocks
f. Cephalic	_____ 6. Finger or toe
g. Cervical	_____ 7. Neck
h. Crural	_____ 8. Arm
i. Cubital	_____ 9. Cheek
j. Digital	_____ 10. Leg
k. Genital	_____ 11. Head
l. Gluteal	_____ 12. Anterior region of elbow

PART E ASSESSMENTS

Match the body regions in column A with the locations in column B. Place the letter of your choice in the space provided. 6

Column A

a. Inguinal
b. Lumbar
c. Mammary
d. Occipital
e. Palmar
f. Pectoral
g. Pedal
h. Perineal
i. Plantar
j. Popliteal
k. Sternal
l. Tarsal

Column B

_____ **1.** Ankle

_____ **2.** Breasts

_____ **3.** Between anus and reproductive organs

_____ **4.** Sole of foot

_____ **5.** Middle of thorax

_____ **6.** Chest

_____ **7.** Posterior region of knee

_____ **8.** Foot

_____ **9.** Inferior posterior region of head

_____ **10.** Abdominal wall near thigh (groin)

_____ **11.** Lower back

_____ **12.** Palm

PART F ASSESSMENTS

ASSESS

CRITICAL THINKING

State the quadrant of the abdominopelvic cavity in which the pain or sound would be located for each of the six conditions listed. In some cases, there may be more than one correct answer, and pain is sometimes referred to another region. This phenomenon, called *referred pain*, occurs when pain is interpreted as originating from some area other than the parts being stimulated. When referred pain is involved in the patient's interpretation of the pain location, the proper diagnosis of the ailment is more challenging. For the purpose of this exercise, assume the pain is interpreted as originating from the organ involved. 3

1. Stomach ulcer _____

2. Appendicitis _____

3. Bowel sounds _____

4. Gallbladder attack _____

5. Kidney stone in left ureter _____

6. Ruptured spleen _____

PART G ASSESSMENTS

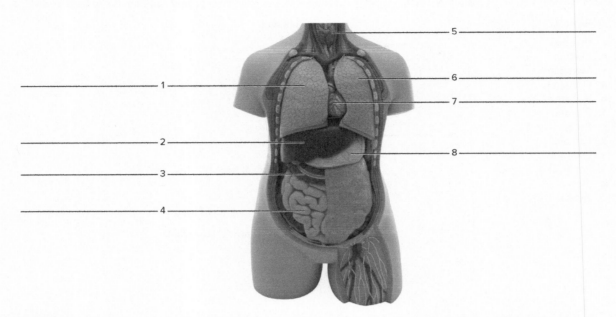

FIGURE 2.9 Using a straight edge, draw lines to divide the torso model in figure 2.9 into its nine abdominopelvic regions. Using the terms provided label the nine abdominopelvic regions and the major organs pictured.

Locate the 16 anatomical terms pertaining to "Body Organization and Terminology." 7

Terms:
Thyroid gland
Right lung
Left lung
Heart
Liver
Large intestine
Small intestine
Right hypochondriac region
Epigastric region
Left hypochondriac region
Right lateral region
Umbilical region
Left lateral region
Right inguinal region
Pubic region
Left inguinal region

LABORATORY EXERCISE

Chemistry of Life

MATERIALS NEEDED

For pH Tests:
Chopped fresh red cabbage
Beaker (250 mL)
Distilled water
Tap water
Vinegar
Baking soda
Laboratory scoop for measuring
Pipets for measuring
Full-range pH test papers
7 assorted common liquids clearly labeled in closed
     bottles on a tray or in a tub
Droppers labeled for each liquid

For Organic Tests:
Test tubes
Test-tube rack
Test-tube clamps
China marker
Hot plate
Beaker for hot water bath (500 mL)
Pipets for measuring
Benedict's solution
Biuret reagent (or 10% NaOH and 1% $CuSO_4$)
Iodine-potassium-iodide (IKI) solution
Sudan IV dye
Egg albumin
10% glucose solution
Clear carbonated soft drink
10% starch solution
Potatoes for potato water
Distilled water
Vegetable oil
Brown paper
Numbered unknown organic samples

⚠ SAFETY

- Review all Laboratory Safety Guidelines in Appendix 1 of your laboratory manual.
- Clean laboratory surfaces before and after laboratory procedures using soap and water.
- Use extreme caution when working with chemicals.

- Wear safety goggles.
- Wear disposable gloves while working with the chemicals.
- Take precautions to prevent chemicals from contacting your skin.
- Do not mix any of the chemicals together unless instructed to do so.
- Clean up any spills immediately and notify the instructor at once.
- Wash your hands before leaving the laboratory.

PURPOSE OF THE EXERCISE

To review the organization of atoms and molecules, types of chemical interactions, and basic categories of organic compounds and to differentiate between types of organic compounds.

LEARNING OUTCOMES

After completing this exercise, you should be able to

1. Associate and illustrate the basic organization of atoms and molecules.

2. Measure pH values of various substances through testing methods.

3. Determine categories of organic compounds with basic colorimetric tests.

4. Discover the organic composition of an unknown solution.

The complexities of the human body arise from the organization and interactions of chemicals. Organisms are made of matter, and the most basic unit of matter is the chemical element. The smallest unit of an element is an atom, and two or more of those can unite to form a molecule. All processes that occur within the body involve chemical reactions—interactions between atoms and molecules. We breathe to supply oxygen to our cells for energy. We eat and drink to bring chemicals into our bodies that our cells need. Water fills and bathes all of our cells and allows an amazing array of reactions to occur, all of which are designed to keep us alive. Chemistry forms the basis of life and thus forms the foundation of anatomy and physiology.

PRACTICE

PROCEDURE A—Matter, Molecules, Bonding, and pH

1. Review section 2.2 titled "Structure of Matter" in chapter 2 of the textbook. Use the Periodic Table of Elements, located in Appendix 2 of the laboratory manual, as a reference.
2. Review figures 3.1, 3.2, and 3.3, which show molecular diagrams, formation of an ionic bond, and formation of a covalent bond, respectively.
3. Complete Part A of Laboratory Report 3.

 PRACTICE

PROCEDURE B—The pH Scale

1. Review the concept heading titled "Acid and Base Concentrations" of section 2.2 in chapter 2 of the textbook.
2. Review figure 3.4, which shows the pH scale. Note the range of the scale and the pH value considered neutral.
3. **Cabbage water tests.** Many tests can determine a pH value, but among the more interesting are colorimetric tests in which an indicator chemical changes color when it reacts. Many plant pigments, especially anthocyanins (which give plants color, from blue to red), can be used as colorimetric pH indicators. One that works well is the red pigment in red cabbage.
   a. Prepare cabbage water to be used as a general pH indicator. Fill a 250 mL beaker to the 100 mL level with chopped red cabbage. Add water to make 150 mL. Place the beaker on a hot plate and simmer the mixture until the pigments come out of the cabbage and the water turns deep purple. Allow the water to cool. (You may proceed to step 4 while you wait for this to finish.)
   b. Label three clean test tubes: one for water, one for vinegar, and one for baking soda.
   c. Place 2 mL of cabbage water into each test tube.
   d. To the first test tube, add 2 mL of distilled water and swirl the mixture. Record the color in Part B of the laboratory report.
   e. Repeat this procedure for test tube 2, adding 2 mL of vinegar and swirling the mixture. Record the results in Part B of the laboratory report.
   f. Repeat this procedure for test tube 3, adding one laboratory scoop of baking soda. Swirl the mixture. Record the results in Part B of the laboratory report.
4. **Testing with pH paper.** Many commercial pH indicators are available. A simple one to use is pH paper which comes in small strips. Don gloves for this procedure so your skin secretions do not contaminate the paper, and to protect you from the chemicals you are testing.
   a. Test the pH of distilled water by dropping one or two drops of distilled water onto a strip of pH paper, and note the color. Compare the color to the color guide on the container. Note the results of the test as soon as color appears, as drying of the paper will change the results. Record the pH value in Part B of the laboratory report.
   b. Repeat this procedure for tap water. Record the results.

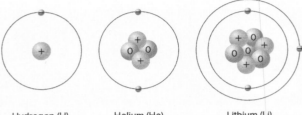

Hydrogen (H)   Helium (He)   Lithium (Li)

FIGURE 3.1 The single electron of a hydrogen atom is located in its first shell. The two electrons of a helium atom fill its first shell. Two of the three electrons of a lithium atom are in the first shell, and one is in the second shell. **APR**

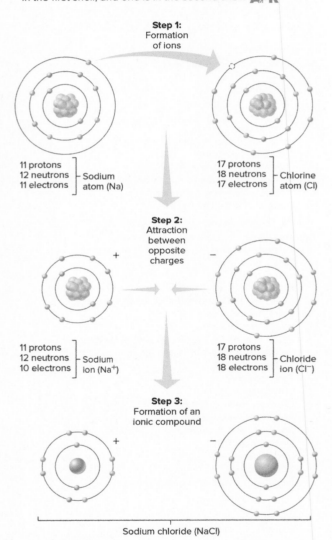

Sodium chloride (NaCl)

FIGURE 3.2 Formation of an ionic bond. **APR**

   c. Individually test the various common substances found on your lab table. Record your results.
   d. Complete Part B of the laboratory report.

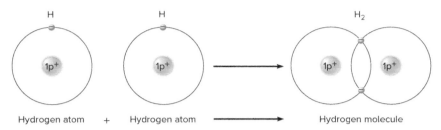

FIGURE 3.3   A hydrogen molecule forms when two hydrogen atoms share a pair of electrons and join by a covalent bond. **APR**

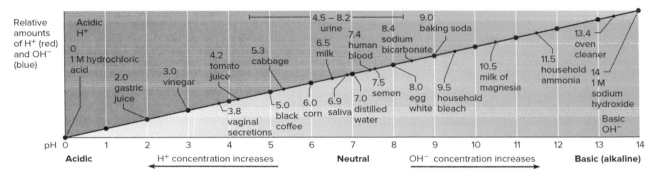

FIGURE 3.4   As the concentration of hydrogen ions (H⁺) increases, a solution becomes more acidic, and the pH value decreases. As the concentration of ions that combine with hydrogen ions (such as hydroxide ions) increases, a solution becomes more basic (alkaline), and the pH value increases. The pH values of some common substances are shown.

PRACTICE

PROCEDURE C—Organic Molecules

In this section, you will perform some simple tests to check for the presence of the following categories of organic molecules (biomolecules): protein, carbohydrate (sugar and starch), and lipid. Specific color changes will occur if the target compound is present. Please note the original color of the indicator being added so you can tell if the color really changed. For example, Biuret reagent and Benedict's solution are both initially blue, so an end color of blue would indicate no change. Use caution when working with these chemicals, and follow all directions. Carefully label the test tubes and avoid contaminating any of your samples. Only the Benedict's test requires heating and a time delay for accurate results. Do not heat any other tubes. To prepare for the Benedict's test, fill a 500 mL beaker half full with water and place it on the hot plate. Turn the hot plate on and bring the water to a boil. To save time, start the water bath before doing the Biuret test.

1. **Biuret test for protein.** In the presence of protein, Biuret reagent reacts with peptide bonds and changes to violet or purple. A pinkish color indicates that shorter polypeptides are present. The color intensity is proportional to the number of peptide bonds; thus, the

intensity reflects the length of polypeptides (amount of protein).

 a. Label six test tubes as follows: 1,2,3,4,5, and 6
 b. To each test tube, add 2 mL of one of the samples as follows:
  1—2 mL distilled water
  2—2 mL egg albumin
  3—2 mL 10% glucose solution
  4—2 mL carbonated soft drink
  5—2 mL 10% starch solution
  6—2 mL potato juice
 c. To each, add 2 mL of Biuret reagent, swirl the tube to mix it, and note the final color. [*Note*: If Biuret reagent is not available, add 2 mL of 10% NaOH (sodium hydroxide) and about 10 drops of 1% CuSO₄ (copper sulfate).] **Be careful—NaOH is very caustic.**
 d. Record your results in Part C of the laboratory report.
 e. Mark a "+" on any tubes with positive results and retain them for comparison while testing your unknown sample. Put remaining test tubes aside so they will not get confused with future trials.

2. **Benedict's test for sugar (monosaccharides).** In the presence of sugar, Benedict's solution changes from its initial blue to a green, yellow, orange, or reddish color, depending on the amount of sugar present. Orange and red indicate greater amounts of sugar.

a. Label six test tubes as follows: 1,2,3,4,5, and 6.

b. To each test tube, add 2 mL of one of the samples as follows:

1—2 mL distilled water

2—2 mL egg albumin

3—2 mL 10% glucose solution

4—2 mL carbonated soft drink

5—2 mL 10% starch solution

6—2 mL potato juice

c. To each, add 2 mL of Benedict's solution, swirl the tube to mix it, and place all tubes into the boiling water bath for 3 to 5 minutes.

d. Note the final color of each tube after heating and record the results in Part C of the laboratory report.

e. Mark a "+" on any tubes with positive results and retain them for comparison while testing your unknown sample. Put remaining test tubes aside so they will not be confused with future trials.

3. **Iodine test for starch.** In the presence of starch, iodine turns a dark purple or blue-black color. Starch is a long chain formed by many glucose units linked together side by side. This regular organization traps the iodine molecules and produces the dark color.

a. Label six test tubes as follows: 1,2,3,4,5, and 6.

b. To each test tube, add 2 mL of one of the samples as follows:

1—2 mL distilled water

2—2 mL egg albumin

3—2 mL 10% glucose solution

4—2 mL carbonated soft drink

5—2 mL 10% starch solution

6—2 mL potato juice

c. To each, add 0.5 mL of IKI (iodine solution) and swirl the tube to mix it.

d. Record the final color of each tube in Part C of the laboratory report.

e. Mark a "+" on any tubes with positive results and retain them for comparison while testing your unknown sample. Put remaining test tubes aside so they will not be confused with future trials.

4. **Tests for lipids.**

a. Label two separate areas of a piece of brown paper as "water" or "oil."

b. Place a drop of water on the area marked "water" and a drop of vegetable oil on the area marked "oil."

c. Let the spots dry several minutes; then record your observations in Part C of the laboratory report. Upon drying, oil leaves a translucent stain (grease spot) on brown paper; water does not leave such a spot.

d. In a test tube, add 2 mL of water and 2 mL of vegetable oil and observe. Shake the tube vigorously; then let it sit for 5 minutes and observe again. Record your observations in Part C of the laboratory report.

e. Sudan IV is a dye that is lipid-soluble but not water-soluble. If lipids are present, then Sudan IV will stain them pink or red. Add a small amount of Sudan IV to the test tube that contains the oil and water. Swirl it; then let it sit for a few minutes. Record your observations in Part C of the laboratory report.

 PRACTICE

PROCEDURE D—Identifying Unknown Compounds

Now you will apply the information you gained with the tests in the previous section. You will retrieve an unknown sample that contains none, one, or any combination of the following types of organic compounds: protein, sugar, starch, or lipid. You will test your sample using each test from the previous section and record your results in Part D of the laboratory report.

1. Label four test tubes (1, 2, 3, 4).

2. Add 2 mL of your unknown sample to each test tube.

3. **Test for protein.** Add 2 mL of Biuret reagent to tube 1. Swirl the tube and observe the color. Record your observations in Part D of the laboratory report.

4. **Test for sugar.** Add 2 mL of Benedict's solution to tube 2. Swirl the tube and place it in a boiling water bath for 3 to 5 minutes. Record your observations in Part D of the laboratory report.

5. **Test for starch.** Add several drops of iodine solution to tube 3. Swirl the tube and observe the color. Record your observations in Part D of the laboratory report.

6. **Test for lipid.** Add 2 mL of water to tube 4. Swirl the tube and note if there is any separation.

7. Add a small amount of Sudan IV to test tube 4, swirl the tube, and record your observations in Part D of the laboratory report.

8. Based on the results of these tests, determine if your unknown sample contains any organic compounds and, if so, what they are. Record and explain your identification in Part D of the laboratory report.

Name _____

Date _____

Section _____

The Ⓐ corresponds to the indicated Learning Outcome(s) found at the beginning of the laboratory exercise.

Chemistry of Life

PART A ASSESSMENTS

Match the terms in column A with the descriptions in column B. Place the letter of your choice in the space provided. Ⓐ

Column A	Column B
a. Atomic number	_____ 1. The number of protons plus the number of neutrons
b. Atomic weight (mass)	_____ 2. A combination of two or more atoms of different elements
c. Base	_____ 3. A small, negatively charged particle that orbits the nucleus
d. Catalyst (enzyme)	_____ 4. An atom that has gained or lost electrons and thus carries an electrical charge
e. Compound	_____ 5. Atoms are held together by sharing electrons
f. Covalent bond	_____ 6. The most fundamental substances of matter
g. Electrolyte	_____ 7. A substance that combines with hydrogen ions
h. Electron	_____ 8. A molecule that influences the rate of chemical reactions but is not consumed in the reaction
i. Elements	_____ 9. Two atoms with the same atomic number but different atomic weights
j. Ion	_____ 10. A substance that releases ions in water
k. Isotopes	_____ 11. The number of protons in an atom
l. Nucleus	_____ 12. The area of an atom where protons and neutrons are located

MOLECULES AND BONDING

After reviewing figures 3.1, 3.2, and 3.3, complete the following:

1. The atomic number of hydrogen is _____, and it has _____ electron(s) in its outer shell. Ⓐ

2. The atomic number of chlorine is _____, and it has _____ electron(s) in its outer shell. Ⓐ

3. Using the style shown in figure 3.2, draw an atom of hydrogen and one of chlorine, clearly indicating the numbers and positions of the protons, neutrons, and electrons. Ⓐ

4. Is the hydrogen atom stable? Ⓐ _____

   Is the chlorine atom stable? Ⓐ _____

5. What type of bond is likely to form the compound HCl (hydrochloric acid)? ⚐ _____

6. Repeat your drawings below but also show and explain how this bond of HCl would form.

   Clearly indicate the positions of protons, neutrons, and electrons. ⚐

PART B ASSESSMENTS

1. Results from cabbage water tests: ⚑

Substance	Cabbage Water	Distilled Water	Vinegar	Baking Soda
Color				
Acid, base, or neutral?				

2. Results from pH paper tests: ⚑

Substance Tested	Distilled Water	Tap Water	Sample 1:_____	Sample 2:_____	Sample 3:_____	Sample 4:_____	Sample 5:_____	Sample 6:_____	Sample 7:_____
pH value									

3. Are the pH values the same for distilled water and tap water? ⚑ _____

4. If not, what might explain this difference? _____

   _____

5. Draw the pH scale here and indicate the following values: 0, 7 (neutral), and 14. Now label the scale with the names and indicate the location of the pH values for each substance you tested.

PART C ASSESSMENTS

1. **Biuret test results for protein.** Enter your results from the Biuret test for protein. A color change to purple indicates that protein is present; pink indicates that short polypeptides are present. /3\

Tube	Contents	Color	Protein Present (+) or Absent (−)
1	Distilled water		
2	Egg albumin		
3	Glucose solution		
4	Soft drink		
5	Starch solution		
6	Potato juice		

2. **Benedict's test results for most sugars (monosaccharides).** Enter your results from the Benedict's test for sugars. A color change to green, yellow, orange, or red indicates that sugar is present. Note the color after the mixture has been heated for 3 to 5 minutes. /3\

Tube	Contents	Color	Sugar Present (+) or Absent (−)
1	Distilled water		
2	Egg albumin		
3	Glucose solution		
4	Soft drink		
5	Starch solution		
6	Potato juice		

3. **Iodine test results for starch.** Enter your results from the iodine test for starch. A color change to dark blue or black indicates that starch is present. /3\

Tube	Contents	Color	Starch Present (+) or Absent (−)
1	Distilled water		
2	Egg albumin		
3	Glucose solution		
4	Soft drink		
5	Starch solution		
6	Potato juice		

4. **Lipid test results.** What did you observe when you allowed the drops of oil and water to dry on the brown paper?

_____

What did you observe when you mixed the oil and water together? _____

_____

What did you observe when you added the Sudan IV dye? /3\ _____

ASSESS

CRITICAL THINKING

If a person were on a low-carbohydrate, high-protein diet, which of the six substances tested would the person want to increase?
_____

Explain your answer.
_____
_____
_____
_____

PART D ASSESSMENTS

1. What is the number of your sample? _____

2. Record the results of your tests on the unknown sample here: **A**

Test Performed	Results
Biuret test for protein	
Benedict's test for sugars	
Iodine test for starch	
Water test for lipid	
Sudan IV test for lipid	

3. Based on these results, what organic compound(s) does your unknown sample contain? **A**
_____

4. Explain your answer. **A**
_____
_____
_____
_____

LABORATORY EXERCISE

Care and Use of the Microscope

PURPOSE OF THE EXERCISE

To become familiar with the major parts of a compound light microscope and their functions and to make use of the microscope to observe small objects.

 ## LEARNING OUTCOMES

After completing this exercise, you should be able to

(1) Locate and identify the major parts of a compound light microscope and differentiate the functions of these parts.

(2) Calculate the total magnification produced by various combinations of eyepiece and objective lenses.

(3) Demonstrate proper use of the microscope to observe and measure small objects.

(4) Prepare a simple microscope slide and sketch the objects you observed.

The human eye cannot perceive objects less than 0.1 mm in diameter, so a microscope is an essential tool for the study of small structures such as cells. The microscope used for this purpose is the *compound light microscope*. It is called

compound because it uses two sets of lenses: an eyepiece, or ocular, lens system and an objective lens system. The eyepiece lens system magnifies, or compounds, the image reaching it after the image is magnified by the objective lens system. Such an instrument can magnify images of small objects up to about 1,000 times.

 ## PRACTICE

PROCEDURE A—Microscope Basics

1. Familiarize yourself with the following list of rules for care of the microscope:
   a. Keep the microscope under its *dustcover* and in a cabinet when it is not being used.
   b. Handle the microscope with great care. It is an expensive and delicate instrument. To move it or carry it, hold it by its *arm* with one hand and support its *base* with the other hand (fig. 4.1).
   c. Always store the microscope with the scanning or lowest-power objective in place. Always start with this objective when using the microscope.
   d. To clean the lenses, rub them gently with *lens paper* or a high-quality cotton swab. If the lenses need additional cleaning, follow the directions in the "Lens-Cleaning Technique" section that follows.
   e. If the microscope has a substage lamp, be sure the electric cord does not hang off the laboratory table where someone might trip over it. The bulb life can be extended if the lamp is cool before the microscope is moved.
   f. Never drag the microscope across the laboratory table.
   g. Never remove parts of the microscope or try to disassemble the eyepiece or objective lenses.
   h. If your microscope is not functioning properly, report the problem to your laboratory instructor immediately.

2. Observe a compound light microscope and study figure 4.1 to learn the names of its major parts. Your microscope might be equipped with a binocular body and two eyepieces (fig. 4.2). The lens system of a compound microscope includes three parts—the condenser, objective lens, and eyepiece (ocular).

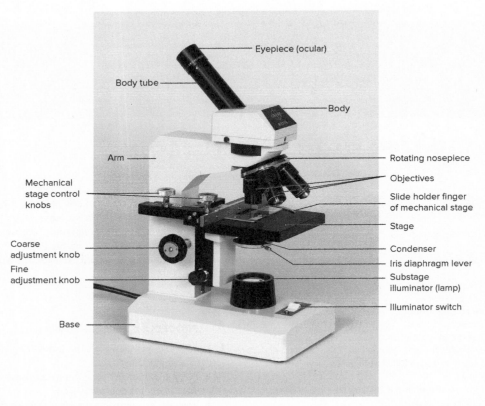

Eyepiece (ocular)

Body tube

Body

Arm

Rotating nosepiece

Mechanical
stage control
knobs

Objectives

Slide holder finger
of mechanical stage

Stage

Coarse
adjustment knob

Condenser

Fine
adjustment knob

Iris diaphragm lever

Substage
illuminator (lamp)

Illuminator switch

Base

FIGURE 4.1 Major parts of a compound light microscope with a monocular (one eyepiece) body and a mechanical stage. Some microscopes are equipped with a binocular (two eyepieces) body. (J and J Photography)

 LENS-CLEANING TECHNIQUE

1. Moisten one end of a high-quality cotton swab with one drop of lens cleaner. Keep the other end dry.

2. Clean the optical surface with the wet end. Dry it with the other end, using a circular motion.

3. Use a hand aspirator to remove lingering dust particles.

4. Start with the scanning objective and work upward in magnification, using a new cotton swab for each objective.

5. When cleaning the eyepiece, do not open the lens unless it is absolutely necessary.

6. Use alcohol for difficult cleaning.

Light enters this system from a *substage illuminator (lamp)* or *mirror* and is concentrated and focused by a *condenser* onto a microscope slide (fig. 4.3). The condenser, which contains a set of lenses, usually is kept in its highest position possible.

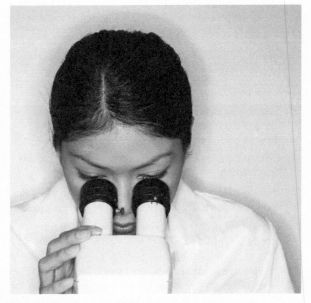

FIGURE 4.2 Scientist using a compound light microscope equipped with a binocular body and two eyepieces. (George Doyle/Getty Images)

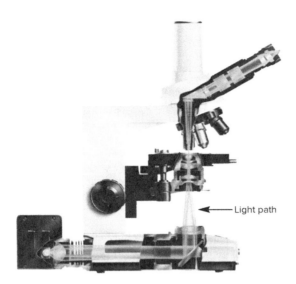

FIGURE 4.3 Microscope, showing the path of light through it. (©Nikon Instruments)

The *iris diaphragm,* located between the light source and the condenser, can be used to increase or decrease the intensity of the light entering the condenser and to control the contrast of the image. Locate the lever that operates the iris diaphragm beneath the *stage* and move it back and forth. Note how this movement causes the size of the opening in the diaphragm to change. (Some microscopes have a revolving plate called a *disc diaphragm* beneath the stage instead of an iris diaphragm. Disc diaphragms have different-sized holes to admit varying amounts of light.) Which way do you move the diaphragm to increase the light intensity? _____ Which way to decrease it? _____

After light passes through a specimen mounted on a microscope slide, it enters an *objective lens system.* This lens projects the light upward into the *body tube,* where it produces a magnified image of the object being viewed.

The *eyepiece (ocular) lens* system then magnifies this image to produce another image, seen by the eye. Typically, the eyepiece lens magnifies the image ten times (10×). Look for the number in the metal of the eyepiece that indicates its power (fig. 4.4). What is the eyepiece power of your microscope? _____

The objective lenses are mounted in a *rotating nosepiece* so that different magnifications can be achieved by rotating any one of several objective lenses into position above the specimen. Commonly, this set of lenses

(a)

(b)

FIGURE 4.4 The powers of this 10× eyepiece (*a*) and this 40× objective (*b*) are marked in the metal. DIN is an international optical standard on quality optics. The 0.65 on the 40× objective is the numerical aperture, a measure of the light-gathering capabilities. ((*a,b*) J and J Photography)

includes a scanning objective (4×), a low-power objective (10×), and a high-power objective, also called a high-dry-power objective (about 40×). Sometimes an oil immersion objective (about 100×) is present. Look for the number printed on each objective lens that indicates its power. What are the objective lens powers of your microscope? _____

To calculate the *total magnification* achieved when using a particular objective, multiply the power of the eyepiece by the power of the objective used. Thus, the 10× eyepiece and the 40× objective produce a total magnification of 10 × 40, or 400×. See a summary of microscope lenses in table 4.1.

TABLE 4.1 Microscope Lenses

Objective Lens Name	Common Objective Lens Magnification	Common Eyepiece Lens Magnification	Total Magnification
Scan	4×	10×	40×
Low-power (LP)	10×	10×	100×
High-power (HP)	40×	10×	400×
Oil immersion	100×	10×	1,000×

Note: If you wish to observe an object under LP, HP, or oil immersion, locate and then center and focus the object first under scan magnification.

3. Complete Part A of Laboratory Report 4.
4. Turn on the substage illuminator and look through the eyepiece. You will see a lighted, circular area called the *field of view.*

   You can measure the diameter of this field of view by focusing the lenses on the millimeter scale of a transparent plastic ruler. To do this, follow these steps:

   a. Place the ruler on the microscope stage in the spring clamp of a slide holder finger on a *mechanical stage* or under the *stage (slide) clips.* (*Note*: If your microscope is equipped with a mechanical stage, it may be necessary to use a short section cut from a transparent plastic ruler. The section should be several millimeters long and can be mounted on a microscope slide for viewing.)

   b. Center the millimeter scale in the beam of light coming up through the condenser, and rotate the scanning objective into position.

   c. While you watch from the side to prevent the lens from touching anything, raise the stage until the objective is as close to the ruler as possible, using the *coarse adjustment knob* (fig. 4.5). (*Note*: The adjustment knobs on some microscopes move the body and objectives downward and upward for focusing.)

   d. Look into the eyepiece and use the *fine adjustment knob* to raise the stage until the lines of the millimeter scale come into sharp focus.

   e. Adjust the light intensity by moving the *iris diaphragm lever* so that the field of view is brightly illuminated but comfortable to your eye. At the same time, take care not to overilluminate the field because transparent objects tend to disappear in bright light.

   f. Position the millimeter ruler so that its scale crosses the greatest diameter of the field of view. In addition, move the ruler so that one of the millimeter marks is against the edge of the field of view.

   g. In millimeters, measure the distance across the field of view.

5. Complete Part B of the laboratory report.
6. Most microscopes are designed to be *parfocal*. This means that when a specimen is in focus with a lower-power objective, it will be in focus (or nearly so) when a higher-power objective is rotated into position. Always

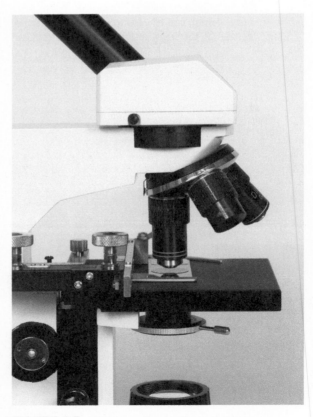

FIGURE 4.5   When you focus using a particular objective, you can prevent it from touching the specimen by watching from the side. (J and J Photography)

center the specimen in the field of view before changing to higher objectives.

Rotate the low-power objective into position, and then look at the millimeter scale of the transparent plastic ruler. If you need to move the low-power objective to sharpen the focus, use the fine adjustment knob.

Adjust the iris diaphragm so that the field of view is properly illuminated. Once again, adjust the millimeter ruler so that the scale crosses the field of view through its greatest diameter, and position the ruler so that a millimeter mark is against one edge of the field. Measure the distance across the field of view in millimeters.

7. Rotate the high-power objective into position while you watch from the side, and then observe the millimeter scale on the plastic ruler. *All focusing using high-power magnification should be done only with the fine adjustment knob. Never use the coarse adjustment knob while observing with high-power magnification.* If you use the coarse adjustment knob with the high-power objective, you can accidentally force the objective into the coverslip and break the slide. This is because the *working distance* (the distance from the objective lens to the slide on the stage) is much shorter when using higher magnifications.

Adjust the iris diaphragm for proper illumination. When using higher magnifications, more illumination usually will help you to view the objects more clearly. Try to measure the distance across the field of view in millimeters.

8. Locate the numeral 4 (or 9) on the plastic ruler and focus on it using the scanning objective. Note how the number appears in the field of view. Move the plastic ruler to the right and note which way the image moves. Slide the ruler away from you and again note how the image moves.

9. Observe a letter *e* slide in addition to the numerals on the plastic ruler. Note the orientation of the letter *e* using the scan, LP, and HP objectives. As you increase magnifications, note that the amount of the letter *e* shown is decreased. Move the slide to the left, then away from you, and note the direction in which the observed image moves. If a *pointer* is visible in your field of view, you can manipulate the pointer to a location (structure) within the field of view by moving the slide or rotating the eyepiece.

10. Examine the slide of the three colored threads using the low-power objective and then the high-power objective. Focus on the location where the three threads cross. By using the fine adjustment knob, determine the order from top to bottom by noting which color is in focus at different depths. The other colored threads will still be visible, but they will be blurred. Be sure to notice whether the stage or the body tube moves up and down with the adjustment knobs of the microscope being used for this depth determination. The vertical depth of the specimen clearly in focus is called the *depth of field*

(*focus*). Whenever specimens are examined, continue to use the fine adjustment focusing knob to determine relative depths of structures clearly in focus within cells, giving a three-dimensional perspective. The depth of field is less at higher magnifications.

 ASSESS

CRITICAL THINKING

What was the sequence of the three colored threads from top to bottom? Explain how you came to that conclusion. **3**

_____

_____

_____

11. Complete Parts C and D of the laboratory report.

 PRACTICE

PROCEDURE B—Slide Preparation

1. Prepare several temporary *wet mounts* using any small, transparent objects of interest, and examine the specimens using the low-power objective and then a high-power objective to observe their details. To prepare a wet mount, follow these steps (fig. 4.6):

   a. Obtain a precleaned microscope slide.

   b. Place a tiny, thin piece of the specimen you want to observe in the center of the slide, and use a medicine dropper to put a drop of water over it. Consult with your instructor if a drop of stain might enhance the image of any cellular structures of your specimen. If the specimen is solid, you might want to tease some of it apart with dissecting needles. In any case, the specimen must be thin enough that light can pass through it. Why is it necessary for the specimen to be so thin?

   _____

   _____

   c. Cover the specimen with a coverslip. Try to avoid trapping bubbles of air beneath the coverslip by slowly lowering it at an angle into the drop of water.

   d. Remove any excess water from the edge of the coverslip with absorbent paper. If your microscope has an inclination joint, do not tilt the microscope while observing wet mounts because the fluid will flow.

   e. Place the slide under the stage (slide) clips or in the slide holder on a mechanical stage, and position the slide so that the specimen is centered in the light beam passing up through the condenser.

   f. Focus on the specimen using the scanning objective first. Next, focus using the low-power objective, and then examine it with the high-power objective.

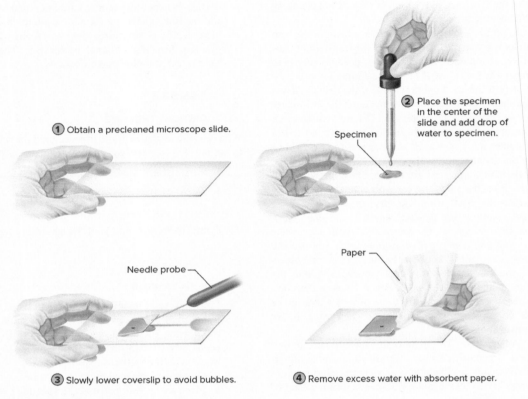

① Obtain a precleaned microscope slide.

② Place the specimen in the center of the slide and add drop of water to specimen.

Specimen

Needle probe —

③ Slowly lower coverslip to avoid bubbles.

Paper —

④ Remove excess water with absorbent paper.

FIGURE 4.6   Steps in the preparation of a wet mount.

2. If an oil immersion objective is available, use it to examine the specimen. To use the oil immersion objective, follow these steps:

   a. Center the object you want to study under the high-power field of view.

   b. Rotate the high-power objective away from the microscope slide, place a small drop of immersion oil on the coverslip, and swing the oil immersion objective into position. To achieve sharp focus, use the fine adjustment knob only.

   c. You will need to open the iris diaphragm more fully for proper illumination. More light is needed because the oil immersion objective covers a very small lighted area of the microscope slide.

   d. The oil immersion objective must be very close to the coverslip to achieve sharp focus, thus care must be taken to avoid breaking the coverslip or damaging the objective lens. For this reason, never lower the objective when you are looking into the eyepiece.

Instead, always raise the objective to achieve focus, or prevent the objective from touching the coverslip by watching the microscope slide and coverslip from the side if the objective needs to be lowered. Usually when using the oil immersion objective, only the fine adjustment knob needs to be used for focusing. *Never switch back to one of the other three objectives while using oil immersion. This could cause damage to the microscope. The other objectives are not designed for oil immersion.*

3. When you have finished working with the microscope, remove the microscope slide from the stage and wipe any oil from the objective lens with lens paper or a high-quality cotton swab. Swing the scanning objective or the low-power objective into position. Wrap the electric cord around the base of the microscope and replace the dustcover.

4. Complete Part E of the laboratory report.

LEARN: ACTIVITY

A stereomicroscope (dissecting microscope) (fig. 4.7) is useful for observing the details of relatively large, opaque specimens. Although this type of microscope achieves less magnification than a compound microscope, it has the advantage of producing a three-dimensional image rather than the flat, two-dimensional image of the compound light microscope. In addition, the image produced by the stereomicroscope is positioned in the same manner as the specimen, rather than being reversed and inverted, as it is by the compound light microscope.

Observe the stereomicroscope. The eyepieces can be pushed apart or together to fit the distance between your eyes. Focus the microscope on the end of your finger. Which way does the image move when you move your finger to the right? _____

When you move it away from you? _____

If the instrument has more than one objective, change the magnification to a higher power. Use the instrument to examine various small, opaque objects available in the laboratory.

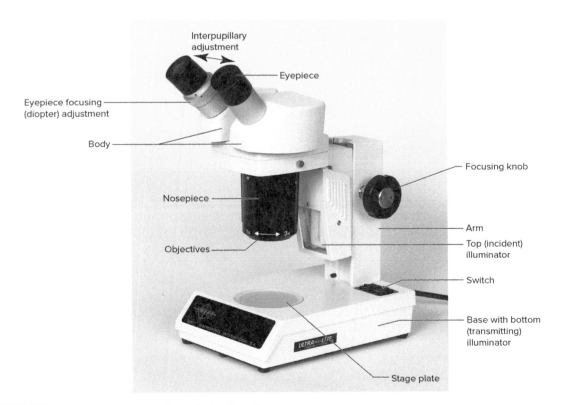

FIGURE 4.7 A stereomicroscope, also called a dissecting microscope. (J and J Photography)

Notes

Name _____

Date _____

Section _____

The Ⓐ corresponds to the indicated Learning Outcome(s) found at the beginning of the laboratory exercise.

Care and Use of the Microscope

♻ PART A ASSESSMENTS

Revisit Procedure A, number 2; then complete the following:

1. What total magnification will be achieved if the 10× eyepiece and the 10× objective are used? Ⓐ2 _____

2. What total magnification will be achieved if the 10× eyepiece and the 100× objective are used? Ⓐ2 _____

♻ PART B ASSESSMENTS

Revisit Procedure A, number 2; then complete the following:

1. Sketch the millimeter scale as it appears under the scanning objective magnification. (The circle represents the field of view through the microscope.)

2. In millimeters, what is the diameter of the scanning field of view? Ⓐ3 _____

3. Microscopic objects often are measured in *micrometers*. A micrometer equals 1/1,000 of a millimeter and is symbolized by µm. In micrometers, what is the diameter of the scanning power field of view? Ⓐ3 _____

4. If a circular object or specimen extends halfway across the scanning field, what is its diameter in millimeters? Ⓐ3 _____

5. In micrometers, what is its diameter? Ⓐ3 _____

♻ PART C ASSESSMENTS

Complete the following:

1. Sketch the millimeter scale as it appears using the low-power objective.

2. What do you estimate the diameter of this field of view to be in millimeters? Ⓐ3

_____

3. How does the diameter of the scanning power field of view compare with that of the low-power field?

_____

4. Why is it more difficult to measure the diameter of the high-power field of view than that of the low-power field? ▲3

_____

_____

5. What change occurred in the intensity of the light in the field of view when you exchanged the low-power objective for the high-power objective? _____

_____

6. Sketch the numeral 4 (or 9) as it appears through the scanning objective of the compound microscope.

7. What has the lens system done to the image of the numeral? (Is it right side up, upside down, or what?) _____

8. When you moved the ruler to the right, which way did the image move? _____

9. When you moved the ruler away from you, which way did the image move? _____

PART D ASSESSMENTS

Match the names of the microscope parts in column A with the descriptions in column B. Place the letter of your choice in the space provided. ▲1

Column A	Column B
a. Adjustment knob (coarse)	_____ **1.** Increases or decreases the light intensity
b. Arm	_____ **2.** Platform that supports a microscope slide
c. Condenser	_____ **3.** Concentrates light onto the specimen
d. Eyepiece (ocular)	_____ **4.** Causes stage (or objective lens) to move upward or downward
e. Field of view	_____ **5.** After light passes through the specimen, it next enters this lens system
f. Iris diaphragm	_____ **6.** Holds a microscope slide in position
g. Nosepiece	_____ **7.** Contains a lens at the top of the body tube
h. Objective lens system	_____ **8.** Serves as a handle for carrying the microscope
i. Stage	_____ **9.** Part to which the objective lenses are attached
j. Stage (slide) clip	_____ **10.** Circular area seen through the eyepiece

PART E ASSESSMENTS

Prepare sketches of the objects you observed using the microscope. For each sketch, include the name of the object, the magnification you used to observe it, and its estimated dimensions in millimeters and micrometers. ▲3 ▲4

Cells

LABORATORY EXERCISE

5

Cell Structure and Function

MATERIALS NEEDED

Textbook
Animal cell model
Clean microscope slides
Coverslips
Flat toothpicks
Medicine dropper
Methylene blue (dilute) or iodine-potassium-
    iodide stain
Prepared microscope slides of human tissues
Compound light microscope
Single-edged razor blade
Plant materials such as leaves, soft stems, fruits, onion
    peel, and vegetables
Cultures of *Amoeba* and *Paramecium*

 SAFETY

- Review all Laboratory Safety Guidelines in Appendix 1 of your laboratory manual.
- Clean laboratory surfaces before and after laboratory procedures.
- Wear disposable gloves for the wet-mount procedures of the cells lining the inside of the cheek.
- Work only with your own materials when preparing the slide of cheek cells. Observe the same precautions as with all body fluids.
- Dispose of laboratory gloves, slides, coverslips, and toothpicks as instructed.
- Use the biohazard container to dispose of items used during the cheek cells procedure.
- Take precautions to prevent cellular stains from contacting your clothes and skin.
- Wash your hands before leaving the laboratory.

PURPOSE OF THE EXERCISE

To review the structure and functions of major cellular components and to observe examples of human cells.

 LEARNING OUTCOMES APR

After completing this exercise, you should be able to

1. Name and locate the components of a cell.
2. Differentiate the functions of cellular components.
3. Prepare a wet mount of cells lining the inside of the cheek; stain the cells; and identify the cell membrane, nucleus, and cytoplasm.
4. Examine cells on prepared slides of human tissues and identify their major components.

Cells are the smallest, most basic units of life. Their arrangement and interactions result in the shape, organization, and construction of the body, and cells are responsible for carrying on its life processes. Under the light microscope, with a properly applied stain to make structures visible, the *cell (plasma) membrane,* the *cytoplasm,* and a *nucleus* are easily seen. The cytoplasm is composed of a clear fluid, the *cytosol,* and numerous *cytoplasmic organelles* suspended in the cytosol.

The cell membrane, composed of lipids and proteins, forms the cell boundary and functions in various methods of membrane transport. Chromatin within the nucleus contains fine strands of DNA and protein. Various cytoplasmic organelles, including mitochondria, endoplasmic reticulum, and Golgi apparatus, provide specialized metabolic functions.

 PRACTICE

PROCEDURE—Cell Structure and Function

1. Review section 3.2 titled "A Composite Cell" in chapter 3 of the textbook.
2. Observe an animal cell model and identify its structures.
3. As a review activity, label figure 5.1 and study figure 5.2.
4. Complete Part A of Laboratory Report 5.
5. Prepare a wet mount of cells lining the inside of the cheek. To do this, follow these steps:
    a. Gently scrape (force is not necessary and should be avoided) the inner lining of your cheek with the broad end of a flat toothpick.

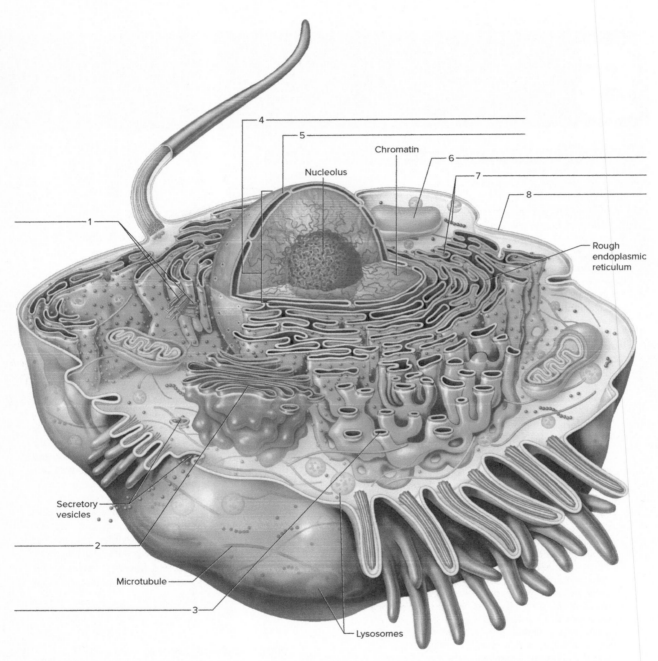

Chromatin

Nucleolus

Rough
endoplasmic
reticulum

Secretory
vesicles

Microtubule

Lysosomes

FIGURE 5.1 Label the structures of this composite cell. The structures are not drawn to scale. **APR**

b. Stir the toothpick in a drop of water on a clean micro-scope slide and dispose of the toothpick as directed by your instructor.

c. Cover the drop with a coverslip.

d. Observe the cheek cells by using the microscope. Compare your image with figure 5.3. To report what you observe, sketch a single cell in the space provided in Part B of the laboratory report.

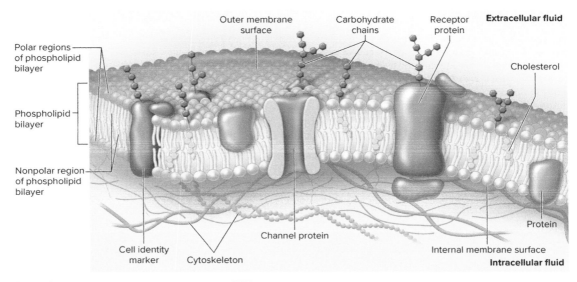

FIGURE 5.2   Structures of the cell membrane. **APR**

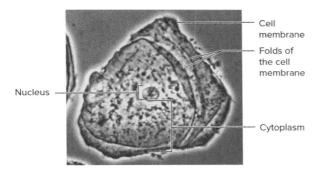

FIGURE 5.3   Iodine-stained cell from inner cheek, as viewed through the compound light microscope using the high-power objective (400×). (©Ed Reschke)

6. Prepare a second wet mount of cheek cells, but this time, add a drop of dilute methylene blue or iodine-potassium-iodide stain to the cells. Cover the liquid with a cover-slip and observe the cells with the microscope. Add to your sketch any additional structures you observe in the stained cells.
7. Answer the questions in Part B of the laboratory report.
8. Observe all safety guidelines for disposal of materials used during this procedure.
9. Using the microscope, observe each of the prepared slides of human tissues. To report what you observe, sketch a single cell of each type in the space provided in Part C of the laboratory report.
10. Complete Parts C and D of the laboratory report.

 ASSESS

CRITICAL THINKING

The cells lining the inside of the cheek are frequently removed for making observations of basic cell structure. The cells are from stratified squamous epithelium. Explain why these cells are used instead of outer body surface tissue.

_____

_____

_____

_____

_____

LEARN: ACTIVITY

Investigate the microscopic structure of various plant materials. To do this, prepare tiny, thin slices of plant specimens using a single-edged razor blade. (*Take care not to injure yourself with the blade.*) Keep the slices in a container of water until you are ready to observe them. To observe a specimen, place it into a drop of water on a clean microscope slide and cover it with a coverslip. Use the microscope and view the specimen using low- and high-power magnifications. Observe near the edges where your section of tissue is most likely to be one cell thick. Add a drop of dilute methylene blue or iodine-potassium-iodide stain, and note if any additional structures become visible. How are the microscopic structures of the plant specimens similar to the human tissues you observed?

_____

How are they different? _____

_____

 LEARN: ACTIVITY

Prepare separate wet mounts of the *Amoeba* and *Paramecium* by putting a drop of culture on a clean glass slide. Gently cover each sample with a clean coverslip. Observe the movements of the *Amoeba* with pseudopodia and the *Paramecium* with cilia. Try to locate cellular components such as the cell membrane, nuclear envelope, nucleus, mitochondria, and contractile vacuoles. Describe the movement of the *Amoeba*.

_____

_____

Describe the movement of the *Paramecium*.

_____

_____

Name _____

Date _____

Section _____

The Ⓐ corresponds to the indicated Learning Outcome(s) found at the beginning of the laboratory exercise.

Cell Structure and Function

PART A ASSESSMENTS

Match the cellular components in column A with the descriptions in column B. Place the letter of your choice in the space provided. **1** **2**

Column A	Column B
a. Chromatin	_____ **1.** Loosely coiled fibers containing protein and DNA within a nucleus
b. Cytoplasm	_____ **2.** Location of ATP production for cellular energy
c. Endoplasmic reticulum	_____ **3.** Small RNA-containing particles for the synthesis of proteins
d. Golgi apparatus	_____ **4.** Membranous sac that stores or transports substances
e. Lysosome	_____ **5.** Dense body of RNA and protein within the nucleus
f. Microtubule	_____ **6.** Slender tubes that provide movement in cilia and flagella
g. Mitochondrion	_____ **7.** Composed of membrane-bound canals and sacs for tubular transport throughout the cytoplasm
h. Nuclear envelope	_____ **8.** Occupies space between the cell membrane and the nucleus
i. Nucleolus	_____ **9.** Flattened, membranous sacs that package a secretion
j. Nucleus	_____ **10.** Membranous sac that contains digestive enzymes
k. Ribosome	_____ **11.** Separates nuclear contents from the cytoplasm
l. Vesicle	_____ **12.** Spherical organelle that contains chromatin and the nucleolus

PART B ASSESSMENTS

Complete the following:

1. Sketch a single cheek cell that has been stained. Label the cellular components you recognize. (The circle represents the field of view through the microscope.) **3**

Magnification _____ ×

2. After comparing the wet mount and the stained cheek cells, describe the advantage gained by staining cells.

_____

_____

3. Are the stained cheek cells nearly the same size and shape? _____ Propose an explanation for your answer.

_____

_____

♻ PART C ASSESSMENTS

Complete the following:

1. Sketch a single cell of each type you observed in the prepared slides of human tissues. Name the tissue, indicate the magnification used, and label the cellular components you recognize. ⒜

_____ ×

Tissue _____

_____ ×

Tissue _____

_____ ×

Tissue _____

_____ ×

Tissue _____

2. What do the various types of cells in these tissues have in common? _____

_____

3. What are the main differences you observed among these cells? _____

_____

PART D ASSESSMENTS

Electron micrographs represent extremely thin slices of cells. The micrograph in figure 5.4 contains a section of a nucleus and some cytoplasm. Compare the organelles shown in this micrograph with organelles of the animal cell model and figure 5.1.

Identify the structures indicated by the arrows in figure 5.4. ⚑

1. _____

2. _____

3. _____

4. _____

5. _____

Answer the following questions after observing the transmission electron micrograph in figure 5.4.

6. What cellular structures were visible in the transmission electron micrograph that were not apparent in the cells you

   observed using the compound light microscope? _____

   _____

7. Before they can be observed by using a transmission electron microscope, cells are sliced into very thin sections. What

   disadvantage does this procedure present in the study of cellular parts? _____

   _____

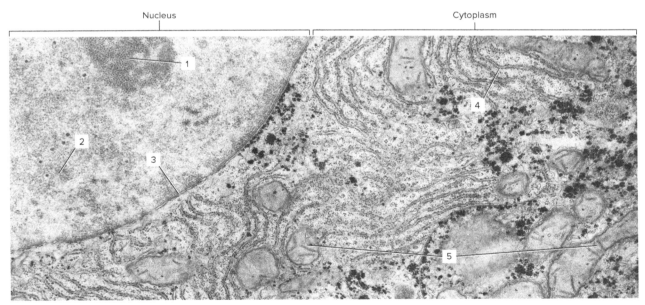

Terms:
Chromatin
Mitochondria
Nuclear envelope
Nucleolus
Rough endoplasmic reticulum

FIGURE 5.4 Transmission electron micrograph of cellular components (23,000×). The view is only a portion of the cell. Identify the numbered cellular structures, using the terms provided. ⚑ **APR** (Keith R. Porter/Science Source)

LABORATORY EXERCISE

6

Movements Through Membranes

MATERIALS NEEDED

For Procedure A—Diffusion:
Textbook
Petri dish
White paper
Forceps
Potassium permanganate crystals
Millimeter ruler (thin and transparent)

For Procedure B—Osmosis:
Thistle tube
Molasses (or Karo dark corn syrup)
Selectively permeable (semipermeable) membrane
    (presoaked dialysis tubing of 1 5/16" or greater
    diameter)
Ring stand and clamp
Beaker
Rubber band
Millimeter ruler

*For Procedure C—Hypertonic, Hypotonic,
    and Isotonic Solutions*:
Test tubes
Marking pen
Test-tube rack
10 mL graduated cylinder
Medicine dropper
Uncoagulated animal blood
Distilled water
0.9% NaCl (aqueous solution)
3.0% NaCl (aqueous solution)
Clean microscope slides
Coverslips
Compound light microscope

For Procedure D—Filtration:
Glass funnel
Filter paper
Support stand and ring
Beaker
Powdered charcoal or ground black pepper
1% glucose (aqueous solution)
1% starch (aqueous solution)
Test tubes
10 mL graduated cylinder
Water bath (boiling water)
Benedict's solution

Iodine-potassium-iodide solution
Medicine dropper

For Alternative Osmosis Activity:
Fresh chicken egg
Beaker
Laboratory balance
Spoon
Vinegar
Corn syrup (Karo)

⚠ SAFETY

- Clean laboratory surfaces before and after
  laboratory procedures.
- Wear disposable gloves when handling chemicals
  and animal blood.
- Wear safety glasses when using chemicals.
- Dispose of laboratory gloves and blood-
  contaminated items as instructed.
- Wash your hands before leaving the laboratory.

PURPOSE OF THE EXERCISE

To demonstrate some of the passive mechanisms by which
substances move through cell membranes.

LEARNING OUTCOMES APR

After completing this exercise, you should be able to

1. Demonstrate the process of diffusion and identify
   examples of diffusion.
2. Explain diffusion by preparing and interpreting a graph.
3. Demonstrate the process of osmosis and identify
   examples of osmosis.
4. Distinguish among hypertonic, hypotonic, and isotonic
   solutions and observe the effects of these solutions on cells.
5. Demonstrate the process of filtration and identify
   examples of filtration.
6. Recognize the anatomical terms pertaining to
   movements through membranes.

A cell membrane functions as a gateway through which chemical substances and small particles may enter or leave a cell. These substances move through the membrane by passive mechanisms that do not require cellular energy, such as diffusion, osmosis, and filtration or by active mechanisms that use cellular energy, such as active transport, phagocytosis, or pinocytosis.

 PRACTICE

PROCEDURE A—Diffusion APR

1. Review the concept of diffusion in section 3.3 titled "Movements Into and Out of the Cell" of chapter 3 in the textbook.
2. To demonstrate *diffusion*, refer to figure 6.1 as you follow these steps:
   a. Place a petri dish, half filled with water, on a piece of white paper that has a millimeter ruler positioned on the paper. Wait until the water surface is still. Allow approximately 3 minutes. *Note*: The petri dish should remain level. A second millimeter ruler may be needed under the petri dish as a shim to obtain a level amount of the water inside the petri dish.
   b. Using forceps, place one crystal of potassium permanganate near the center of the petri dish and near the millimeter ruler (fig. 6.1).
   c. Measure the radius of the purple circle at 1-minute intervals for 10 minutes and record the results in Part A of Laboratory Report 6.
3. Complete Part A of the laboratory report.

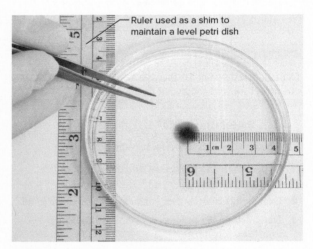

FIGURE 6.1   To demonstrate diffusion, place one crystal of potassium permanganate in the center of a petri dish containing water. Place the crystal near the millimeter ruler (positioned under the petri dish).
(©J and J Photography)

 LEARN: ACTIVITY

Repeat the demonstration of diffusion using a petri dish filled with ice-cold water and a second dish filled with very hot water. At the same moment, add a crystal of potassium permanganate to each dish and observe the circle as before. What difference do you note in the rate of diffusion in the two dishes? How do you explain this difference?

_____

_____

_____

_____

_____

_____

_____

 PRACTICE

PROCEDURE B—Osmosis APR

1. Review the concept of osmosis in section 3.3 titled "Movements Into and Out of the Cell" of chapter 3 in the textbook.
2. To demonstrate *osmosis*, refer to figure 6.2 as you follow these steps:
   a. One person plugs the tube end of a thistle tube with a finger.
   b. Another person then fills the bulb with molasses until it is about to overflow at the top of the bulb. Allow the molasses to enter the first centimeter of the stem, leaving the rest filled with trapped air.
   c. Cover the bulb opening with a single-thickness piece of moist selectively permeable (semipermeable) membrane. Dialysis tubing that has been soaked for 30 minutes can easily be cut open because it becomes pliable.
   d. Tightly secure the membrane in place with several wrappings of a rubber band.
   e. Immerse the bulb end of the tube in a beaker of water. If leaks are noted, repeat the procedures.
   f. Support the upright portion of the tube with a clamp on a support stand. Folded paper between the stem and the clamp will protect the thistle tube stem from breakage.
   g. Mark the meniscus level of the molasses in the tube. *Note*: The best results will occur if the mark of the molasses is a short distance up the stem of the thistle tube when the experiment starts.
   h. Measure the level changes after 10 and 30 minutes and record the results in Part B of the laboratory report.
3. Complete Part B of the laboratory report.

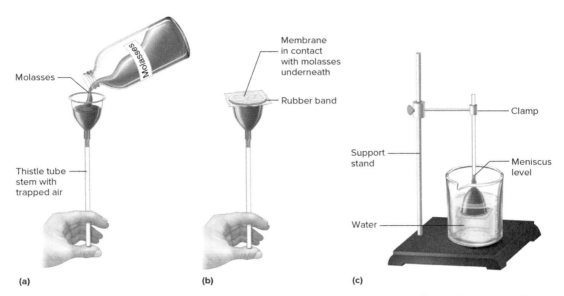

FIGURE 6.2 (*a*) Fill the bulb of the thistle tube with molasses; (*b*) tightly secure a piece of selectively permeable (semipermeable) membrane over the bulb opening; and (*c*) immerse the bulb in a beaker of water. *Note*: These procedures require the participation of two people.

 ALTERNATIVE PROCEDURE

Eggshell membranes possess selectively permeable properties. To demonstrate osmosis using a natural membrane, soak a fresh chicken egg in vinegar for about 24 hours to remove the shell. Use a spoon to carefully handle the delicate egg. Place the egg in a hypertonic solution (corn syrup) for about 24 hours. Remove the egg, rinse it, and using a laboratory balance weigh the egg to establish a baseline weight. Place the egg in a hypotonic solution (distilled water). Remove the egg and weigh it every 15 minutes for an elapsed time of 75 minutes. Explain any weight changes that were noted during this experiment.

_____

_____

_____

_____

_____

 PRACTICE

PROCEDURE C—Hypertonic, Hypotonic, and Isotonic Solutions APR

1. Review the concept of osmosis in section 3.3 titled "Movements Into and Out of the Cell" of chapter 3 in the textbook.

2. To demonstrate the effect of *hypertonic, hypotonic,* and *isotonic* solutions on animal cells, follow these steps:
   a. Place three test tubes in a rack and mark them *tube 1, tube 2,* and *tube 3.* (*Note:* One set of tubes can be used to supply test samples for the entire class.)
   b. Using 10 mL graduated cylinders, add 3 mL of distilled water to tube 1; add 3 mL of 0.9% NaCl to tube 2; and add 3 mL of 3.0% NaCl to tube 3.
   c. Place 3 drops of fresh, uncoagulated animal blood into each of the tubes, and gently mix the blood with the solutions. Wait 5 minutes.
   d. Using three separate medicine droppers, remove a drop from each tube and place the drops on three separate microscope slides marked *1, 2,* and *3.*
   e. Cover the drops with coverslips and observe the blood cells, using the high power of the microscope.
3. Complete Part C of the laboratory report.

 ALTERNATIVE PROCEDURE

Various substitutes for blood can be used for Procedure C. Onion cells, cucumber cells, and cells lining the inside of the cheek represent three possible options.

PRACTICE

PROCEDURE D—Filtration

1. Review the concept of filtration in section 3.3 titled "Movements Into and Out of the Cell" of chapter 3 in the textbook.
2. To demonstrate *filtration*, follow these steps:
   a. Place a glass funnel in the ring of a support stand over an empty beaker. Fold a piece of filter paper in half and then in half again. Open one thickness of the filter paper to form a cone. Wet the cone, and place it in the funnel. The filter paper is used to demonstrate how movement across membranes is limited by the size of the molecules, but it does not represent a working model of biological membranes.
   b. Prepare a mixture of 5 cc (approximately 1 teaspoon) powdered charcoal (or ground black pepper) and equal amounts of 1% glucose solution and 1% starch solution in a beaker. Pour some of the mixture into the funnel until it nearly reaches the top of the filter-paper cone. Care should be taken to prevent the mixture from spilling over the top of the filter paper. Collect the filtrate in the beaker below the funnel (fig. 6.3).
   c. Test some of the filtrate in the beaker for the presence of glucose. To do this, place 1 mL of filtrate in a clean test tube and add 1 mL of Benedict's solution. Place the test tube in a water bath of boiling water for 2 minutes and then allow the liquid to cool slowly. If the color of the solution changes to green, yellow, or red, glucose is present (fig. 6.4).
   d. Test some of the filtrate in the beaker for the presence of starch. To do this, place a few drops of filtrate in a test tube and add 1 drop of iodine-potassium-iodide solution. If the color of the solution changes to blue-black, starch is present.
   e. Observe any charcoal in the filtrate.
3. Complete Parts D and E of the laboratory report.

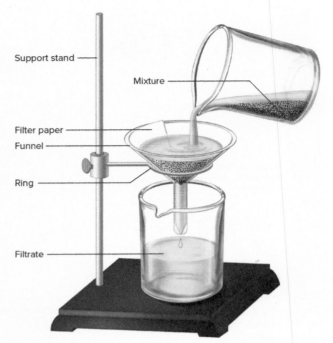

FIGURE 6.3 Apparatus used to illustrate filtration.

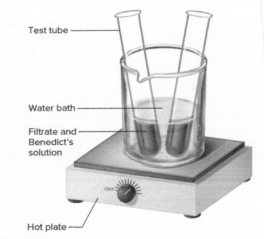

FIGURE 6.4 Heat the filtrate and Benedict's solution in a boiling water bath for 2 minutes.

Name _____

Date _____

Section _____

The corresponds to the indicated Learning Outcome(s) found at the beginning of the laboratory exercise.

Movements Through Membranes

PART A ASSESSMENTS

Complete the following:

1. Enter data for changes in the movement of the potassium permanganate. **1**

Elapsed Time	Radius of Purple Circle in Millimeters
Initial	_____
1 minute	_____
2 minutes	_____
3 minutes	_____
4 minutes	_____
5 minutes	_____
6 minutes	_____
7 minutes	_____
8 minutes	_____
9 minutes	_____
10 minutes	_____

2. Prepare a graph that illustrates the diffusion distance of potassium permanganate in 10 minutes. **2**

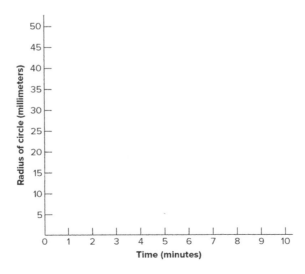

3. Interpret your graph. /2\ _____

_____

4. Define *diffusion*. _____

_____

ASSESS

CRITICAL THINKING

By answering yes or no, indicate which of the following provides an example of diffusion. /1\

1. A perfume bottle is opened, and soon the odor can be sensed in all parts of the room. _____

2. A tea bag is dropped into a cup of hot water, and, without being stirred, all of the liquid becomes the color of the tea leaves. _____

3. A person blows air molecules into a balloon by forcefully exhaling. _____

PART B ASSESSMENTS

Complete the following:

1. What was the change in the level of molasses in 10 minutes? /3\ _____

2. What was the change in the level of molasses in 30 minutes? /3\ _____

3. How do you explain this change? /3\ _____

_____

4. Define *osmosis*. _____

_____

ASSESS

CRITICAL THINKING

By answering yes or no, indicate which of the following involves osmosis. /3\

1. A fresh potato is peeled, weighed, and soaked in a strong salt solution. The next day, it is discovered that the potato has lost weight. _____

2. Air molecules escape from a punctured tire as a result of high pressure inside. _____

3. Plant seeds soaked in water swell and become several times as large as before soaking. _____

PART C ASSESSMENTS

Complete the following:

1. In the spaces, sketch a few blood cells from each of the test tubes, and indicate the magnification. Then, draw an arrow(s) indicating the movement of water (into the cell, out of the cell, or both equally). Ⓐ

Tube 1　　　　　　　　　　　Tube 2　　　　　　　　　　　Tube 3
(distilled water)　　　　　　　　(0.9% NaCl)　　　　　　　　(3.0% NaCl)

_____× 　　　　　　_____× 　　　　　　_____×

2. Based on your results, which tube contained a solution hypertonic to the blood cells? _____

   Give the reason for your answer. Ⓐ _____

   _____

3. Which tube contained a solution hypotonic to the blood cells? _____

   Give the reason for your answer. Ⓐ _____

   _____

4. Which tube contained a solution isotonic to the blood cells? _____

   Give the reason for your answer. Ⓐ _____

   _____

 PART D ASSESSMENTS

Complete the following:

1. Which of the substances in the mixture you prepared passed through the filter paper into the filtrate? Ⓢ _____

   _____

2. What evidence do you have for your answer to question 1? Ⓢ _____

   _____

3. What force was responsible for the movement of substances through the filter paper? Ⓢ _____

   _____

4. What substances did not pass through the filter paper? Ⓢ _____

5. What factor prevented these substances from passing through? 🔺 _____
_____

6. Define *filtration*. _____
_____

⟳ ASSESS

CRITICAL THINKING

By answering yes or no, indicate which of the following involves filtration: 🔺

1. Oxygen molecules move into a cell and carbon dioxide molecules leave a cell because of differences in the concentrations of these substances on either side of the cell membrane. _____

2. Blood pressure forces water molecules and small, dissolved solutes from the blood outward through the thin wall of a blood capillary. _____

3. Noninstant coffee is made using a coffeemaker. _____

7

Cell Cycle

MATERIALS NEEDED

Textbook
Models of animal cells during mitosis
Microscope slides of whitefish mitosis (blastula)
Compound light microscope
Microscope slide of human chromosomes from
    leukocytes in mitosis
Oil immersion objective

PURPOSE OF THE EXERCISE

To review the phases in the cell cycle and to observe cells in various phases of their life cycles.

LEARNING OUTCOMES APR

After completing this exercise, you should be able to

1. Describe the cell cycle and locate structures involved with the process.

2. Identify and sketch the phases (stages) in the life cycle of a particular cell.

3. Arrange into a correct sequence a set of models or drawings of cells in various phases of their life cycles.

4. Recognize the anatomical terms pertaining to the cell cycle.

The cell cycle consists of the series of changes a cell undergoes from the time it is formed until it divides. Typically, a newly formed diploid cell with 46 chromosomes grows to a certain size and then divides to form two new cells *(daughter cells)*, each with 46 chromosomes. This cell division process involves two major steps: (1) division of the cell's nuclear parts, *mitosis*, and (2) division of the cell's cytoplasm, *cytokinesis*. Before the cell divides, it must synthesize biochemicals and other contents. This period of preparation is called *interphase*. The extensive period of interphase is divided into three phases. The S phase, when

DNA synthesis occurs, is between two gap phases (G1 and G2), when cell growth occurs and cytoplasmic organelles duplicate. Eventually, some specialized cells such as skeletal muscle cells and most nerve cells cease further cell division but remain alive.

A special type of cell division, called *meiosis,* occurs in the reproductive system to produce haploid gametes with 23 chromosomes. Meiosis is not included in this laboratory exercise.

PRACTICE

PROCEDURE—Cell Cycle

1. Review section 3.4 titled "The Cell Cycle" in chapter 3 of the textbook.

2. As a review activity, study the various phases of the cell's life cycle represented in figures 7.1–7.3.

3. Label the structures indicated in figure 7.4.

4. Using figure 7.1 as a guide, observe the animal mitosis models, and review the major events in a cell's life cycle represented by each of them. Be sure you can arrange these models in correct sequence if their positions are changed. The acronym IPMAT can help you arrange the correct order of phases in the cell cycle. This includes interphase followed by the four phases of mitosis. Cytokinesis overlaps anaphase and telophase.

5. Complete Part A of Laboratory Report 7.

6. Obtain a slide of the whitefish mitosis (blastula).

   a. Examine the slide using the high-power objective of a microscope. The tissue on this slide was obtained from a developing embryo (blastula) of a fish, and many of the embryonic cells are undergoing mitosis. The chromosomes of these dividing cells are darkly stained (see fig. 7.3).

   b. Search the tissue for cells in various phases of cell division. There are several sections on the slide. If you cannot locate different phases in one section, examine the cells of another section because the phases occur randomly on the slide.

   c. Each time you locate a cell in a different phase, sketch it in an appropriate circle in Part B of the laboratory report.

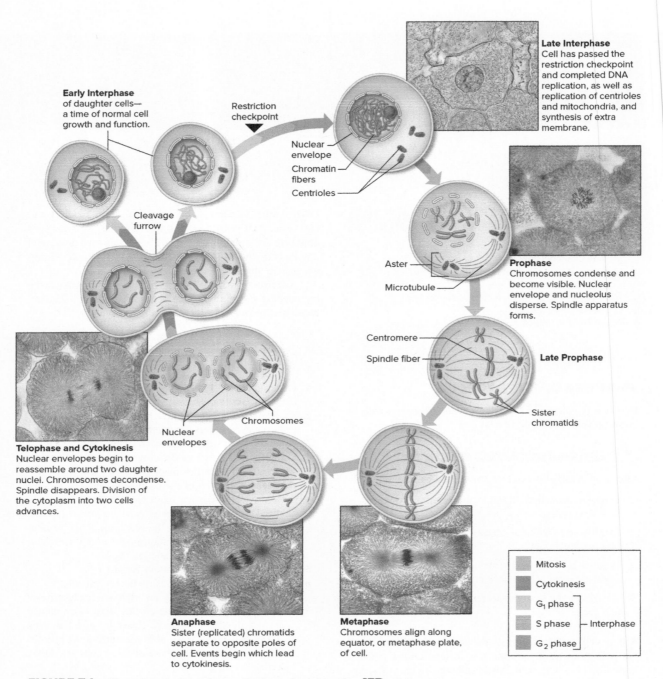

Late Interphase
Cell has passed the restriction checkpoint and completed DNA replication, as well as replication of centrioles and mitochondria, and synthesis of extra membrane.

Early Interphase
of daughter cells—
a time of normal cell growth and function.

Restriction checkpoint

Nuclear envelope
Chromatin fibers
Centrioles

Cleavage furrow

Aster
Microtubule

Prophase
Chromosomes condense and become visible. Nuclear envelope and nucleolus disperse. Spindle apparatus forms.

Centromere
Spindle fiber

Late Prophase

Sister chromatids

Telophase and Cytokinesis
Nuclear envelopes begin to reassemble around two daughter nuclei. Chromosomes decondense. Spindle disappears. Division of the cytoplasm into two cells advances.

Chromosomes
Nuclear envelopes

Anaphase
Sister (replicated) chromatids separate to opposite poles of cell. Events begin which lead to cytokinesis.

Metaphase
Chromosomes align along equator, or metaphase plate, of cell.

Mitosis
Cytokinesis
G_1 phase ⎤
S phase ⎬ Interphase
G_2 phase ⎦

FIGURE 7.1 The cell cycle: interphase, mitosis, and cytokinesis. **APR** (©Ed Reschke)

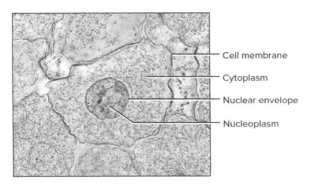

FIGURE 7.2   Cell in interphase (400×). The nucleoplasm contains a fine network of chromatin. (©Ed Reschke)

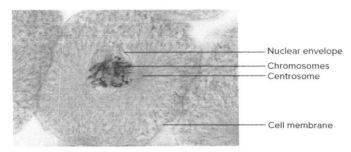

FIGURE 7.3   Cell in prophase (400×). (©Ed Reschke)

ASSESS

CRITICAL THINKING

Which phase (stage) of the cell cycle was the most numerous in the blastula? _____

Explain your answer. _____
_____
_____

7. Complete Parts C and D of the laboratory report.

LEARN: ACTIVITY

Using the oil immersion objective of a microscope, see if you can locate some human chromosomes by examining a prepared slide of human chromosomes from leukocytes. The cells on this slide were cultured in a special medium and were stimulated to undergo mitosis. The mitotic process was arrested in metaphase by exposing the cells to a chemical called colchicine, and the cells were caused to swell osmotically. As a result of this treatment, the chromosomes were spread apart. A complement of human chromosomes should be visible when they are magnified about 1,000×. Each chromosome is double-stranded and consists of two chromatids joined by a common centromere (fig. 7.5).

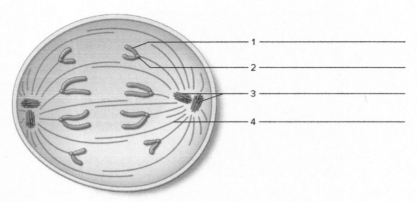

FIGURE 7.4   Label the structures indicated in the dividing cell during anaphase. ⚛

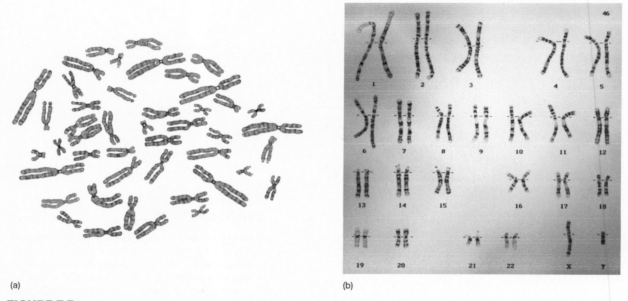

(a)

(b)

FIGURE 7.5   (*a*) A complement of human chromosomes of a female (1,000×). (*b*) A *karyotype* can be constructed by arranging the homologous chromosome pairs together in a chart. A karyotype can aid in the diagnosis of genetic conditions and abnormalities. The completed karyotype indicates a normal male.   **APR**   (James Cavallini/Science Source; Randy Allbritton/Getty Images)

Name _____

Date _____

Section _____

The 🅐 corresponds to the indicated Learning Outcome(s) found at the beginning of the laboratory exercise.

Cell Cycle

♻ PART A ASSESSMENTS

Complete the table by listing the major events of each phase in the second column. 🅐

Phase	Major Events Occurring
Interphase (G_1, S, and G_2)	
Mitosis Prophase	
Metaphase	
Anaphase	
Telophase	
Cytokinesis	

PART B ASSESSMENTS

Sketch an interphase cell and cells in different phases (stages) of mitosis to illustrate the whitefish cell's life cycle. Label the major cellular structures represented in the sketches and indicate cytokinesis locations. (The circles represent fields of view through the microscope.)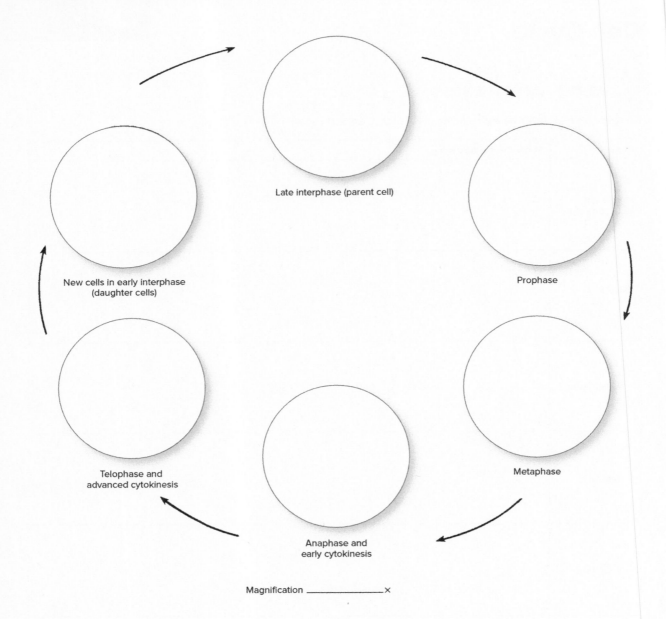

Late interphase (parent cell)

Prophase

New cells in early interphase
(daughter cells)

Telophase and
advanced cytokinesis

Anaphase and
early cytokinesis

Metaphase

Magnification _____×

PART C ASSESSMENTS

1. Identify the mitotic phase represented by each of the micrographs in figure 7.6a–d. **3**

   a. _____   c. _____

   b. _____   d. _____

2. Identify the structures indicated by numbers in figure 7.6 by placing the correct numbers in the spaces provided. **1**

   _____ Cell membrane   _____ Metaphase plate

   _____ Centrosome   _____ Nuclear envelope

   _____ Cleavage furrow   _____ Spindle fiber

(a)

(b)

(c)

(d)

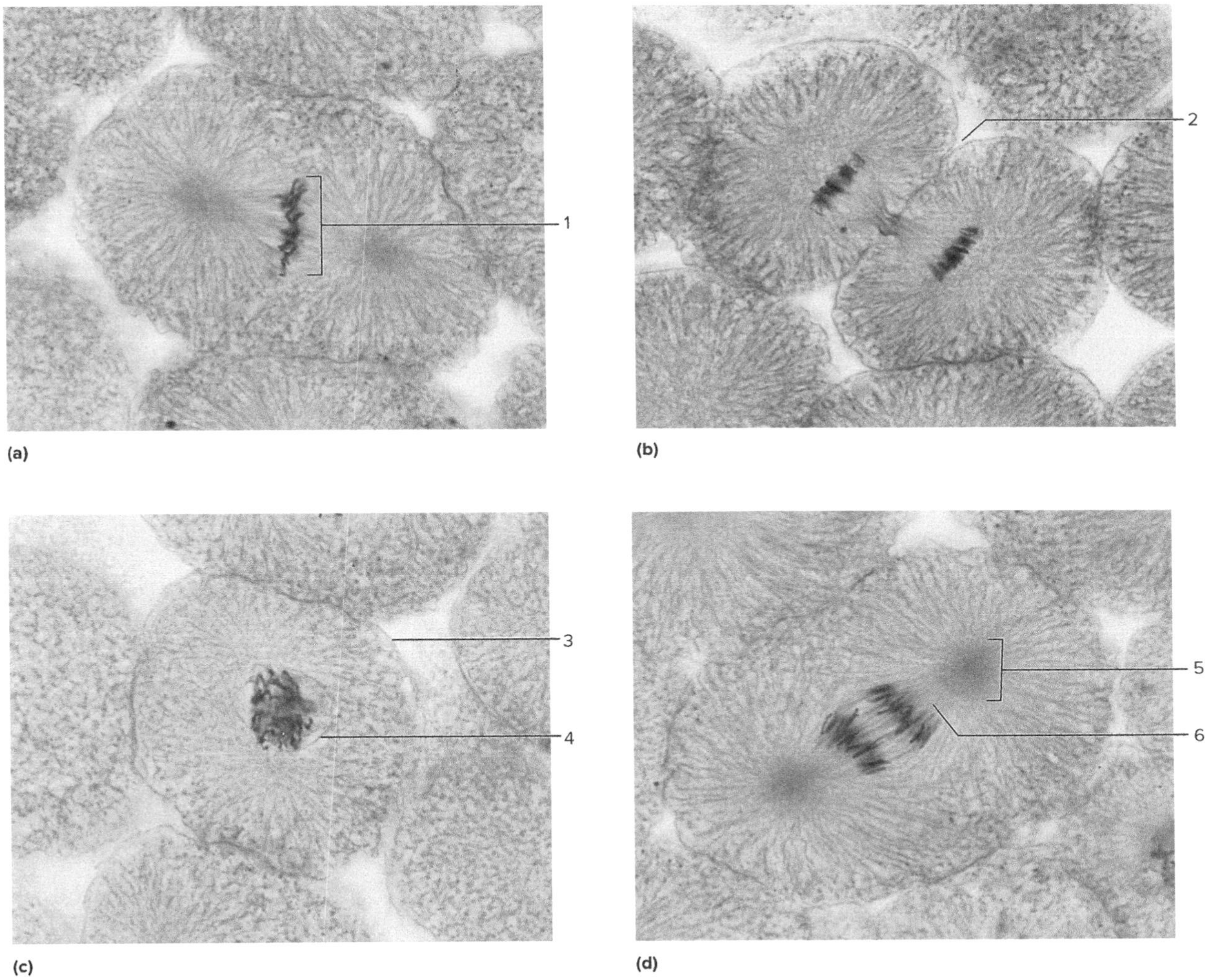

FIGURE 7.6   Identify the mitotic phase and structures of the cell in each of these micrographs of the whitefish blastula (250×). ((a,b,c,d) ©Ed Reschke)

Tissues

LABORATORY EXERCISE

8

Epithelial Tissues

MATERIALS NEEDED

Textbook
Compound light microscope
Prepared slides of the following epithelial tissues:
  Simple squamous epithelium (lung)
  Simple cuboidal epithelium (kidney)
  Simple columnar epithelium (small intestine)
  Pseudostratified (ciliated) columnar epithelium
    (trachea)
  Stratified squamous epithelium (esophagus)
  Transitional epithelium (urinary bladder)
Colored pencils

PURPOSE OF THE EXERCISE

To review the characteristics of epithelial tissues and to observe examples.

 LEARNING OUTCOMES APR

After completing this exercise, you should be able to

(1) Differentiate the special characteristics of each type of epithelial tissue.

(2) Sketch and label the characteristics of epithelial tissues that you were able to observe.

(3) Indicate a location and function of each type of epithelial tissue.

(4) Identify the major types of epithelial tissues on microscope slides.

A tissue is composed of a layer or group of cells similar in size, shape, and function. A study of tissues is called *histology*. Within the human body, there are four major types of tissues: (1) *epithelial*, which cover the body's external and internal surfaces and compose most glands; (2) *connective*, which bind and support parts; (3) *muscle*, which make

movement possible; and (4) *nervous*, which conduct impulses from one part of the body to another and help to control and coordinate body activities.

Epithelial tissues are tightly packed single (simple) to multiple (stratified) layers of cells that provide protective barriers. The underside of this tissue contains an acellular basement membrane layer composed of adhesive cellular secretions and collagen through which the epithelial cells anchor to underlying connective tissue. The cells readily divide and lack blood vessels. Epithelial cells always have a free (apical) surface exposed to the outside or to an open space internally and a basal surface that attaches to the basement membrane. The functions of epithelial cells include protection, filtration, secretion, and absorption. Many epithelial cell shapes are used to name and identify the variations. Many of the prepared slides contain more than the tissue to be studied, therefore be certain that your view matches the correct tissue. Also be aware that stained colors of all tissues might vary.

 PRACTICE

PROCEDURE—Epithelial Tissues APR

1. Review section 5.2 titled "Epithelial Tissues" in chapter 5 of the textbook.
2. Complete Part A of Laboratory Report 8.
3. Use the microscope to observe the prepared slides of types of epithelial tissues. As you observe each tissue, look for its special distinguishing features as described in the textbook, such as cell size, shape, and arrangement. Compare your prepared slides of epithelial tissues to the micrographs in figure 8.1. As you observe each type of epithelial tissue, prepare a labeled sketch of a representative portion of the tissue in Part B of the laboratory report.
4. Complete Part B of the laboratory report.
5. Test your ability to recognize each type of epithelial tissue. To do this, have a laboratory partner select one of the prepared slides, cover its label, and focus the microscope on the tissue. Then see if you can correctly identify the tissue. 4

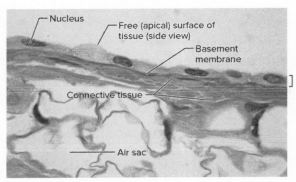

(a) Simple squamous epithelium (side view) (from lung) (250×)

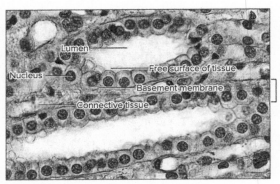

(b) Simple cuboidal epithelium (from kidney) (165×)

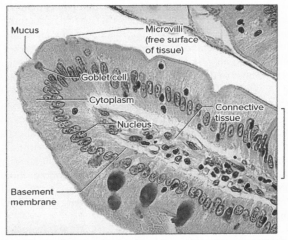

(c) Simple columnar epithelium (nonciliated) (from intestine) (400×)

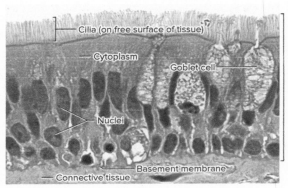

(d) Pseudostratified columnar epithelium with cilia (from trachea) (1,000×)

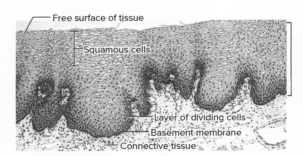

(e) Stratified squamous epithelium (nonkeratinized) (from esophagus) (100×)

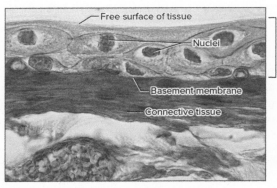

(f) Transitional epithelium (stretched) (from urinary bladder) (675×)

FIGURE 8.1 Micrographs of epithelial tissues *(a–f)*. *Note*: A bracket to the right of a micrograph indicates the tissue layer.
APR *((a,d,f)* ©Ed Reschke; *(b,c)* Victor P.Eroschenko; *(e)* Al Telser/McGraw-Hill Education)

Name _____

Date _____

Section _____

The Ⓐ corresponds to the indicated Learning Outcome(s) found at the beginning of the laboratory exercise.

Epithelial Tissues

♻ PART A ASSESSMENTS

Match the tissues in column A with the characteristics in column B. Place the letter of your choice in the space provided.
(Some answers may be used more than once.) Ⓐ1 Ⓐ3

Column A	Column B
a. Simple columnar epithelium	_____ **1.** Consists of several layers of cells, allowing an expandable lining
	_____ **2.** Commonly possesses cilia that move dust and mucus out of the respiratory airways
b. Simple cuboidal epithelium	_____ **3.** Single layer of flattened cells
	_____ **4.** Nuclei located at different levels within a single row of aligned cells
c. Simple squamous epithelium	_____ **5.** Forms walls of capillaries and air sacs of lungs
	_____ **6.** Appears layered (stratified) but is a single layer of cells (simple)
d. Pseudostratified columnar epithelium	_____ **7.** Deeper cells cuboidal, or columnar; older cells flattened nearest the free surface
	_____ **8.** Forms inner lining of urinary bladder
	_____ **9.** Lines kidney tubules and ducts of salivary glands
e. Stratified squamous epithelium	_____ **10.** Forms lining of stomach and intestines
	_____ **11.** Elongated cells with elongated nuclei located near basement membrane
f. Transitional epithelium	_____ **12.** Forms lining of oral cavity, esophagus, anal canal, and vagina

PART B ASSESSMENTS

In the space that follows, sketch a few cells of each type of epithelium you observed. For each sketch, label the major characteristics, indicate the magnification used, write an example of a location in the body, and provide a function.
1 2 3

Simple squamous epithelium (_____×) Location: _____ Function: _____	Simple cuboidal epithelium (_____ ×) Location: _____ Function: _____
Simple columnar epithelium (_____×) Location: _____ Function: _____	Pseudostratified columnar epithelium with cilia (_____×) Location: _____ Function: _____
Stratified squamous epithelium (_____ ×) Location: _____ Function: _____	Transitional epithelium (_____×) Location: _____ Function: _____

ASSESS

CRITICAL THINKING

As a result of your observations of epithelial tissues, which one(s) provide(s) the best protection? Explain your answer. 3

_____

_____

_____

LEARN: ACTIVITY

Use colored pencils to differentiate various cellular structures in Part B. Select a different color for a nucleus, cytoplasm, cell membrane, basement membrane, goblet cell, and cilia whenever visible.

Connective Tissues

MATERIALS NEEDED

Textbook
Compound light microscope
Prepared slides of the following:
  Areolar tissue
  Adipose tissue
  Reticular connective tissue
  Dense connective tissue (regular type)
  Elastic connective tissue
  Hyaline cartilage
  Elastic cartilage
  Fibrocartilage
  Bone (compact, ground, cross section)
  Blood (human smear)
Colored pencils

PURPOSE OF THE EXERCISE

To review the characteristics of connective tissues and to observe examples of the major types.

LEARNING OUTCOMES APR

After completing this exercise, you should be able to

1. Differentiate the special characteristics of each of the major types of connective tissue.

2. Sketch and label the characteristics of connective tissues that you were able to observe.

3. Indicate a location and function of each type of connective tissue.

4. Identify the major types of connective tissues on microscope slides.

Connective tissues contain a variety of cell types and occur in all regions of the body. They bind structures together, provide support and protection, fill spaces, store fat, and produce blood cells.

Connective tissue cells are often widely scattered in an abundance of extracellular matrix. The matrix consists of fibers and a ground substance of various densities and consistencies. Many of the prepared slides contain more than the tissue to be studied, so be certain that your view matches the correct tissue. Additional study of bone and blood will be found in Laboratory Exercises 12 and 37.

PRACTICE

PROCEDURE—Connective Tissues APR

1. Review section 5.3 titled "Connective Tissues" in chapter 5 of the textbook.
2. Complete Part A of Laboratory Report 9.
3. Use a microscope to observe the prepared slides of various connective tissues. As you observe each tissue, look for its special distinguishing features as described in the textbook. Compare your prepared slides of connective tissues to the micrographs in figure 9.1. As you observe each type of connective tissue, prepare a labeled sketch of a representative portion of the tissue in Part B of the laboratory report.
4. Complete Part B of the laboratory report.
5. Test your ability to recognize each of these connective tissues by having a laboratory partner select a slide, cover its label, and focus the microscope on this tissue. Then see if you can correctly identify the tissue. A

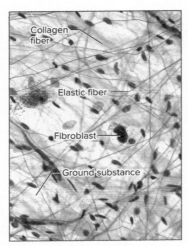

(a) Areolar tissue (from beneath the skin) (800×)

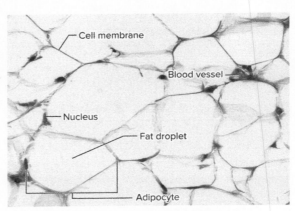

(b) Adipose tissue (from subcutaneous layer) (400×)

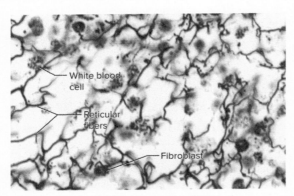

(c) Reticular connective tissue (from spleen) (1,000×)

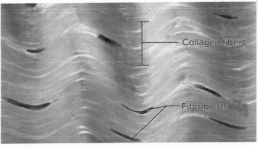

(d) Dense regular connective tissue (from tendon) (500×)

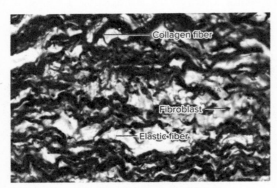

(e) Elastic connective tissue (from artery wall) (160×)

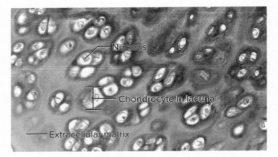

(f) Hyaline cartilage (from costal cartilage of ribs) (160×)

FIGURE 9.1   Micrographs of connective tissues *(a–j)*.   **APR**   ((*a, j*) Al Telser/McGraw-Hill Education; (*b*) Alvin Telser/McGraw-Hill Education; (*c, e, f, g, i*) ©Ed Reschke; (*d*) Ed Reschke/Photolibrary/Getty Images; (*h*) Victor P. Eroschenko)

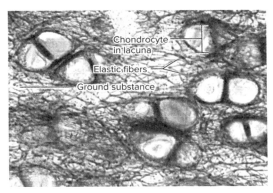

(g) Elastic cartilage (from ear) (400×)

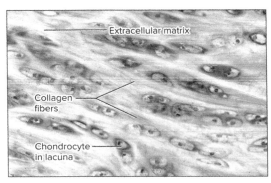

(h) Fibrocartilage (from intervertebral discs) (400×)

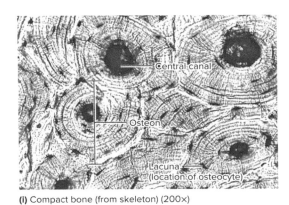

(i) Compact bone (from skeleton) (200×)

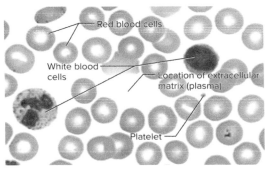

(j) Blood (400×)

FIGURE 9.1   *Continued.* APR

Notes

Name _____

Date _____

Section _____

LABORATORY
REPORT

9

The ⬚ corresponds to the indicated Learning Outcome(s) found at the beginning of the laboratory exercise.

Connective Tissues

PART A ASSESSMENTS

Match the tissues in column A with the characteristics in column B. Place the letter of your choice in the space provided. (Some answers may be used more than once.) ⬚ ⬚

Column A	Column B
a. Adipose tissue	_____ **1.** Forms framework of outer ear
b. Areolar tissue	_____ **2.** Functions as heat insulator beneath skin
c. Blood	_____ **3.** Contains large amounts of fluid and transports nutrients, wastes, and gases
d. Bone (compact)	_____ **4.** Cells in solid matrix arranged around central canal
e. Dense connective tissue (regular)	_____ **5.** Binds skin and fills spaces between organs
f. Elastic cartilage	_____ **6.** Main tissue of tendons and ligaments
g. Elastic connective tissue	_____ **7.** Provides stored energy supply in fat droplets in cytoplasm
h. Fibrocartilage	_____ **8.** Forms the ends of many long bones
i. Hyaline cartilage	_____ **9.** Pads between vertebrae that are shock absorbers
j. Reticular connective tissue	_____ **10.** Matrix contains collagen fibers and mineral salts
	_____ **11.** Occurs in some ligament attachments between vertebrae and larger artery walls
	_____ **12.** Forms supporting tissue in walls of liver and spleen

PART B ASSESSMENTS

In the space that follows, sketch a small section of each of the types of connective tissues you observed. For each sketch, label the major characteristics, indicate the magnification used, write an example of a location in the body, and provide a function.
Ⓐ Ⓐ Ⓐ

Areolar tissue (_____×)
Location: _____
Function: _____

Adipose tissue (_____×)
Location: _____
Function: _____

Reticular connective tissue (_____×)
Location: _____
Function: _____

Dense connective tissue (regular type) (_____×)
Location: _____
Function: _____

Elastic connective tissue (_____×)
Location: _____
Function: _____

Hyaline cartilage (_____×)
Location: _____
Function: _____

Elastic cartilage (____×) Location: _____ Function: _____	Fibrocartilage (____×) Location: _____ Function: _____
Bone (compact) (____×) Location: _____ Function: _____	Blood (____×) Location: _____ Function: _____

ASSESS

CRITICAL THINKING

Abdominal impact injuries in the region of the LUQ often involve the spleen. Explain the structural tissue characteristics within the spleen that make it so vulnerable to serious injury.

_____
_____
_____
_____

LEARN: ACTIVITY

Use colored pencils to differentiate various cellular structures in Part B. Select a different color for the cells, fibers, and ground substance whenever visible.

LABORATORY EXERCISE

10

Muscle and Nervous Tissues

MATERIALS NEEDED

Textbook
Compound light microscope
Prepared slides of the following:
    Skeletal muscle tissue
    Smooth muscle tissue
    Cardiac muscle tissue
    Nervous tissue (spinal cord smear and/or
        cerebellum)
Colored pencils

PURPOSE OF THE EXERCISE

To review the characteristics of muscle and nervous tissues
and to observe examples of these tissues.

 LEARNING OUTCOMES APR

After completing this exercise, you should be able to

1. Differentiate the special characteristics of each type
   of muscle tissue and nervous tissue.

2. Sketch and label the characteristics of the different
   muscle and nervous tissues that you were able to
   observe.

3. Indicate an example location and function of each
   type of muscle tissue and nervous tissue.

4. Identify three types of muscle tissues and nervous
   tissues on microscope slides.

Muscle tissues are characterized by the presence of
elongated cells or muscle fibers that can contract in
response to specific stimuli. As they shorten, these fibers pull
at their attached ends and cause body parts to move. The three
types of muscle tissues are *skeletal*, *smooth*, and *cardiac*.

Nervous tissues occur in the brain, spinal cord, and
peripheral nerves. They consist of *neurons* (nerve cells), the
impulse-conducting cells of the nervous system, and *neuroglia*, which perform supportive and protective functions
for neurons.

 PRACTICE

PROCEDURE—Muscle and Nervous Tissues APR

1. Review section 5.5 titled "Muscle Tissues" and section 5.6
   titled "Nervous Tissues" in chapter 5 of the textbook.
2. Complete Part A of Laboratory Report 10.
3. Using microscope, observe each of the types of muscle
   tissues on the prepared slides. Look for the special features of each type, as described in the textbook. Compare
   your prepared slides of muscle tissues to the micrographs
   in figure 10.1 and the muscle tissue characteristics in
   table 10.1. As you observe each type of muscle tissue, prepare a labeled sketch of a representative portion of the
   tissue in Part B of the laboratory report.
4. Observe the prepared slide of nervous tissue and identify neurons (nerve cells), neuron cellular processes,
   and neuroglia. Compare your prepared slide of nervous tissue to the micrograph in figure 10.1. Prepare a
   labeled sketch of a representative portion of the tissue in
   Part B of the laboratory report. APR

TABLE 10.1 Muscle Tissue Characteristics

Characteristic	Skeletal Muscle	Smooth Muscle	Cardiac Muscle
Appearance of cells	Unbranched and relatively parallel	Spindle-shaped	Branched and connected in complex networks
Striations	Present and obvious	Absent	Present but faint
Nucleus	Multinucleated	Uninucleated	Uninucleated (usually)
Intercalated discs	Absent	Absent	Present
Control	Voluntary (usually)	Involuntary	Involuntary

5. Complete Part B of the laboratory report.
6. Test your ability to recognize each of these muscle and nervous tissues by having your laboratory partner select a slide, cover its label, and focus the microscope on this tissue. Then see if you correctly identify the tissue. 4

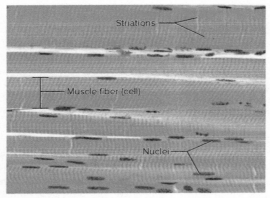

(a) Skeletal muscle (400×)

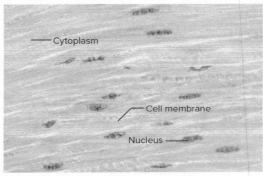

(b) Smooth muscle (from small intestine) (400×)

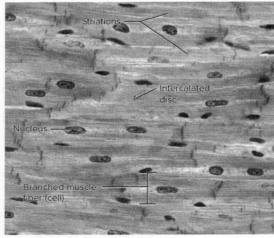

(c) Cardiac muscle (from heart) (400×)

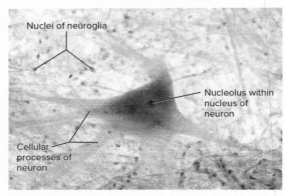

(d) Nervous tissue (350×)

FIGURE 10.1   Micrographs of muscle and nervous tissues *(a–d)*. **APR** *((a–d)* Al Telser/McGraw-Hill Education)

LABORATORY
REPORT

Name _____

Date _____

Section _____

The ⃞A corresponds to the indicated Learning Outcome(s) found at the beginning of the laboratory exercise.

10

Muscle and Nervous Tissues

PART A ASSESSMENTS

Match the tissues in column A with the characteristics in column B. Place the letter of your choice in the space provided. (Some answers may be used more than once.) ⃞1 ⃞3

Column A	Column B
a. Cardiac muscle	_____ **1.** Coordinates, regulates, and integrates body functions
	_____ **2.** Contains intercalated discs
	_____ **3.** Muscle that lacks striations
	_____ **4.** Striated and involuntary
b. Nervous tissue	_____ **5.** Striated and voluntary
	_____ **6.** Contains neurons and neuroglia
	_____ **7.** Muscle attached to bones
c. Skeletal muscle	_____ **8.** Muscle that composes heart
	_____ **9.** Moves food through the digestive tract
	_____ **10.** Conducts impulses along cellular processes
	_____ **11.** Muscle under conscious control
d. Smooth muscle	_____ **12.** Muscle of blood vessels and urinary bladder

PART B ASSESSMENTS

In the space that follows, sketch a few cells or fibers of each of the three types of muscle tissues and of nervous tissue as they appear through the microscope. For each sketch, label the major structures of the cells or fibers, indicate the magnification used, write an example of a location in the body, and provide a function. 🔺 🔺 🔺

Skeletal muscle tissue (_____x)	Smooth muscle tissue (_____x)
Location: _____	Location: _____
Function: _____	Function: _____
Cardiac muscle tissue (_____x)	Nervous tissue (_____x)
Location: _____	Location: _____
Function: _____	Function: _____

🌀 **LEARN: ACTIVITY**

Use colored pencils to differentiate various cellular structures in Part B.

Integumentary System

LABORATORY EXERCISE

11

Integumentary System

MATERIALS NEEDED

Textbook
Skin model
Hand magnifier or dissecting microscope
Forceps
Microscope slide and coverslip
Compound light microscope
Prepared microscope slide of human scalp or axilla
Prepared slide of dark (heavily pigmented) human skin
Prepared slide of thick skin (plantar or palmar)
Tattoo slide

PURPOSE OF THE EXERCISE

To observe the structures and tissues of the integumentary system and to review the functions of these parts.

 LEARNING OUTCOMES APR

After completing this exercise, you should be able to

① Locate and name the structures of the integumentary system.

② Describe the major functions of these structures.

③ Distinguish the locations and tissues among the epidermis, dermis, and subcutaneous layers.

④ Identify and sketch the layers of the skin and associated structures observed on the prepared slide.

The integumentary system includes the skin, hair, nails, sebaceous glands, and sweat glands. These structures provide a protective covering for deeper tissues, aid in regulating body temperature, retard water loss, house sensory receptors, synthesize various chemicals, and excrete small quantities of wastes.

The skin consists of two distinct layers. The outer layer, the *epidermis,* consists of keratinized stratified squamous epithelium. The inner layer, the *dermis,* consists of a thicker layer of mainly dense connective tissue. Beneath the dermis is the *subcutaneous layer* (not considered a true layer of the skin), composed of adipose and areolar connective tissues.

 PRACTICE

PROCEDURE—Integumentary System

1. Review section 6.1 entitled "Layers of the Skin" and section 6.2 entitled "Accessory Structures of the Skin: Epidermal Derivatives" in chapter 6 of the textbook.

2. As a review activity, label figures 11.1 and 11.2, and study figure 11.3. Locate as many of these structures as possible on a skin model.

3. Complete Part A of Laboratory Report 11.

4. Use the hand magnifier or dissecting microscope and proceed as follows:
   a. Observe the skin, hair, and nails on your hand.
   b. Compare the type and distribution of hairs on the front and back of your forearm.

5. Use low-power magnification of the compound light microscope and proceed as follows:
   a. Pull out a single hair with forceps and mount it on a microscope slide under a coverslip.
   b. Observe the root and shaft of the hair and note the scalelike parts that make up the shaft.

6. Complete Part B of the laboratory report.

7. As vertical sections of human skin are observed, remember that the lenses of the microscope invert and reverse images. It is important to orient the position of the epidermis, dermis, and subcutaneous (hypodermis) layers using scan magnification before continuing with additional observations. Compare all of your skin observations to figure 11.3. Use low-power magnification of the compound light microscope and proceed as follows:
   a. Observe the prepared slide of human scalp or axilla.
   b. Locate the epidermis, dermis, and subcutaneous layer; a hair follicle; an arrector pili muscle; a sebaceous gland; and a sweat gland.
   c. Focus on the epidermis with high power and locate the stratum corneum, stratum granulosum, stratum spinosum, and stratum basale. Note how the shapes of the cells in these layers differ.
   d. Observe the dense connective tissue (irregular type) that makes up the bulk of the dermis along with some areolar connective tissue.
   e. Observe the adipose tissue that composes most of the subcutaneous layer along with some areolar connective tissue.

8. Observe the prepared slide of dark (heavily pigmented) human skin with low-power magnification. The pigment is most abundant in the epidermis. Focus on this region with the high-power objective. The pigment-producing cells, or melanocytes, are located among the deeper layers of epidermal cells. Differences in skin color are primarily due to the amount of pigment (melanin) produced by these cells.

 ## ASSESS

CRITICAL THINKING

Explain the advantage for melanin granules being located in the deep layer of the epidermis.

_____

_____

_____

_____

_____

9. Observe the prepared slide of thick skin from the palm of a hand or the sole of a foot (fig. 11.2). Locate the stratum lucidum. Note how the stratum corneum compares to your observation of the human scalp.
10. Complete Part C of the laboratory report.
11. Using low-power magnification, locate a hair follicle sectioned longitudinally through its bulblike base. Also locate a sebaceous gland close to the follicle and find a sweat gland (fig. 11.3). Observe the detailed structure of these parts with high-power magnification.
12. Complete Parts D and E of the laboratory report.

 ## LEARN: ACTIVITY

Observe a vertical section of human skin through a tattoo, using low-power magnification. Note the location of the dispersed ink granules within the upper portion of the dermis. From a thin vertical section of a tattoo, it is not possible to determine the figure or word of the entire tattoo as seen on the surface of the skin because only a small, thin segment of the image is in view at any one time. Compare this to the location of melanin granules found in dark (heavily pigmented) skin. Suggest reasons why a tattoo is permanent but a suntan is not.

_____

_____

_____

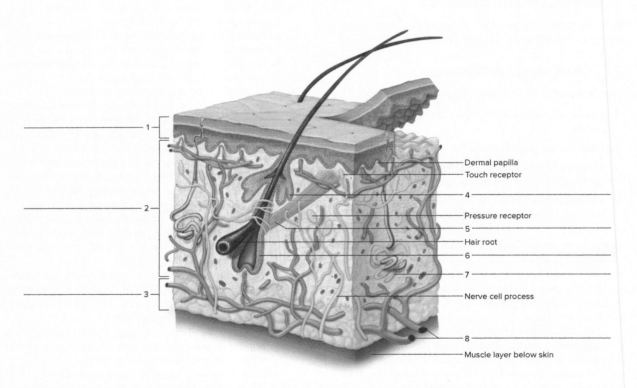

FIGURE 11.1   Label this vertical section of the skin and subcutaneous layer. APR

FIGURE 11.2 Label the features associated with this hair follicle. ⒜ APR

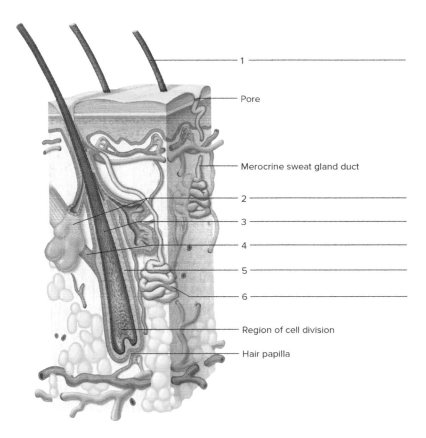

Pore

Merocrine sweat gland duct

1

2

3

4

5

6

Region of cell division

Hair papilla

FIGURE 11.3 Features of human skin are indicated in these micrographs (*a–e*). APR ((*a*) ©Victor B. Eichler, Ph.D.; (*b*) Al Telser/McGraw-Hill Education; (*c*) Image Source/Getty Images; (*d*) Al Telser/McGraw-Hill Education (*e*) doc-stock/Alamy Stock Photo)

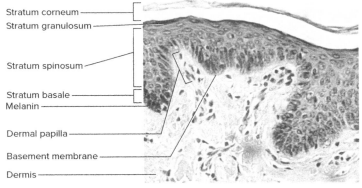

Stratum corneum
Stratum granulosum
Stratum spinosum
Stratum basale
Melanin
Dermal papilla
Basement membrane
Dermis

(a) Epidermis with melanin (400×)

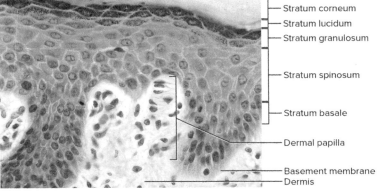

Stratum corneum
Stratum lucidum
Stratum granulosum
Stratum spinosum
Stratum basale
Dermal papilla
Basement membrane
Dermis

(b) Strata of thick skin (500×)

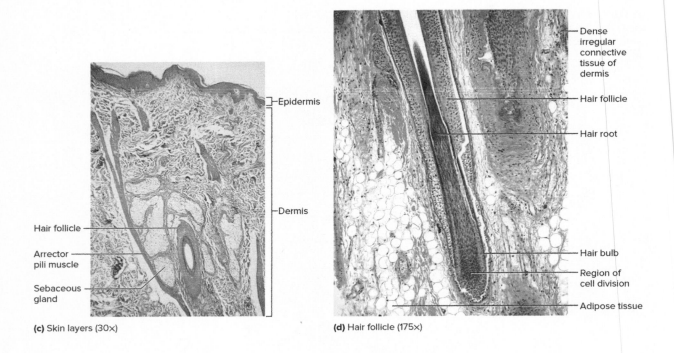

(c) Skin layers (30×)

Epidermis

Dermis

Hair follicle

Arrector pili muscle

Sebaceous gland

Dense irregular connective tissue of dermis

Hair follicle

Hair root

Hair bulb

Region of cell division

Adipose tissue

(d) Hair follicle (175×)

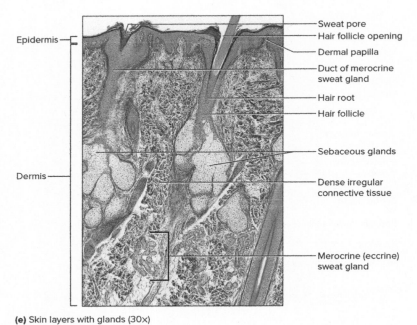

Epidermis

Dermis

Sweat pore

Hair follicle opening

Dermal papilla

Duct of merocrine sweat gland

Hair root

Hair follicle

Sebaceous glands

Dense irregular connective tissue

Merocrine (eccrine) sweat gland

(e) Skin layers with glands (30×)

FIGURE 11.3   *Continued.* **APR**

Name _____

Date _____

Section _____

The Ⓐ corresponds to the indicated Learning Outcome(s) found at the beginning of the laboratory exercise.

Integumentary System

PART A ASSESSMENTS

Match the structures in column A with the descriptions and functions in column B. Place the letter of your choice in the space provided. Ⓐ1 Ⓐ2

Column A	Column B
a. Apocrine sweat gland	_____ 1. Oily secretion that helps to waterproof body surface
b. Arrector pili muscle	_____ 2. Outermost layer of epidermis
c. Dermis	_____ 3. Becomes active at puberty in axillary and groin regions
d. Epidermis	_____ 4. Epidermal pigment
e. Hair follicle	_____ 5. Inner layer of skin
f. Keratin	_____ 6. Responds to elevated body temperature
g. Melanin	_____ 7. General name of entire superficial layer of the skin
h. Merocrine sweat gland	_____ 8. Gland that secretes an oily mixture
i. Sebaceous gland	_____ 9. Tough protein of nails and hair
j. Sebum	_____ 10. Cell division and deepest layer of epidermis
k. Stratum basale	_____ 11. Tubelike part that contains the root of the hair
l. Stratum corneum	_____ 12. Causes hair to stand on end and goose bumps to appear

PART B ASSESSMENTS

Complete the following:

1. How does the skin of your palm differ from that on the back (posterior) of your hand? Ⓐ1 _____

_____

_____

2. Describe the differences you observed in the type and distribution of hair on the front (anterior) and back (posterior) of your forearm. Ⓐ1 _____

_____

3. Explain how a hair is formed. Ⓐ2 _____

_____

4. What cells produce the pigment in hair? Ⓐ2 _____

PART C ASSESSMENTS

Complete the following:

1. Distinguish the locations and tissues among the epidermis, dermis, and subcutaneous layer. 3 _____

_____

_____

2. How do the cells of stratum corneum and stratum basale differ? 3 _____

_____

_____

3. State the specific location of melanin observed in dark (heavily pigmented) skin. 3 _____

_____

4. What special qualities, due to the presence of fibers, does the connective tissue of the dermis have? 3 _____

_____

PART D ASSESSMENTS

Complete the following:

1. What part of the hair extends from between the bulb and the shaft? 1 _____

2. In which layer of skin are sebaceous glands found? 1 _____

3. How are sebaceous glands associated with hair follicles and what do they secrete? 1 _____

_____

4. The ducts of apocrine sweat glands open into _____ . 1

PART E ASSESSMENTS

Using the scanning objective, sketch a vertical section of human skin. Label the skin layers and a hair follicle, a sebaceous gland, and a sweat gland. 4

Skeletal System

LABORATORY EXERCISE

12

Bone Structure and Classification

MATERIALS NEEDED

Textbook
Prepared microscope slide of ground compact bone
Human bone specimens, including long, short, flat,
    irregular, and sesamoid types
Human long bone, sectioned longitudinally
Fresh animal bones, sectioned longitudinally and
    transversely
Compound light microscope
Dissecting microscope
Fresh chicken bones (radius and ulna from wings)
Vinegar or dilute hydrochloric acid

 SAFETY

- Wear disposable gloves for handling fresh bones
  and for the demonstration of a bone soaked in
  vinegar or dilute hydrochloric acid.
- Wash your hands before leaving the laboratory.

PURPOSE OF THE EXERCISE

To review the way bones are classified and to examine the
structure of a long bone.

LEARNING OUTCOMES APR

After completing this exercise, you should be able to

1. Locate the major structures of a long bone.
2. Distinguish between compact and spongy bone.
3. Differentiate the special characteristics of compact
   bone tissue.

4. Arrange five groups of bones based on their shapes and
   identify an example for each group.
5. Describe the functions of various structures of a bone.

A bone represents an organ of the skeletal system. As such,
it is composed of a variety of tissues, including bone tissue, cartilage, dense connective tissue, blood, and nervous tissue. Bones are not only alive but also multifunctional. They
support and protect softer tissues, provide points of attachment
for muscles, house blood-producing cells, and store inorganic
salts.

Bones are classified according to their shapes as long,
short, flat, or irregular. Although the bones of the skeleton
vary greatly in size and shape, they have much in common
structurally and functionally.

 PRACTICE

PROCEDURE—Bone Structure and Classification

1. Review section 5.3 titled "Connective Tissues" in chapter
   5 of the textbook and section 7.2 titled "Bone Shape and
   Structure" in chapter 7 of the textbook.
2. As a review activity, label figures 12.1 and 12.2.
3. Examine the microscopic structure of bone tissue by
   observing a prepared microscope slide of ground compact bone. Use figure 12.3, a micrograph of bone tissue,
   to locate the following features:

   osteon (Haversian system)—cylinder-shaped unit
   central canal (Haversian canal)—contains blood vessels
       and nerves
   lamella—concentric ring of matrix around central canal
   lacuna—small chamber for an osteocyte
   bone extracellular matrix—collagen and calcium
       phosphate
   canaliculus—minute tube containing cellular process

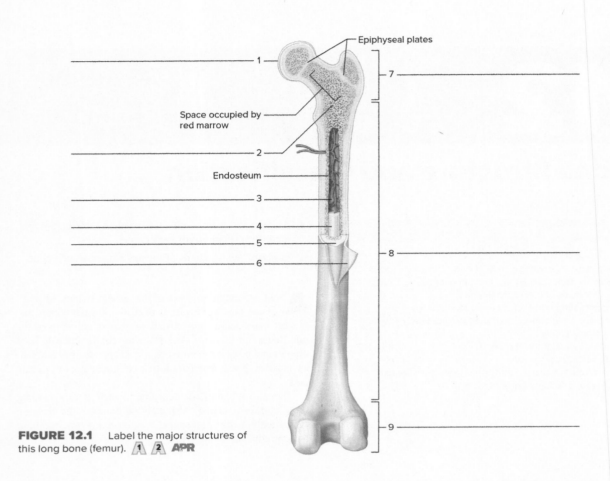

Epiphyseal plates

Space occupied by red marrow

Endosteum

1

2

3

4

5

6

7

8

9

FIGURE 12.1   Label the major structures of this long bone (femur). Ⓐ Ⓐ **APR**

![recycle icon] **ASSESS**

CRITICAL THINKING

Explain how bone cells embedded in a solid ground substance obtain nutrients and eliminate wastes. ③

_____
_____
_____
_____
_____

4. Observe the individual bone specimens and arrange them into groups according to the following shapes and examples: ④

   long–femur; humerus; phalanges
   short–carpals; tarsals; sesamoid (round) bones
   flat–ribs; some cranial bones, such as frontal bone
   irregular–vertebrae; some facial bones, such as sphenoid

5. Complete Part A of Laboratory Report 12.

6. Examine the sectioned long bones and locate the following:

   epiphysis
      proximal–nearest limb attachment to torso
      distal–farthest from limb attachment to torso
   epiphyseal plate–growth zone of hyaline cartilage
   articular cartilage–on ends of epiphyses
   diaphysis–shaft between epiphyses
   metaphysis–expanded portion of bone between
      diaphysis and epiphysis
   periosteum–strong membrane around bone
      (except articular cartilage) of dense irregular
      connective tissue
   compact bone–forms diaphysis and epiphyseal surfaces
   spongy bone–within epiphyses
      trabeculae–a structural lattice in spongy bone
   medullary cavity–hollow chamber
   endosteum–thin membrane of reticular connective
      tissue that lines the medullary cavity
   yellow marrow–occupies medullary cavity and stores
      adipose tissue
   red marrow–occupies spongy bone in some epiphyses
      and flat bones and produces blood cells

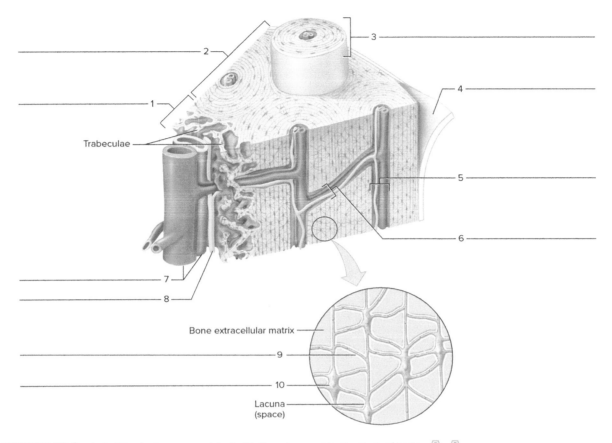

Trabeculae

Bone extracellular matrix

Lacuna
(space)

FIGURE 12.2   Label the features associated with the microscopic structure of bone.

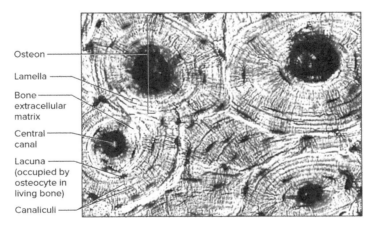

Osteon

Lamella

Bone
extracellular
matrix

Central
canal

Lacuna
(occupied by
osteocyte in
living bone)

Canaliculi

FIGURE 12.3   Micrograph of ground compact bone tissue (200×).   APR   (©Victor B. Eichler)

7. Use the dissecting microscope to observe the compact bone and spongy bone of the sectioned specimens. Also, examine the marrow in the medullary cavity and the spaces within the spongy bone of the fresh specimen.

8. Complete Parts B and C of the laboratory report.

 LEARN: ACTIVITY

Examine a fresh chicken bone and a chicken bone that has been soaked for several days in vinegar or overnight in dilute hydrochloric acid. Wear disposable gloves for handling these bones. This acid treatment removes the inorganic salts from the bone extracellular matrix. Rinse the bones in water and note the texture and flexibility of each (fig. 12.4*a*). The bone becomes soft and flexible without the support of the inorganic salts with calcium.

Examine the specimen of chicken bone that has been exposed to high temperature (baked at 121°C [250°F] for 2 hours). This treatment removes the protein and other organic substances from the bone extracellular matrix (fig. 12.4*b*). The bone becomes brittle and fragile without the benefit of the collagen fibers. A living bone with a combination of the qualities of inorganic and organic substances possesses tensile strength.

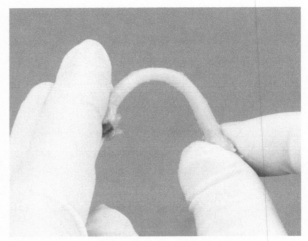

(a)

(b)

FIGURE 12.4 Results of fresh chicken bone demonstration: (*a*) soaked in vinegar; (*b*) baked in oven.
((*a,b*) ©J & J Photography)

Name _____

Date _____

Section _____

The 🄰 corresponds to the indicated Learning Outcome(s) found at the beginning of the laboratory exercise.

Bone Structure and Classification

🔄 PART A ASSESSMENTS

Complete the following statements. (*Note*: Questions 1-6 pertain to bone classification by shape.)

1. A bone that is platelike is classified as a(an) _____ bone. 🄰

2. The bones of the wrist are examples of _____ bones. 🄰

3. The bone of the thigh is an example of a(an) _____ bone. 🄰

4. Vertebrae are examples of _____ bones. 🄰

5. The patella (kneecap) is a special type of short bone called a _____ bone. 🄰

6. The bones of the skull that form a protective covering of the brain are examples of _____ bones. 🄰

7. Distinguish between the epiphysis and the diaphysis of a long bone. 🄰 _____

   _____

8. Describe where cartilage is found on the surface of a long bone. 🄰 _____

   _____

9. Describe where the periosteum is found on the surface of a long bone. 🄰 _____

   _____

🔄 PART B ASSESSMENTS

Complete the following:

1. Distinguish between the locations and tissues of the periosteum and those of the endosteum. 🄰 _____

   _____

   _____

2. What structural differences did you note between compact bone and spongy bone? 🄰 _____

   _____

   _____

3. How are these structural differences related to the locations and functions of these two types of bone? 🄰 _____

   _____

   _____

4. From your observations, how does the marrow in the medullary cavity compare with the marrow in the spaces of the spongy bone? 🄰 _____

   _____

   _____

PART C ASSESSMENTS

Identify the structures indicated in figure 12.5.

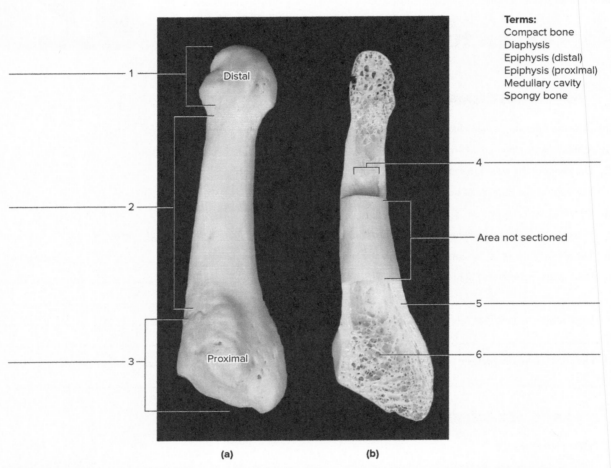

Terms:
Compact bone
Diaphysis
Epiphysis (distal)
Epiphysis (proximal)
Medullary cavity
Spongy bone

Distal

Proximal

1

2

3

4

Area not sectioned

5

6

(a)   (b)

FIGURE 12.5 Identify the structures indicated in (*a*) the unsectioned long bone (fifth metatarsal) and (*b*) the partially sectioned long bone, using the terms provided. ▲ ▲ (©J and J Photography)

13

Organization of the Skeleton

MATERIALS NEEDED

Textbook
Human skeleton, articulated
Human skeleton, disarticulated
Radiographs (X rays) of skeletal structures

PURPOSE OF THE EXERCISE

To review the organization of the skeleton, the major bones of the skeleton, and the terms used to describe skeletal structures.

LEARNING OUTCOMES APR

After completing this exercise, you should be able to

1. Distinguish between the axial skeleton and the appendicular skeleton.
2. Locate and label the major bones of the human skeleton.
3. Associate the terms used to describe skeletal structures and locate examples of such structures on the human skeleton.

The skeleton can be separated into two major portions: (1) the *axial skeleton,* which consists of the bones and cartilages of the head, neck, and trunk, and (2) the *appendicular skeleton,* which consists of the bones of the limbs and those that anchor the limbs to the axial skeleton. The bones that anchor the limbs include the pectoral and pelvic girdles. Men and women, although variation can exist, possess the same total bone number of 206.

PRACTICE

PROCEDURE—Organization of the Skeleton

1. Review section 7.5 titled "Skeletal Organization" in chapter 7 of the textbook. (Pronunciations of the names for major skeletal structures are included within the narrative of chapter 7.)
2. As a review activity, label figure 13.1.

3. Examine the articulated human skeleton and locate the following parts. As you locate the following bones, note the number of each in the skeleton. Palpate as many of the corresponding bones in your skeleton as possible.

axial skeleton

skull	
cranium	(8)
face	(14)
middle ear bone	(6)
hyoid bone	(1)
vertebral column	
vertebra	(24)
sacrum	(1)
coccyx	(1)
thoracic cage	
rib	(24)
sternum	(1)

appendicular skeleton

pectoral (shoulder) girdle	
scapula	(2)
clavicle	(2)
upper limbs	
humerus	(2)
radius	(2)
ulna	(2)
carpal	(16)
metacarpal	(10)
phalanx	(28)
pelvic girdle	
hip bone (coxal bone; pelvic bone; innominate bone)	(2)
lower limbs	
femur	(2)
tibia	(2)
fibula	(2)
patella	(2)
tarsal	(14)
metatarsal	(10)
phalanx	(28)
Total	**206 bones**

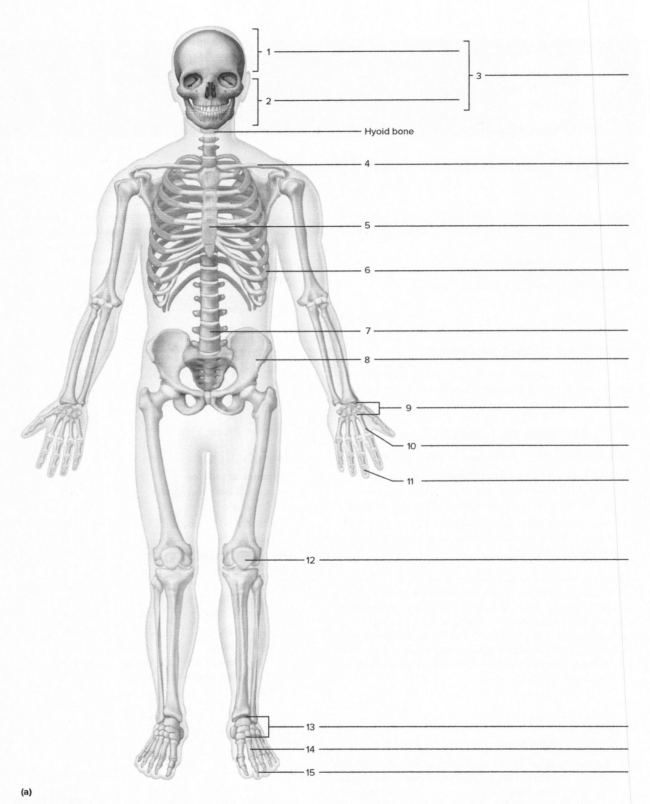

Hyoid bone

(a)

FIGURE 13.1   Label the major bones of the skeleton: (*a*) anterior view; (*b*) posterior view. The axial portion is shown in orange; the appendicular portions are shown in yellow.   APR

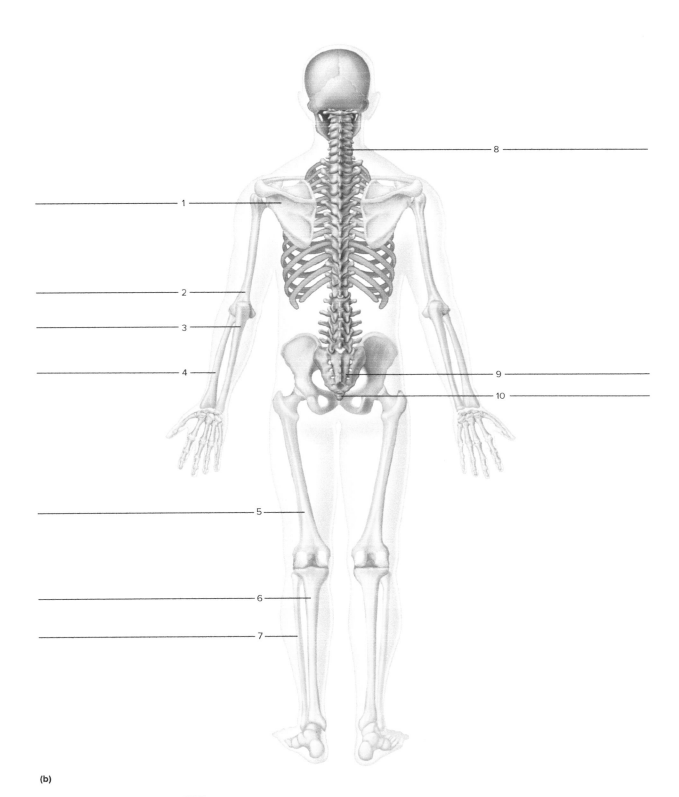

(b)

FIGURE 13.1   *Continued.* APR

4. Study table 7.4 in chapter 7 of the textbook. Bone features (bone markings) can be grouped together in a category of *projections, articulations, depressions,* or *openings.* Within each category more specific examples occur. Each of the bones listed in this section represents only one example of a location in the human body. Locate an example of each of the following features on the example, bone from a disarticulated skeleton, noting the size, shape, and location in the human skeleton:

<u>**Projections: sites for a tendon or ligament attachment**</u>

    crest (ridgelike)—hip bone

    epicondyle (superior to condyle)—femur

    line (linea) (slightly raised ridge)—femur

    process (prominent)—vertebra

    protuberance (outgrowth)—skull (occipital)

    ramus (extension)—hip bone

    spine (thornlike)—scapula

    trochanter (large)—femur

    tubercle (small, knoblike)—humerus

    tuberosity (rough elevation)—radius

<u>**Articulations: where bones connect at a joint**</u>

    condyle (rounded process)—skull (occipital)

    facet (nearly flat)—vertebra

    head (expanded end)—femur

<u>**Depressions: recessed areas in bones**</u>

    fossa (shallow basin)—humerus

    fovea (tiny pit)—femur

<u>**Openings: open spaces in bones**</u>

    fissure (slit)—skull (orbit)

    foramen (hole)—vertebra

    meatus (tubelike passageway)—skull (temporal)

    sinus (cavity)—skull (maxilla)

 ASSESS

CRITICAL THINKING

Locate and name the largest foramen in the skull.

_____

_____

_____

Locate and name the largest foramen in the skeleton.

_____

_____

_____

5. Complete Parts A–D of Laboratory Report 13.

 LEARN: ACTIVITY

Images on radiographs (X rays) are produced by allowing X rays from an X-ray tube to pass through a body part and to expose photographic film positioned on the opposite side of the part. The image that appears on the film after it is developed reveals the presence of parts with different densities. Bone, for example, is very dense tissue and is a good absorber of X rays. Thus, bone generally appears light on the film. Air-filled spaces, on the other hand, absorb almost no X-rays and appear as dark areas on the film. Liquids and soft tissues absorb intermediate quantities of X rays, so they usually appear in various shades of gray.

Examine the available radiographs of skeletal structures by holding each film in front of a light source. Identify as many of the bones and features as you can.

Name _____

Date _____

Section _____

The ⓐ corresponds to the indicated Learning Outcome(s) found at the beginning of the laboratory exercise.

Organization of the Skeleton

♻ PART A ASSESSMENTS

Complete the following statements:

1. The two divisions of the skeleton are the _____ skeleton and the appendicular skeleton. ⓐ

2. The extra bones that sometimes develop between the flat bones of the skull are called _____ bones. ⓑ

3. The cranium and facial bones compose the _____. ⓐ

4. The _____ bone supports the tongue. ⓑ

5. The _____ at the inferior end of the sacrum is composed of four fused vertebrae. ⓑ

6. Most ribs are attached anteriorly to the _____. ⓑ

7. The thoracic cage is composed of _____ pairs of ribs. ⓑ

8. The scapulae and clavicles together form the _____ girdle. ⓐ

9. The humerus, radius, and _____ articulate to form the elbow joint. ⓑ

10. The wrist is composed of eight bones called _____. ⓑ

11. The hip bones are attached posteriorly to the _____. ⓑ

12. The _____ covers the anterior surface of the knee. ⓑ

13. The bones that articulate with the distal ends of the tibia and fibula are called _____. ⓑ

14. All finger and toe bones are called _____. ⓑ

♻ PART B ASSESSMENTS

Match the terms in column A with the definitions in column B. Place the letter of your choice in the space provided. ⓒ

Column A	Column B
a. Condyle	_____ 1. Small, nearly flat articular surface
b. Crest	_____ 2. Shallow basin
c. Facet	_____ 3. Rounded process
d. Foramen	_____ 4. Opening or hole
e. Fossa	_____ 5. Knoblike rough elevation
f. Line	_____ 6. Ridgelike projection
g. Tuberosity	_____ 7. Slightly raised ridge

PART C ASSESSMENTS

Match the terms in column A with the definitions in column B. Place the letter of your choice in the space provided.

Column A		Column B
a. Fovea	_____	**1.** Tubelike passageway
b. Head	_____	**2.** Tiny pit or depression
c. Meatus	_____	**3.** Small, knoblike process
d. Sinus	_____	**4.** Thornlike projection
e. Spine	_____	**5.** Rounded enlargement at end of bone
f. Trochanter	_____	**6.** Air-filled cavity within bone
g. Tubercle	_____	**7.** Relatively large process

PART D ASSESSMENTS

1. Identify the bones indicated in figure 13.2.

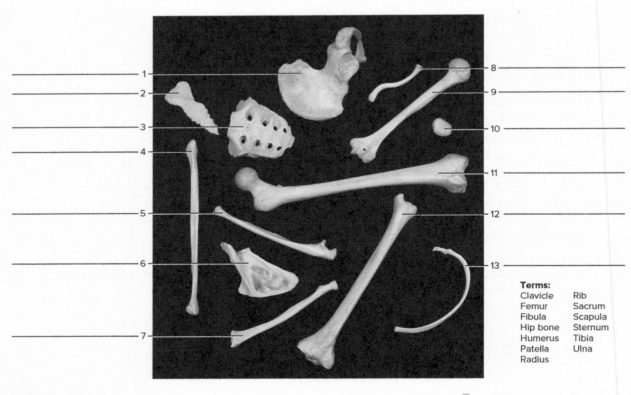

Terms:

Clavicle	Rib
Femur	Sacrum
Fibula	Scapula
Hip bone	Sternum
Humerus	Tibia
Patella	Ulna
Radius	

FIGURE 13.2   Identify the bones in this random arrangement, using the terms provided. (©J & J Photography)

2. List any of the bones shown in figure 13.2 that are included as part of the axial skeleton. _____

_____

14

Skull

MATERIALS NEEDED

Textbook
Human skull, articulated
Human skull, disarticulated (Beauchene)
Human skull, sagittal section
Colored pencils
Fetal skull

PURPOSE OF THE EXERCISE

To examine the structure of the human skull and to identify the bones and major features of the skull.

 LEARNING OUTCOMES APR

After completing this exercise, you should be able to

1. Distinguish between the cranium and the facial skeleton.
2. Locate and label the bones of the skull and their major features.
3. Locate and label the major sutures of the cranium.
4. Locate and label the sinuses of the skull.

A human skull consists of twenty-two bones that, except for the lower jaw, are firmly interlocked along sutures. Eight of these immovable bones make up the braincase, or cranium, and thirteen more immovable bones and the mandible form the facial skeleton.

 PRACTICE

PROCEDURE—Skull APR

1. Review section 7.6 titled "Skull" in chapter 7 of the textbook.
2. As a review activity, label figures 14.1–14.5.
3. Study the bones with paranasal sinuses (fig. 14.6).

4. Examine the **cranial bones** of the articulated human skull and the sectioned skull. Also observe the corresponding disarticulated bones. Locate the following bones and features in the laboratory specimens and, at the same time, palpate as many of these bones and features in your skull as possible.

frontal bone   **(1)**
  supraorbital foramen
  frontal sinus

parietal bone   **(2)**
  sagittal suture
  coronal suture

occipital bone   **(1)**
  lambdoid suture
  foramen magnum
  occipital condyle

temporal bone   **(2)**
  squamous suture
  external acoustic meatus
  mandibular fossa
  mastoid process
  styloid process
  carotid canal
  jugular foramen
  internal acoustic meatus
  zygomatic process

sphenoid bone   **(1)**
  sella turcica
  greater and lesser wings
  sphenoidal sinus

ethmoid bone   **(1)**
  cribriform plate
  perpendicular plate
  superior nasal concha
  middle nasal concha
  ethmoidal sinus
  crista galli

5. Complete Parts A and B of Laboratory Report 14.

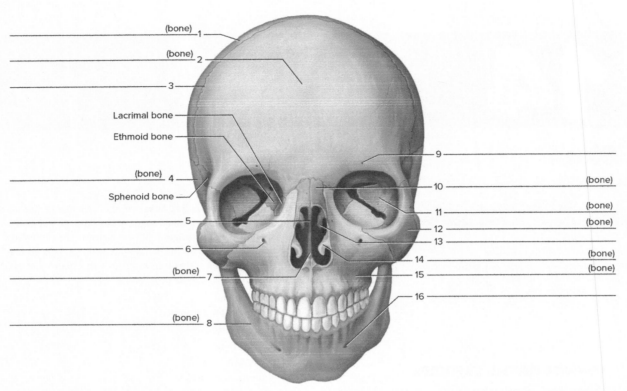

(bone) 1
(bone) 2
3
Lacrimal bone
Ethmoid bone
(bone) 4
Sphenoid bone
5
6
(bone) 7
(bone) 8

9
10 (bone)
11 (bone)
12 (bone)
13
14 (bone)
15 (bone)
16

FIGURE 14.1   Label the anterior bones and features of the skull. (If the line lacks the word *bone*, label the particular feature of that bone.) 🅰 **2** **APR**

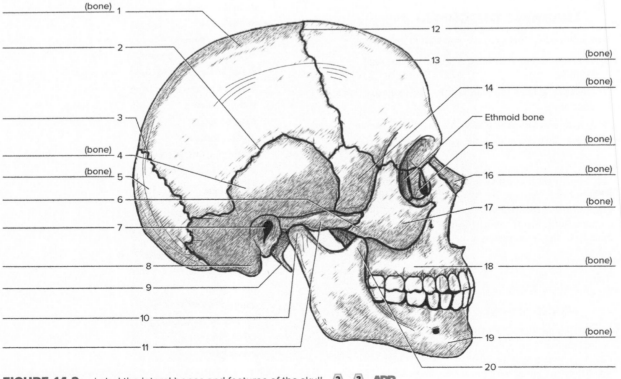

(bone) 1
2
3
(bone) 4
(bone) 5
6
7
8
9
10
11

12
13 (bone)
14 (bone)
Ethmoid bone
15 (bone)
16 (bone)
17 (bone)
18 (bone)
19 (bone)
20

FIGURE 14.2   Label the lateral bones and features of the skull. 🅰 **2** 🅰 **3** **APR**

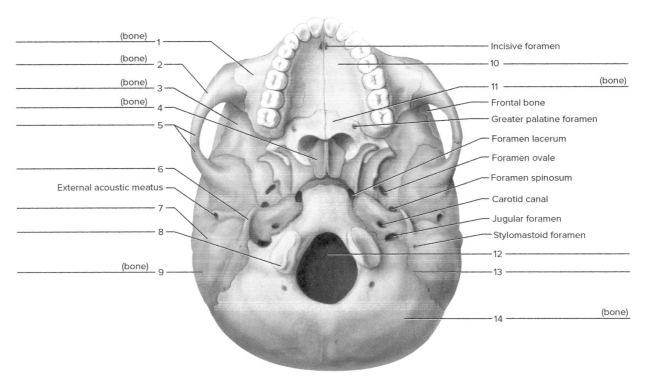

(bone) 1
(bone) 2
(bone) 3
(bone) 4
5
6
External acoustic meatus
7
8
(bone) 9

Incisive foramen
10
11 (bone)
Frontal bone
Greater palatine foramen
Foramen lacerum
Foramen ovale
Foramen spinosum
Carotid canal
Jugular foramen
Stylomastoid foramen
12
13
14 (bone)

FIGURE 14.3 Label the inferior bones and features of the skull. [2] APR

Anterior

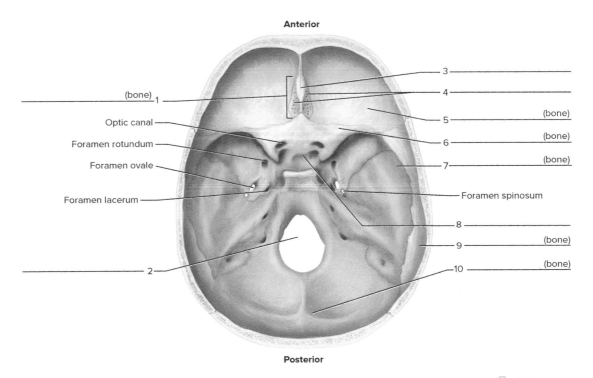

(bone) 1
Optic canal
Foramen rotundum
Foramen ovale
Foramen lacerum
2

3
4
5 (bone)
6 (bone)
7 (bone)
Foramen spinosum
8
9 (bone)
10 (bone)

Posterior

FIGURE 14.4 Label the bones and features of the floor of the cranial cavity as viewed from above. [2] APR

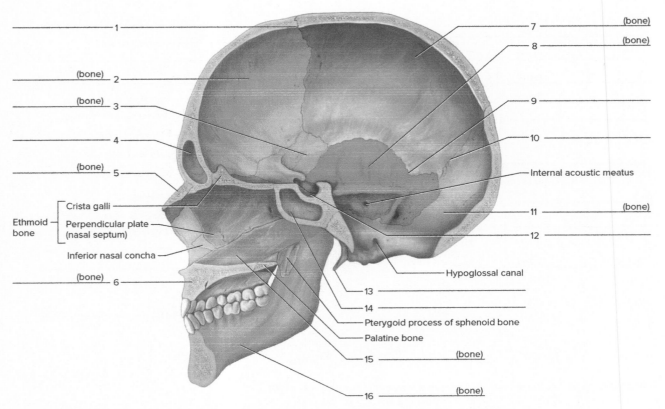

1 _____

(bone) 2 _____

(bone) 3 _____

4 _____

(bone) 5 _____

Crista galli _____

Ethmoid bone — Perpendicular plate (nasal septum) _____

Inferior nasal concha _____

(bone) 6 _____

7 _____ (bone)

8 _____ (bone)

9 _____

10 _____

Internal acoustic meatus

11 _____ (bone)

12 _____

Hypoglossal canal

13 _____

14 _____

Pterygoid process of sphenoid bone

Palatine bone

15 _____ (bone)

16 _____ (bone)

FIGURE 14.5   Label the bones and features of the sagittal section of the skull.   2 3 4 APR

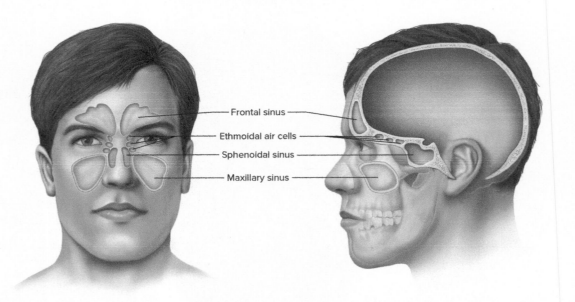

Frontal sinus

Ethmoidal air cells

Sphenoidal sinus

Maxillary sinus

FIGURE 14.6   Paranasal sinuses are located in the frontal bone, ethmoid bone, sphenoid bone, and both maxillary bones.

6. Examine the **facial bones** of the articulated and sectioned skulls and the corresponding disarticulated bones. Locate the following:

maxilla   (2)
    maxillary sinus
    palatine process
    alveolar process
    alveolar arch
palatine bone   (2)
zygomatic bone   (2)
    temporal process
    zygomatic arch (formed by temporal and zygomatic
      processes)
lacrimal bone   (2)
nasal bone   (2)
vomer   (1)
inferior nasal concha   (2)
mandible   (1)
    ramus
    mandibular condyle
    coronoid process
    alveolar border
    mandibular foramen
    mental foramen

7. Study the skull bones of a disarticulated skull (fig. 14.7).
8. Complete Part C of the laboratory report.

 LEARN: ACTIVITY

Use colored pencils to differentiate the bones illustrated in figure 14.2. Select a different color for each bone. This activity should help you locate various bones shown in a lateral view of a skull. You can check your work by referring to the corresponding figure in the textbook, presented in full color.

9. Study table 7.7 of chapter 7 in the textbook. Locate the following features of the human skull:

carotid canal
foramen lacerum
foramen magnum
foramen ovale
foramen rotundum
foramen spinosum
greater palatine foramen
hypoglossal canal
incisive foramen
inferior orbital fissure

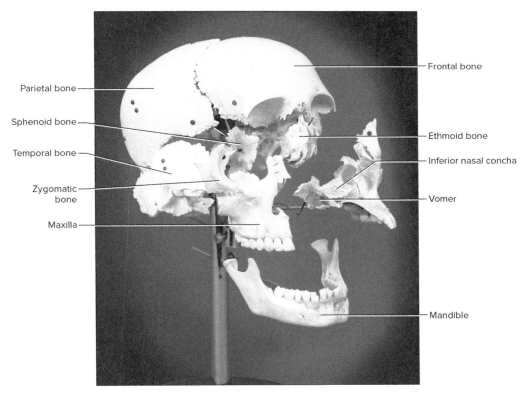

FIGURE 14.7 Bones of a disarticulated skull. (©J and J Photography)

infraorbital foramen
internal acoustic meatus
jugular foramen
mandibular foramen
mental foramen
optic canal
stylomastoid foramen
superior orbital fissure
supraorbital foramen

10. Complete Parts D and E of the laboratory report.

ASSESS

CRITICAL THINKING

Examine the inside of the cranium on a sectioned skull. What area appears to be the weakest area? Explain your answer.

_____

_____

_____

_____

LEARN: ACTIVITY

Examine the fetal skeleton and skull (figs. 14.8 and 14.9). The skull is incompletely developed and the cranial bones are separated by fibrous membranes. These membranous areas are called as fontanels, or "soft spots." The fontanels close as the cranial bones grow together. The posterior and lateral fontanels usually close during the first year after birth, whereas the anterior fontanel may not close until the middle or end of the second year. What other features characterize the fetal skull?

_____

_____

_____

_____

_____

_____

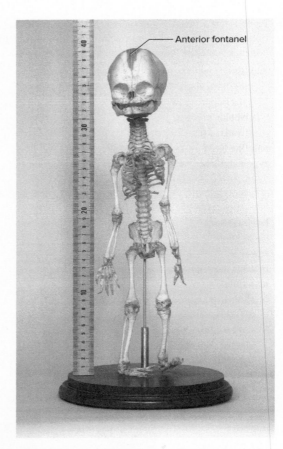

FIGURE 14.8   Anterior view of a fetal skeleton next to a metric scale. The gestational age of this skeleton is 7–8 months. (©J & J Photography)

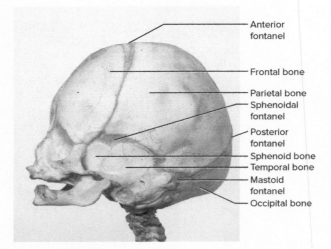

FIGURE 14.9   Lateral view of a fetal skull. (©J & J Photography)

Name _____

Date _____

Section _____

The Ⓐ corresponds to the indicated Learning Outcome(s) found at the beginning of the laboratory exercise.

Skull

🔁 PART A ASSESSMENTS

Complete the following statements:

1. Name six cranial bones that are visible on a lateral view of a skull. Ⓐ _____

   _____

2. The _____ suture joins the frontal bone to the parietal bones. Ⓐ

3. The parietal bones are firmly interlocked along the midline by the _____ suture. Ⓐ

4. The _____ suture joins the parietal bones to the occipital bone. Ⓐ

5. The temporal bones are joined to the parietal bones along the _____ sutures. Ⓐ

6. Name the three cranial bones that contain sinuses. Ⓐ Ⓐ _____

7. Name a facial bone that contains a sinus. Ⓐ Ⓐ _____

🔁 PART B ASSESSMENTS

Match the bones in column A with the features in column B. Place the letter of your choice in the space provided. (Some answers are used more than once.) Ⓐ

Column A	Column B
a. Ethmoid bone	_____ **1.** Forms sagittal, coronal, squamous, and lambdoid sutures
	_____ **2.** Cribriform plate
b. Frontal bone	_____ **3.** Crista galli
	_____ **4.** External acoustic meatus
	_____ **5.** Foramen magnum
c. Occipital bone	_____ **6.** Mandibular fossa
	_____ **7.** Mastoid process
d. Parietal bone	_____ **8.** Middle nasal concha
	_____ **9.** Occipital condyle
e. Sphenoid bone	_____ **10.** Sella turcica
	_____ **11.** Styloid process
f. Temporal bone	_____ **12.** Supraorbital foramen

PART C ASSESSMENTS

Match the bones in column A with the characteristics in column B. Place the letter of your choice in the space provided. 2

Column A		Column B
a. Inferior nasal concha	_____	**1.** Forms bridge of nose
	_____	**2.** Only movable bone in the facial skeleton
b. Lacrimal bone	_____	**3.** Contains coronoid process
	_____	**4.** Creates prominence of cheek inferior and lateral to the eye
c. Mandible	_____	**5.** Contains sockets of upper teeth
d. Maxilla	_____	**6.** Forms inferior portion of nasal septum
e. Nasal bone	_____	**7.** Forms anterior portion of zygomatic arch
	_____	**8.** Scroll-shaped bone
f. Palatine bone	_____	**9.** Forms anterior roof of mouth
	_____	**10.** Contains mental foramen
g. Vomer	_____	**11.** Forms posterior roof of mouth
h. Zygomatic bone	_____	**12.** Small medial bone of each orbit

 PART D ASSESSMENTS

Match the passageways in column A with the structures transmitted through them in column B. Place the letter of your choice in the space provided. 2

Column A		Column B
a. Foramen magnum	_____	**1.** Maxillary division of trigeminal nerve
b. Foramen ovale	_____	**2.** Nerve fibers of spinal cord
c. Foramen rotundum	_____	**3.** Optic nerve
d. Incisive foramen	_____	**4.** Vagus and accessory nerves
e. Internal acoustic meatus	_____	**5.** Nasopalatine nerves
f. Jugular foramen	_____	**6.** Mandibular division of trigeminal nerve
g. Optic canal	_____	**7.** Vestibular and cochlear nerves

PART E ASSESSMENTS

Identify the numbered bones and features of the skulls indicated in figures 14.10–14.13. 2 3

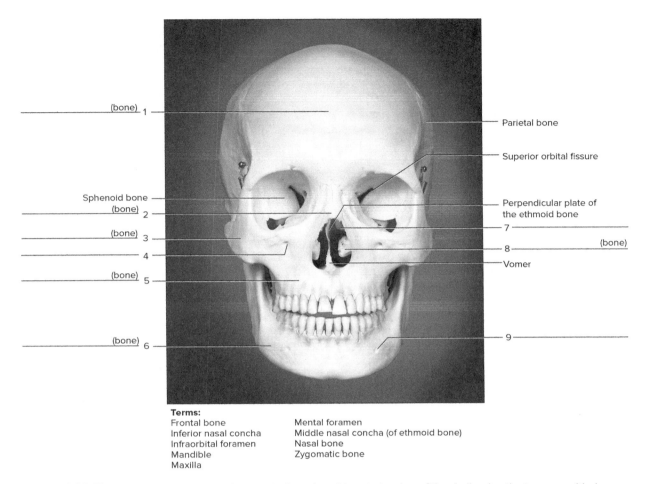

Terms:

Frontal bone

Inferior nasal concha

Infraorbital foramen

Mandible

Maxilla

Mental foramen

Middle nasal concha (of ethmoid bone)

Nasal bone

Zygomatic bone

FIGURE 14.10   Identify the bones and features indicated on this anterior view of the skull, using the terms provided.

(J and J Photography)

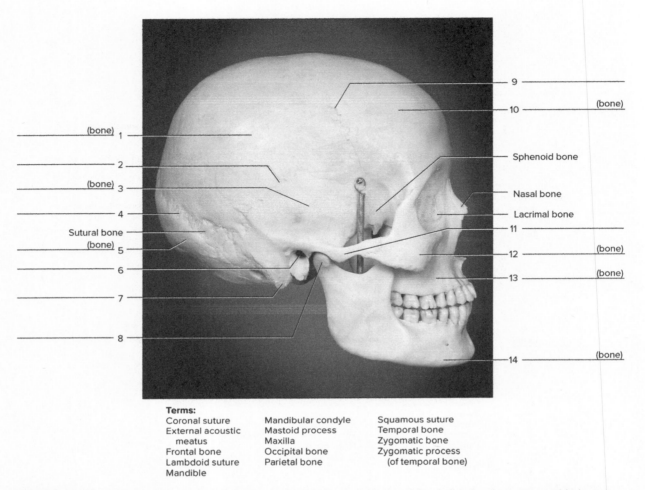

(bone) 1

2

(bone) 3

4

Sutural bone
(bone) 5

6

7

8

9

10   (bone)

Sphenoid bone

Nasal bone

Lacrimal bone

11

12   (bone)

13   (bone)

14   (bone)

Terms:

Coronal suture
External acoustic
   meatus
Frontal bone
Lambdoid suture
Mandible

Mandibular condyle
Mastoid process
Maxilla
Occipital bone
Parietal bone

Squamous suture
Temporal bone
Zygomatic bone
Zygomatic process
   (of temporal bone)

FIGURE 14.11   Identify the bones and features indicated on this lateral view of the skull, using the terms provided.
(©J and J Photography)

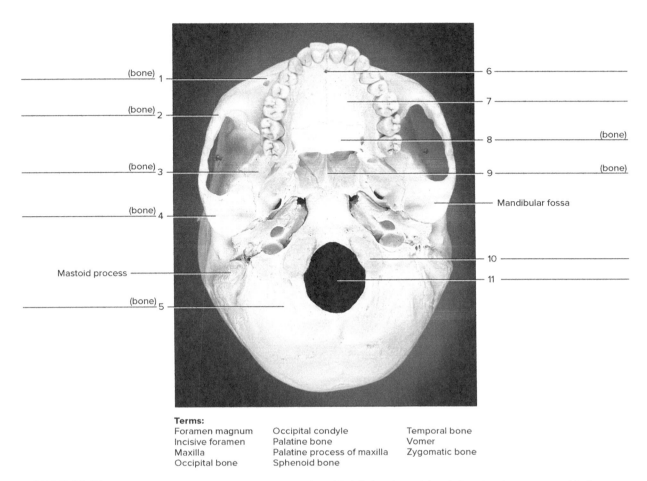

(bone) 1

(bone) 2

(bone) 3

(bone) 4

Mastoid process

(bone) 5

6

7

8 (bone)

9 (bone)

Mandibular fossa

10

11

Terms:

Foramen magnum	Occipital condyle	Temporal bone
Incisive foramen	Palatine bone	Vomer
Maxilla	Palatine process of maxilla	Zygomatic bone
Occipital bone	Sphenoid bone	

FIGURE 14.12   Identify the bones and features indicated on this inferior view of the skull, using the terms provided.
(J and J Photography)

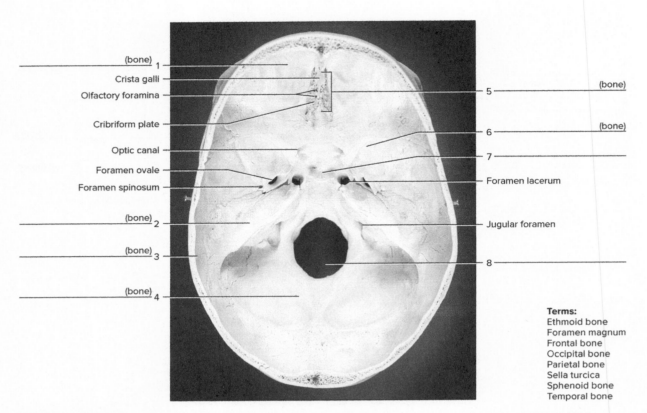

(bone) 1
Crista galli
Olfactory foramina
Cribriform plate
Optic canal
Foramen ovale
Foramen spinosum
(bone) 2
(bone) 3
(bone) 4

(bone) 5
(bone) 6
7
Foramen lacerum
Jugular foramen
8

Terms:
Ethmoid bone
Foramen magnum
Frontal bone
Occipital bone
Parietal bone
Sella turcica
Sphenoid bone
Temporal bone

FIGURE 14.13   Identify the bones and features on this floor of the cranial cavity of a skull, using the terms provided. (©J & J Photography)

15

Vertebral Column and Thoracic Cage

MATERIALS NEEDED

Textbook
Human skeleton, articulated
Samples of cervical, thoracic, and lumbar vertebrae
Human skeleton, disarticulated

PURPOSE OF THE EXERCISE

To examine the vertebral column and the thoracic cage of the human skeleton and to identify the bones and major features of these parts.

 LEARNING OUTCOMES APR

After completing this exercise, you should be able to

1. Identify the major features of the vertebral column.

2. Locate the features of a vertebra.

3. Distinguish the cervical, thoracic, and lumbar vertebrae and locate the sacrum and coccyx.

4. Identify the structures and functions of the thoracic cage.

5. Distinguish between true and false ribs.

The *vertebral column,* consisting of twenty-six bones, extends from the skull to the pelvis and forms the vertical axis of the human skeleton. The vertebral column includes seven cervical vertebrae, twelve thoracic vertebrae, five lumbar vertebrae, one sacrum of five fused vertebrae, and one coccyx of usually four fused vertebrae. To help you to remember the number of cervical, thoracic, and lumbar vertebrae from superior to inferior, consider this saying: breakfast at 7, lunch at 12, and dinner at 5. These vertebrae are separated from one another by cartilaginous intervertebral discs and are held together by ligaments.

The *thoracic cage* surrounds the thoracic and upper abdominal cavities. It includes the ribs, the thoracic vertebrae, the sternum, and the costal cartilages. The thoracic cage protects organs in the thoracic and upper abdominal cavities, supports the pectoral girdle and arm, and aids breathing.

 PRACTICE

PROCEDURE A—Vertebral Column

1. Review section 7.7 titled "Vertebral Column" in chapter 7 of the textbook.

2. As a review activity, label figures 15.1–15.4.

3. Examine the vertebral column of the human skeleton and locate the following bones and features. At the same time, locate as many of the corresponding bones and features in your skeleton as possible.

cervical vertebrae	**(7)**
atlas (C1)	
axis (C2)	
vertebra prominens (C7)	
thoracic vertebrae	**(12)**
lumbar vertebrae	**(5)**
sacrum	**(1)**
coccyx	**(1)**

intervertebral discs (fibrocartilage)
vertebral canal (contains spinal cord)
cervical curvature
thoracic curvature
lumbar curvature
sacral curvature
intervertebral foramina
    (passageway for spinal nerves)

 ASSESS

CRITICAL THINKING

Note the four curvatures of the vertebral column. What functional advantages exist with curvatures for skeletal structure instead of a straight vertebral column?

_____
_____
_____
_____

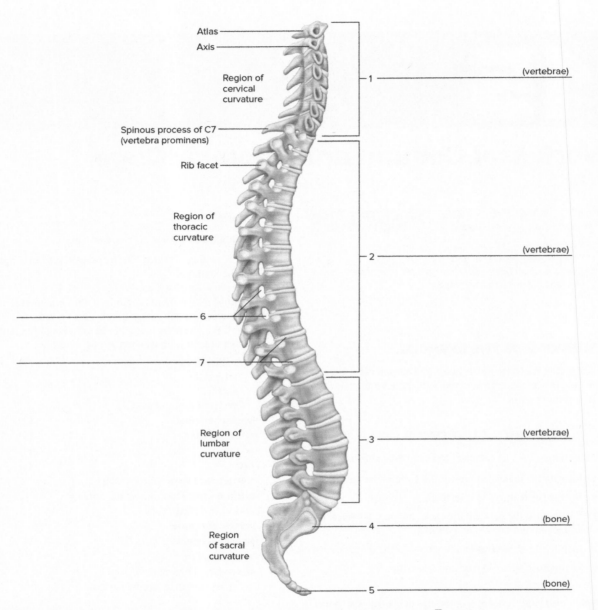

Atlas
Axis
Region of cervical curvature
Spinous process of C7 (vertebra prominens)
Rib facet
Region of thoracic curvature
6
7
Region of lumbar curvature
Region of sacral curvature

1 ———— (vertebrae)
2 ———— (vertebrae)
3 ———— (vertebrae)
4 ———— (bone)
5 ———— (bone)

FIGURE 15.1   Label the bones and features of the vertebral column (right lateral view). Ⓐ APR

(a) Atlas (C1)

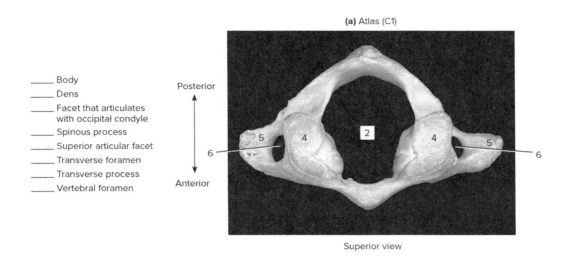

_____ Body
_____ Dens
_____ Facet that articulates
       with occipital condyle
_____ Spinous process
_____ Superior articular facet
_____ Transverse foramen
_____ Transverse process
_____ Vertebral foramen

Posterior ↕ Anterior

Superior view

(b) Axis (C2)

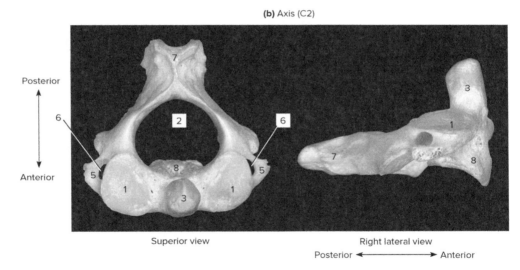

Posterior ↕ Anterior

Superior view

Right lateral view

Posterior ←——→ Anterior

FIGURE 15.2   Label the superior features of (a) the atlas and the superior and right lateral features of (b) the axis by placing the correct numbers in the spaces provided. 2 3 APR

(©J & J Photography)

Superior views   Right lateral views

_____ Body
_____ Inferior vertebral notch
_____ Lamina
_____ Pedicle
_____ Spinous process
_____ Superior articular process
_____ Transverse foramen
_____ Transverse process
_____ Vertebral foramen

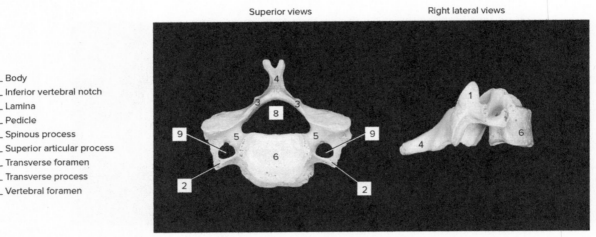

(a) Cervical vertebra

Posterior ↑

Anterior ↓

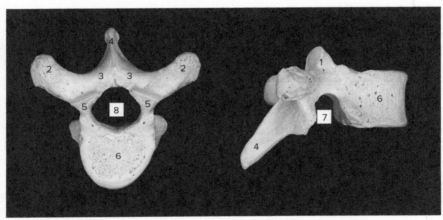

(b) Thoracic vertebra

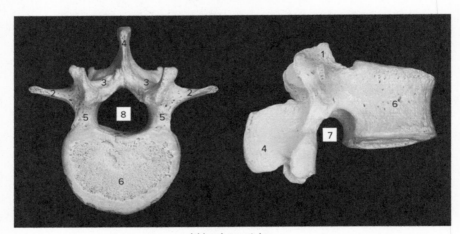

(c) Lumbar vertebra

Posterior ◄————► Anterior

FIGURE 15.3   Label the superior and right lateral features of the (*a*) cervical, (*b*) thoracic, and (*c*) lumbar vertebrae by placing the correct numbers in the spaces provided. **A** **3** **APR**

(©J & J Photography)

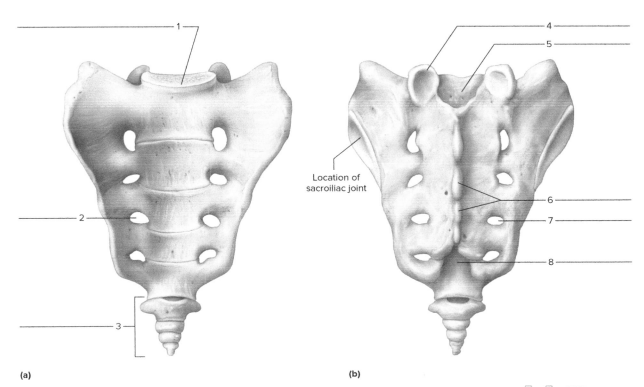

(a)   **(b)**

FIGURE 15.4   Label the coccyx and the features of the sacrum: (*a*) anterior view; (*b*) posterior view.

4. Compare the available samples of cervical, thoracic, and lumbar vertebrae by noting differences in size and shape and by locating the following features:

   vertebral foramen
   body
   pedicles
   laminae
   spinous process
   transverse processes
   superior articular processes
   inferior articular processes
   inferior vertebral notch
   transverse foramina
   facets
   dens of axis

5. Examine the sacrum and coccyx. Locate the following features:

   sacrum
      superior articular process
      posterior sacral foramen
      anterior sacral foramen
      sacral promontory
      sacral canal
      tubercles
      median sacral crest
      sacral hiatus

   coccyx

6. Complete Parts A and B of Laboratory Report 15.

PRACTICE

PROCEDURE B—Thoracic Cage

1. Review section 7.8 titled "Thoracic Cage" in chapter 7 of the textbook.
2. As a review activity, label figure 15.5.
3. Study a typical rib with articulation sites with a thoracic vertebra (fig. 15.6)
4. Examine the thoracic cage of the human skeleton and locate the following bones and features:

   rib
      head
      tubercle
      neck
      shaft
      anterior (sternal) end
      facets
      true ribs (pairs 1–7)
      false ribs (pairs 8–12) (includes floating ribs) floating ribs (pairs 11–12)

   costal cartilages (hyaline cartilage)

   sternum
      jugular (suprasternal) notch
      clavicular notch
      manubrium
      sternal angle
      body
      xiphoid process

5. Complete Parts C and D of the laboratory report.

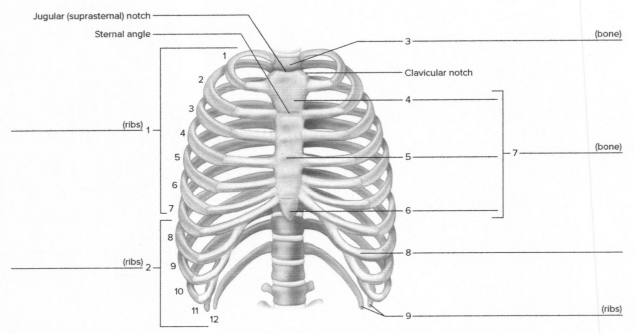

Jugular (suprasternal) notch
Sternal angle
Clavicular notch
(bone)
(ribs) 1
(ribs) 2
1
2
3
4
5
6
7
8
9
10
11
12
3
4
5
6
7
8
9
(bone)
(bone)
(ribs)

FIGURE 15.5   Label the bones and features of the thoracic cage (anterior view). 🅰 🅰 APR

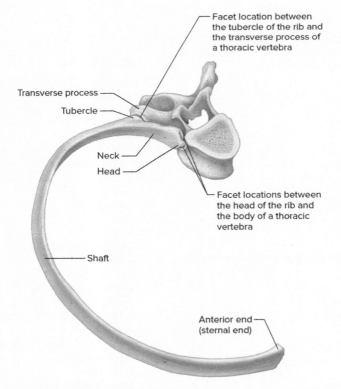

Facet location between
the tubercle of the rib and
the transverse process of
a thoracic vertebra

Transverse process
Tubercle
Neck
Head

Facet locations between
the head of the rib and
the body of a thoracic
vertebra

Shaft

Anterior end
(sternal end)

FIGURE 15.6   Superior view of a typical rib with the articulation sites with a thoracic vertebra.

Name _____

Date _____

Section _____

The Ⓐ corresponds to the indicated Learning Outcome(s) found at the beginning of the laboratory exercise.

Vertebral Column and Thoracic Cage

♻ PART A ASSESSMENTS

Complete the following statements:

1. The vertebral column encloses and protects the _____. Ⓐ

2. The number of separate bones in the vertebral column of an adult is _____. Ⓐ

3. The _____ of the vertebrae support the weight of the head and trunk. Ⓐ

4. The intervertebral foramina provide passageways for _____. Ⓐ

5. Transverse foramina of cervical vertebrae serve as passageways for _____ leading to the brain. Ⓐ

6. The first vertebra is also called the _____. Ⓐ

7. The second vertebra is also called the _____. Ⓐ

8. When the head is moved from side to side, the first vertebra pivots around the _____ of the second vertebra. Ⓐ

9. The _____ vertebrae have the largest and strongest bodies. Ⓐ

10. The typical number of vertebrae that fuse in the adult to form the sacrum is _____. Ⓐ

11. The upper, anterior margin of the sacrum that projects forward is called the _____. Ⓐ

12. An opening called the _____ exists at the tip of the sacral canal. Ⓐ

♻ PART B ASSESSMENTS

Based on your observations, compare typical cervical, thoracic, and lumbar vertebrae in relation to the characteristics indicated in the table. The table is partly completed. For your responses, consider characteristics such as size, shape, presence or absence, and unique features. Ⓐ Ⓐ

Vertebra	Number	Size	Body	Spinous Process	Transverse Foramina
Cervical	7		Smallest	C2 through C6 are forked (bifid)	
Thoracic		Intermediate			
Lumbar					Absent

PART C ASSESSMENTS

Complete the following statements:

1. The last two pairs of ribs that have no cartilaginous attachments to the sternum are sometimes called

   _____ ribs. **5**

2. There are _____ pairs of true ribs. **5**

3. Costal cartilages are composed of _____ tissue. **4**

4. The manubrium articulates with the _____ on its superior border. **4**

5. List three general functions of the thoracic cage. **4** _____

   _____

 PART D ASSESSMENTS

Identify the bones and features indicated in the radiograph of the neck in figure 15.7. **1 2**

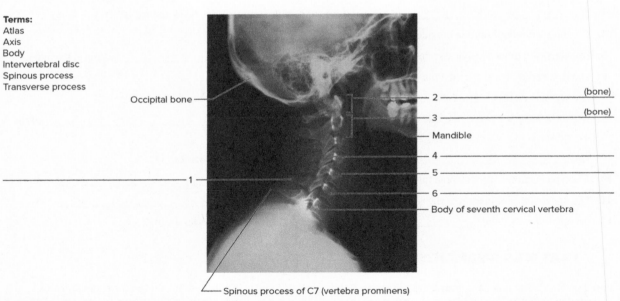

Terms:
Atlas
Axis
Body
Intervertebral disc
Spinous process
Transverse process

Occipital bone

2 _____ (bone)

3 _____ (bone)

Mandible

4 _____

5 _____

6 _____

Body of seventh cervical vertebra

1

Spinous process of C7 (vertebra prominens)

FIGURE 15.7   Identify the bones and features indicated in this radiograph of the neck (lateral view), using the terms provided. **APR** (Stockbyte/Getty Images)

LABORATORY EXERCISE

16

Pectoral Girdle and Upper Limb

MATERIALS NEEDED

Textbook
Human skeleton, articulated
Human skeleton, disarticulated
Colored pencils

PURPOSE OF THE EXERCISE

To examine the bones of the pectoral girdle and upper limb and to identify the major features of these bones.

LEARNING OUTCOMES APR

After completing this exercise, you should be able to

1. Locate and identify the bones of the pectoral girdle and their major features.

2. Locate and identify the bones of the upper limb and their major features.

3. Recognize the anatomical terms pertaining to the pectoral girdle and upper limb.

The pectoral girdle (shoulder girdle) consists of two clavicles and two scapulae. These parts support the upper limbs and serve as attachments for various muscles that move these limbs.

Each upper limb includes a humerus, a radius, an ulna, eight carpals, five metacarpals, and fourteen phalanges. These bones form the framework of the arm, forearm, and hand. They also function as parts of levers when the limbs are moved.

PRACTICE

PROCEDURE A—Pectoral Girdle

1. Review section 7.9 entitled "Pectoral Girdle" in chapter 7 of the textbook.
2. As a review activity, label figures 16.1 and 16.2.
3. Examine the bones of the pectoral girdle and locate the following features. At the same time, locate as many of

the corresponding surface bones and features of your own skeleton as possible.

clavicle
  medial (sternal) end
  lateral (acromial) end

scapula
  spine
  acromion process
  coracoid process
  glenoid cavity
  borders
    superior border
    medial (vertebral) border
    lateral (axillary) border
  fossae
    supraspinous fossa
    infraspinous fossa

ASSESS

CRITICAL THINKING

Why is the clavicle the bone that is most commonly fractured in the human body?

_____

_____

_____

_____

_____

4. Complete Part A of Laboratory Report 16.

PRACTICE

PROCEDURE B—Upper Limb

1. Review section 7.10 entitled "Upper Limb" in chapter 7 of the textbook.
2. As a review activity, label figures 16.3–16.6.

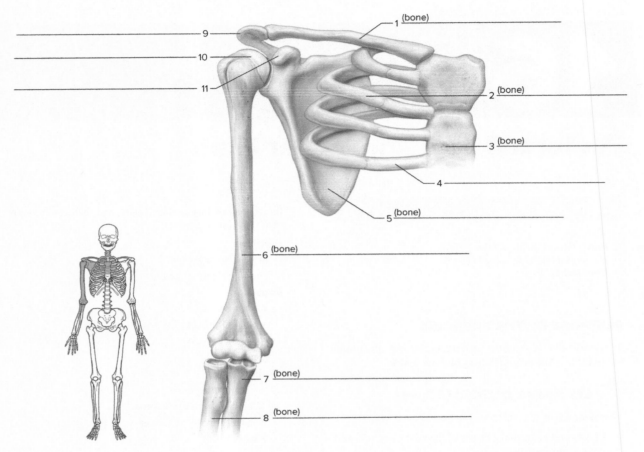

1 _____ (bone)

9 _____

10 _____

11 _____

2 _____ (bone)

3 _____ (bone)

4 _____

5 _____ (bone)

6 _____ (bone)

7 _____ (bone)

8 _____ (bone)

FIGURE 16.1   Label the bones and features of the right shoulder and upper limb (anterior view). A A APR

3. Examine the following bones and features of the upper limb:

humerus
 proximal features
  head
  greater tubercle
  lesser tubercle
  anatomical neck
  surgical neck
  intertubercular sulcus
 shaft
  deltoid tuberosity
 distal features
  capitulum
  trochlea
  medial epicondyle

  lateral epicondyle
  coronoid fossa
  olecranon fossa
radius
 head
 radial tuberosity
 styloid process
 ulnar notch
ulna
 trochlear notch (semilunar notch)
 radial notch
 olecranon process
 coronoid process
 styloid process
 head

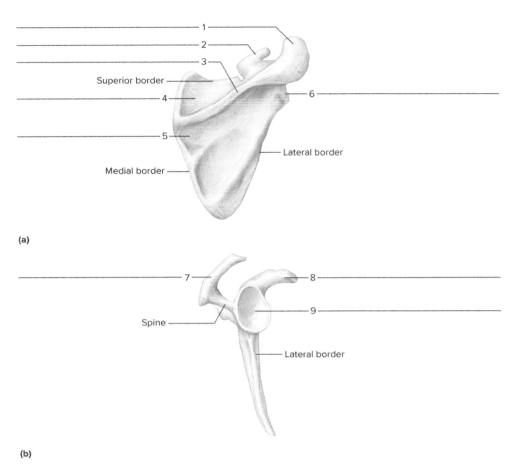

1 _____
2 _____
3 _____
Superior border _____
4 _____
5 _____
Medial border _____
Lateral border
6 _____

(a)

7 _____
8 _____
9 _____
Spine _____
Lateral border

(b)

FIGURE 16.2   Label (a) the posterior surface of the right scapula and (b) the lateral aspect of the right scapula. Ⓐ **APR**

carpal bones
   proximal row (listed lateral to medial)
     scaphoid
     lunate
     triquetrum
     pisiform
   distal row (listed medial to lateral)
     hamate
     capitate
     trapezoid
     trapezium
metacarpal bones
phalanges
   proximal phalanx
   middle phalanx
   distal phalanx
4. Complete Parts B–E of the laboratory report.

The following mnemonic device will help you learn the eight carpals:

So Long Top Part
Here Comes The Thumb

The first letter of each word corresponds to the first letter of a carpal. This device arranges the carpals in order for the proximal, transverse row of four bones from lateral to medial, followed by the distal, transverse row from medial to lateral, which ends nearest the thumb. This arrangement assumes the hand is in the anatomical position.

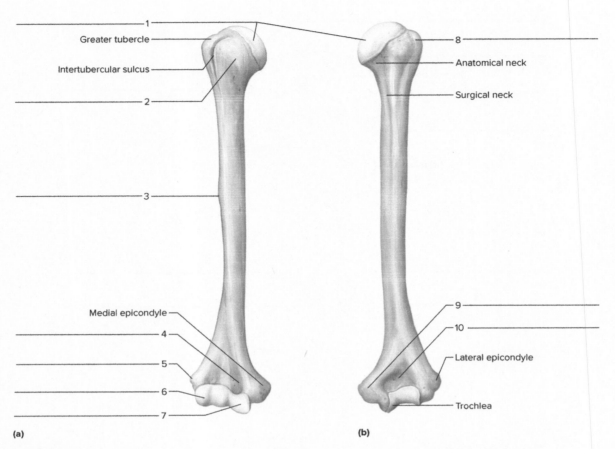

Greater tubercle

Intertubercular sulcus

1

2

3

Medial epicondyle

4

5

6

7

(a)

8

Anatomical neck

Surgical neck

9

10

Lateral epicondyle

Trochlea

(b)

FIGURE 16.3   Label the (a) anterior surface and (b) posterior surface of the right humerus. 🄰 APR

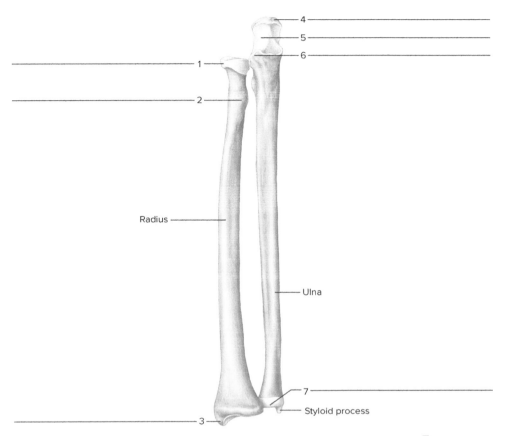

Radius

Ulna

Styloid process

FIGURE 16.4   Label the major anterior features of the right radius and ulna (anterior view). APR

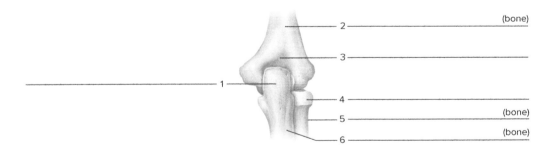

FIGURE 16.5   Label the bones and features of the right elbow, posterior view.

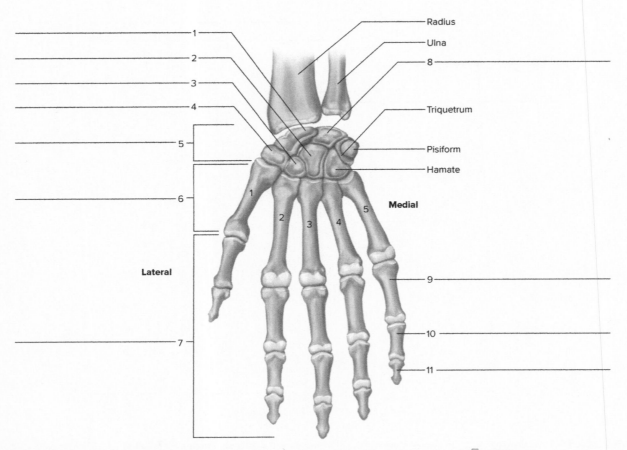

FIGURE 16.6 Label the bones and groups of bones in this anterior view of the right hand. **2** **APR**

LEARN: ACTIVITY

Use different colored pencils to distinguish the individual bones in figure 16.6.

Name _____

Date _____

Section _____

The 🄰 corresponds to the indicated Learning Outcome(s) found at the beginning of the laboratory exercise.

Pectoral Girdle and Upper Limb

♻ PART A ASSESSMENTS

Complete the following statements:

1. The pectoral girdle is an incomplete ring because it is open in the back between the _____ . 🄰

2. The medial end of a clavicle articulates with the _____ of the sternum. 🄰

3. The lateral end of a clavicle articulates with the _____ process of the scapula. 🄰

4. The _____ is a bone that serves as a brace between the sternum and the scapula. 🄰

5. The _____ divides the posterior side of the scapula into unequal portions. 🄰

6. The lateral tip of the shoulder is the _____ process of the scapula. 🄰

7. Near the lateral end of the scapula, the _____ process curves anteriorly and inferiorly from the clavicle. 🄰

8. The glenoid cavity of the scapula articulates with the _____ of the humerus. 🄰

♻ PART B ASSESSMENTS

Match the bones in column A with the bones and features in column B. Place the letter of your choice in the space provided. 🄰

Column A	Column B
a. Carpals	_____ 1. Capitate
	_____ 2. Coronoid fossa
	_____ 3. Deltoid tuberosity
b. Humerus	_____ 4. Greater tubercle
	_____ 5. Five palmar bones
c. Metacarpals	_____ 6. Fourteen bones in digits
	_____ 7. Intertubercular sulcus
d. Phalanges	_____ 8. Lunate
	_____ 9. Olecranon fossa
e. Radius	_____ 10. Radial tuberosity
	_____ 11. Trapezium
f. Ulna	_____ 12. Trochlear notch

PART C ASSESSMENTS

Identify the bones and features indicated in the radiographs of figures 16.7–16.9. 🔺1 🔺2

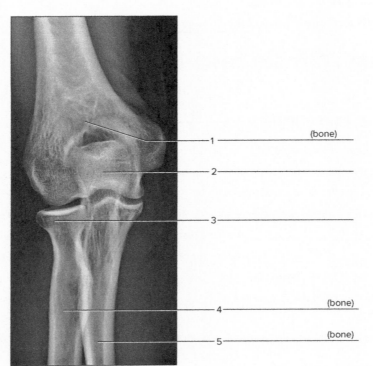

1 _____ (bone)

2 _____

3 _____

4 _____ (bone)

5 _____ (bone)

Terms:
Head of radius
Humerus
Olecranon process
Radius
Ulna

FIGURE 16.7 Identify the bones and features indicated on this radiograph of the right elbow (anterior view), using the terms provided. **APR** (©Dale Butler)

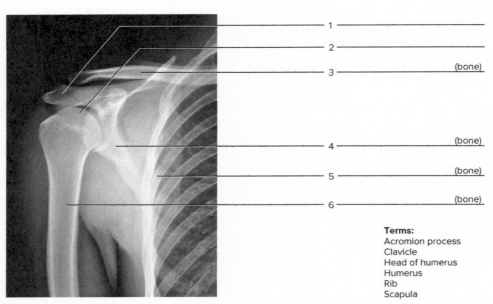

1 _____

2 _____

3 _____ (bone)

4 _____ (bone)

5 _____ (bone)

6 _____ (bone)

Terms:
Acromion process
Clavicle
Head of humerus
Humerus
Rib
Scapula

FIGURE 16.8 Identify the bones and features indicated on this radiograph of the right shoulder (anterior view), using the terms provided. **APR** (sanyanwuji/Getty Images)

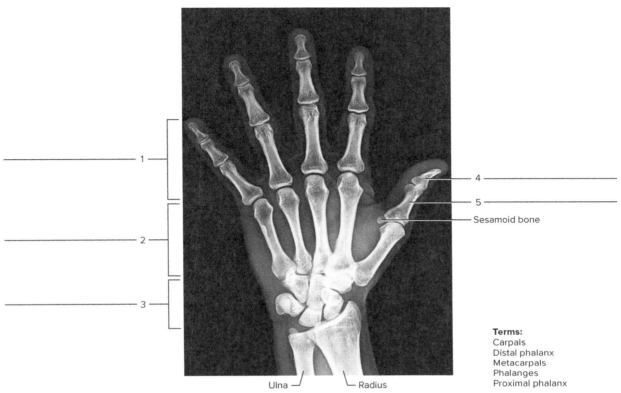

1

2

3

4

5

Sesamoid bone

Ulna

Radius

Terms:
Carpals
Distal phalanx
Metacarpals
Phalanges
Proximal phalanx

FIGURE 16.9   Identify the bones indicated on this radiograph of the right hand (anterior view), using the terms provided. **APR** (©Dale Butler)

PART D ASSESSMENTS

Complete the labeling of the bones of the hand in figure 16.10.

_____ Capitate
_____ Distal phalanges
_____ Hamate
_____ Lunate
_____ Metacarpals
_____ Middle phalanges
____4____ Pisiform
_____ Proximal phalanges
_____ Scaphoid
_____ Trapezium
_____ Trapezoid
____3____ Triquetrum

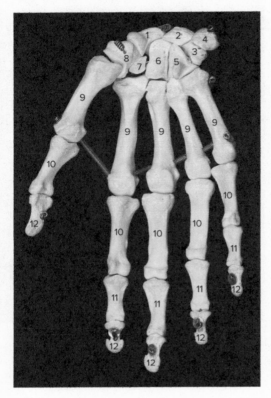

FIGURE 16.10   Label the bones numbered on this anterior view of the right hand by placing the correct numbers in the spaces provided. (© J & J Photography)

LABORATORY EXERCISE

17

Pelvic Girdle and Lower Limb

MATERIALS NEEDED

Textbook
Human skeleton, articulated
Human skeleton, disarticulated
Male and female pelves
Colored pencils

PURPOSE OF THE EXERCISE

To examine the bones of the pelvic girdle and lower limb and to identify the major features of these bones.

LEARNING OUTCOMES APR

After completing this exercise, you should be able to

1. Locate and identify the bones of the pelvic girdle and their major features.

2. Locate and identify the bones of the lower limb and their major features.

3. Recognize the anatomical terms pertaining to the pelvic girdle and lower limb.

The pelvic girdle includes two coxal (hip) bones that articulate with each other anteriorly at the pubic symphysis and posteriorly with the sacrum. Together, the pelvic girdle, sacrum, and coccyx constitute the pelvis. The pelvis, in turn, provides support for the trunk of the body and provides attachments for the lower limbs.

The bones of the lower limb form the framework of the thigh, leg, and foot. Each limb includes a femur, a patella, a tibia, a fibula, seven tarsals, five metatarsals, and fourteen phalanges.

PRACTICE

PROCEDURE A—Pelvic Girdle

1. Review section 7.11 entitled "Pelvic Girdle" in chapter 7 of the textbook.
2. As a review activity, label figures 17.1 and 17.2.

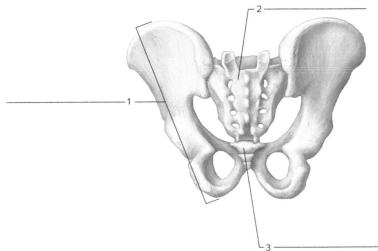

FIGURE 17.1 Label the bones of the pelvis (posterior view).

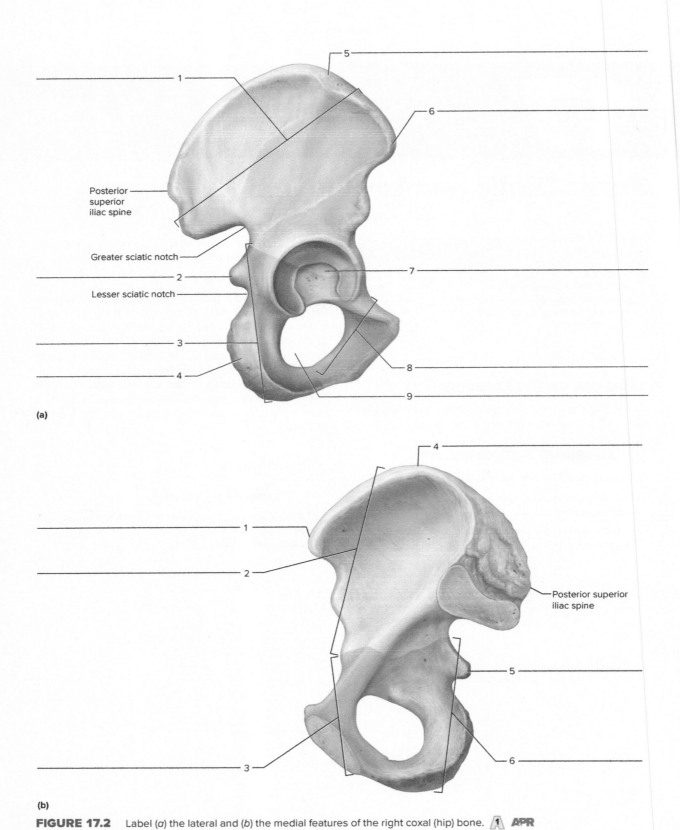

Posterior
superior
iliac spine

Greater sciatic notch

Lesser sciatic notch

(a)

Posterior superior
iliac spine

(b)

FIGURE 17.2   Label (*a*) the lateral and (*b*) the medial features of the right coxal (hip) bone.   APR

3. Examine the bones of the pelvic girdle and locate the following:

coxal bone (hip bone; pelvic bone; innominate bone)
   acetabulum
   ilium
   iliac crest
   iliac fossa
   sacroiliac joint
   anterior superior iliac spine
   posterior superior iliac spine
   greater sciatic notch
   lesser sciatic notch
   ischium
   ischial tuberosity
   ischial spine
   pubis
   pubic symphysis
   pubic arch
   obturator foramen

4. Complete Part A of Laboratory Report 17.

 ASSESS

CRITICAL THINKING

Examine the male and female pelves and study figure 17.3. Look for major differences between them. Note especially the flare of the iliac bones, the angle of the pubic arch, the distance between the ischial spines and ischial tuberosities, and the curve and width of the sacrum. In what ways are the differences you observed related to the function of the female pelvis as a birth canal?

_____

_____

_____

_____

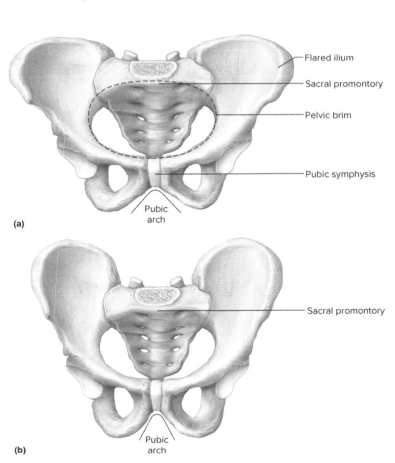

(a)

(b)

FIGURE 17.3 The female pelvis is usually wider in all diameters and roomier than that of the male pelvis. (*a*) Female pelvis. (*b*) Male pelvis. **APR**

PRACTICE

PROCEDURE B—Lower Limb

1. Review section 7.12 entitled "Lower Limb" in chapter 7 of the textbook.
2. As a review activity, label figures 17.4–17.7.
3. Examine the bones of the lower limb and locate each of the following:

femur
    proximal features
      head
      fovea capitis
      neck
      greater trochanter
      lesser trochanter
    shaft
      gluteal tuberosity
      linea aspera
    distal features
      lateral epicondyle
      medial epicondyle
      lateral condyle
      medial condyle

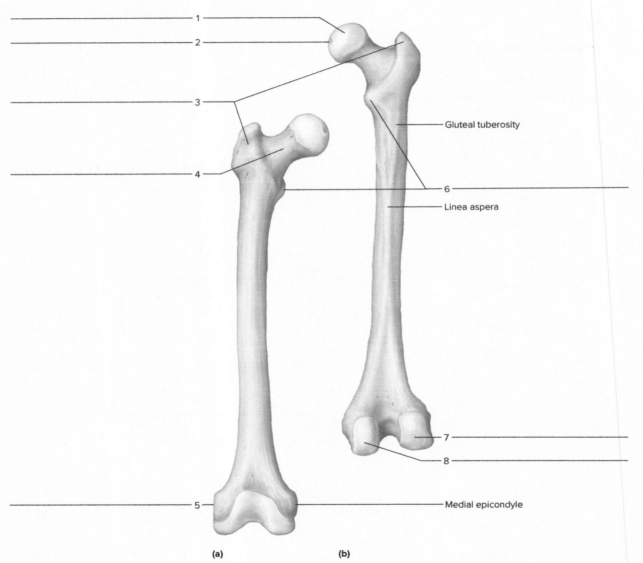

(a)   (b)

FIGURE 17.4 Label the features of (*a*) the anterior surface and (*b*) the posterior surface of the right femur. **2** **APR**

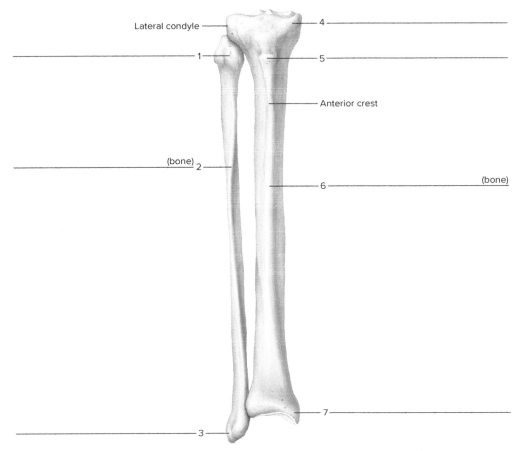

Lateral condyle

4

1

5

Anterior crest

(bone) 2

6 (bone)

7

3

FIGURE 17.5 Label the bones and features of the right tibia and fibula in this anterior view. 🅰 **APR**

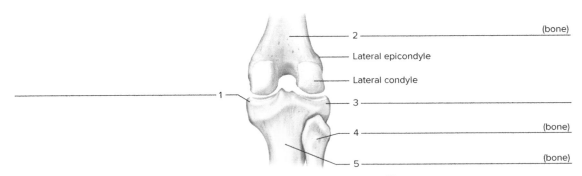

2 (bone)

Lateral epicondyle

Lateral condyle

1

3

4 (bone)

5 (bone)

FIGURE 17.6 Label the bones and features of the right knee, posterior view. 🅰 **APR**

patella
tibia
    medial condyle
    lateral condyle
    tibial tuberosity
    anterior crest (border)
    medial malleolus
fibula
    head
    lateral malleolus
tarsal bones
    talus
    calcaneus
    navicular
    cuboid

lateral cuneiform
intermediate cuneiform
medial cuneiform
metatarsal bones
phalanges
    proximal phalanx
    middle phalanx
    distal phalanx

4. Complete Parts B–E of the laboratory report.

LEARN: ACTIVITY

Use different colored pencils to distinguish the individual bones in figure 17.7.

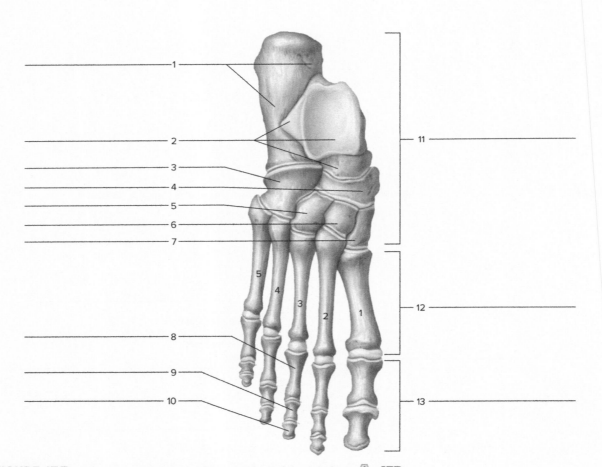

FIGURE 17.7   Label the bones of the superior surface of the right foot. 2 APR

Name _____

Date _____

Section _____

The 🅐 corresponds to the indicated Learning Outcome(s) found at the beginning of the laboratory exercise.

Pelvic Girdle and Lower Limb

PART A ASSESSMENTS

Complete the following statements:

1. The pelvic girdle consists of two _____ . 🅐

2. The head of the femur articulates with the _____ of the coxal (hip) bone. 🅐

3. The _____ is the largest portion of the coxal (hip) bone. 🅐

4. The distance between the _____ represents the shortest diameter of the pelvic outlet. 🅐

5. The pubic bones come together anteriorly to form a cartilaginous joint called the _____ . 🅐

6. The _____ is the portion of the ilium that causes the prominence of the coxal (hip) bone. 🅐

7. When a person sits, the _____ of the ischium supports the weight of the body. 🅐

8. The angle formed by the pubic bones below the pubic symphysis is called the _____ . 🅐

9. The _____ is the largest foramen in the skeleton. 🅐

10. The ilium joins the sacrum at the _____ joint. 🅐

PART B ASSESSMENTS

Match the bones in column A with the features in column B. Place the letter of your choice in the space provided. 🅑

Column A	Column B
a. Femur	_____ 1. Middle phalanx
	_____ 2. Lesser trochanter
b. Fibula	_____ 3. Medial malleolus
	_____ 4. Fovea capitis
c. Metatarsals	_____ 5. Calcaneus
	_____ 6. Lateral cuneiform
d. Patella	_____ 7. Tibial tuberosity
	_____ 8. Talus
e. Phalanges	_____ 9. Linea aspera
	_____ 10. Lateral malleolus
f. Tarsals	_____ 11. Sesamoid bone
g. Tibia	_____ 12. Five bones that form the instep

PART C ASSESSMENTS

Identify the bones and features indicated in the radiographs of figures 17.8–17.10. 🅰 🅰

Terms:
Head of femur
Ilium
Obturator foramen
Pubic symphysis
Pubis
Sacrum

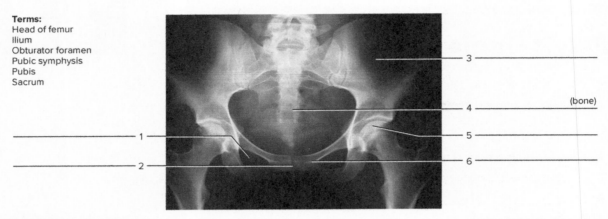

3 ————————

4 ———— (bone)

1 ————

5 ————————

2 ————

6 ————————

FIGURE 17.8   Identify the bones and features indicated on this radiograph of the anterior view of the pelvic region, using the terms provided. **APR**   (©Dale Butler)

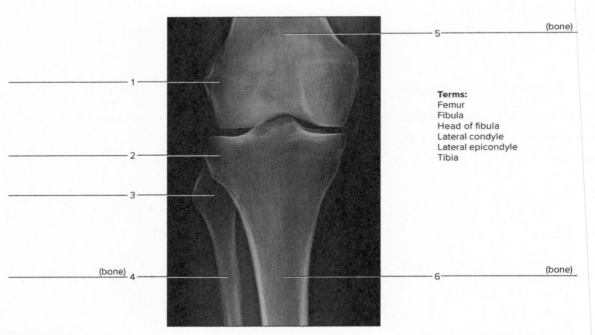

5 ———— (bone)

1 ————

2 ————

3 ————

Terms:
Femur
Fibula
Head of fibula
Lateral condyle
Lateral epicondyle
Tibia

(bone) 4 ————

6 ———— (bone)

FIGURE 17.9   Identify the bones and features indicated in this radiograph of the right knee (anterior view), using the terms provided. **APR**   ©Dale Butler

Terms:
Calcaneus
Distal phalanx
Metatarsal
Proximal phalanx
Talus
Tibia

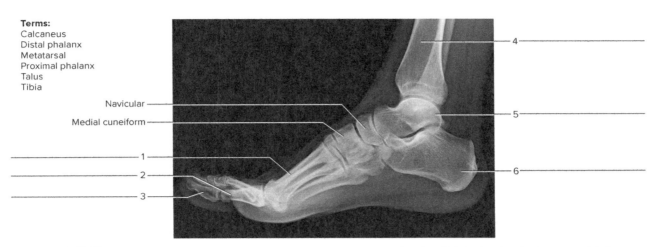

Navicular

Medial cuneiform

4

5

6

1

2

3

FIGURE 17.10   Identify the bones indicated on this radiograph of the right foot (medial view), using the terms provided.
(Courtesy Dale Butler)

 PART D ASSESSMENTS

Identify the bones of the foot in figure 17.11. /2/

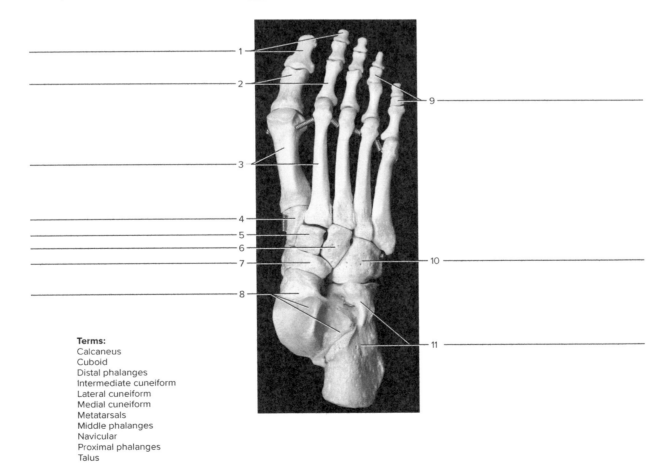

1
2
9
3
4
5
6
7
8
10
11

Terms:
Calcaneus
Cuboid
Distal phalanges
Intermediate cuneiform
Lateral cuneiform
Medial cuneiform
Metatarsals
Middle phalanges
Navicular
Proximal phalanges
Talus

FIGURE 17.11   Identify the bones indicated on this superior view of the right foot, using the terms provided.
(©J and J Photography)

18

Joint Structure and Movements

MATERIALS NEEDED

Textbook
Human skull
Human skeleton, articulated
Models of synovial joints (shoulder, elbow, hip, and knee)
Fresh animal joint (knee joint preferred)
Radiographs of major joints

⚠ SAFETY

- Wear disposable gloves when handling the fresh animal joint.
- Wash your hands before leaving the laboratory.

PURPOSE OF THE EXERCISE

To examine examples of the three types of joints, to identify the major features of these joints, and to review the types of movements produced at synovial joints.

LEARNING OUTCOMES APR

After completing this exercise, you should be able to

1. Identify key structural and functional characteristics of fibrous, cartilaginous, and synovial joints.
2. Locate examples of each type of joint.
3. Locate examples of each of the six types of synovial joints.
4. Demonstrate the types of movements that occur at synovial joints.
5. Examine the structure of the shoulder, elbow, hip, and knee joints.

Joints are junctions between bones. Although they vary considerably in structure, they can be classified according to the type of tissue that binds the bones together. Thus, the three groups of *structural joints* can be

identified as *fibrous joints, cartilaginous joints,* and *synovial joints.* Fibrous joints are filled with dense fibrous connective tissue, cartilaginous joints are filled with a type of cartilage, and synovial joints contain synovial fluid inside a joint cavity.

Functional joints can also be classified by the degree of movement allowed: *synarthroses* are immovable, *amphiarthroses* allow slight movement, and *diarthroses* allow free movement. Movements occurring at freely movable synovial joints are due to the contractions of skeletal muscles. In each case, the type of movement depends on the type of joint involved and the way in which the muscles are attached to the bones on either side of the joint.

PRACTICE

PROCEDURE A—Types of Joints APR

1. Review section 8.1 titled "Types of Joints" in chapter 8 of the textbook.
2. Examine the human skull and articulated skeleton to locate examples of the following types of joints:

fibrous joints

　suture (synarthrotic joint)—skull
　gomphosis (synarthrotic joint)—tooth sockets
　syndesmosis (amphiarthrotic joint)—tibiofibular
　　joint

cartilaginous joints

　synchondrosis (synarthrotic joint)—epiphyseal
　　plates
　symphysis (amphiarthrotic joint)—intervertebral
　　discs; pubic symphysis

3. Complete Part A of Laboratory Report 18.
4. Locate examples of the following types of synovial joints (diarthrotic joints) in the skeleton and in your own body. Experiment with each joint to experience its range of movements.

synovial joints

　ball-and-socket joint—hip; shoulder
　condylar (ellipsoidal) joint—radiocarpal;
　　metacarpophalangeal

plane (gliding) joint—intercarpal; intertarsal

hinge joint—elbow; knee; interphalangeal

pivot joint—radioulnar at elbow; dens at atlas

saddle joint—base of thumb with trapezium

5. Examine models of the shoulder, elbow, hip, and knee joints. Locate the major knee joint structures of figure 18.1 that are visible on the knee joint model.

6. Complete Parts B and C of the laboratory report.

LEARN: ACTIVITY

Examine a longitudinal section of a fresh synovial animal knee joint. Locate the dense connective tissue that forms the joint capsule and the hyaline cartilage that forms the articular cartilage on the ends of the bones. Locate the synovial membrane on the inner surface of the joint capsule. Does the joint have any semilunar cartilages (menisci)?

_____

What is the function of such cartilages?

_____

_____

_____

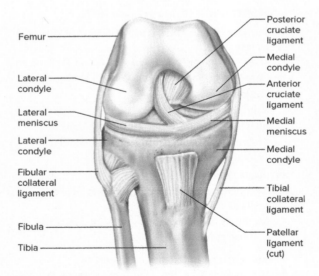

FIGURE 18.1   Anterior view of right bent knee (patella removed). **APR**

Labels:
- Femur
- Lateral condyle
- Lateral condyle
- Lateral meniscus
- Lateral condyle
- Fibular collateral ligament
- Fibula
- Tibia
- Posterior cruciate ligament
- Medial condyle
- Anterior cruciate ligament
- Medial meniscus
- Medial condyle
- Tibial collateral ligament
- Patellar ligament (cut)

PRACTICE

PROCEDURE B—Joint Movements

1. Review section 8.2 titled "Types of Joint Movements" in chapter 8 of the textbook and study figures 18.2 and 18.3.

2. Most joints and body parts are extended and/or adducted when the body is positioned in anatomical position. Skeletal muscle action involves the more movable end (*insertion*) being pulled toward the more stationary end (*origin*). In the limbs, the origin is usually proximal to the insertion; in the trunk, the origin is usually medial to the insertion. Use these concepts as reference points as you move joints. Move various parts of your body to demonstrate the synovial joint movements described in tables 18.1 and 18.2.

3. Have your laboratory partner do some of the movements and see if you can correctly identify the movements made.

ASSESS

CRITICAL THINKING

Describe a body position that can exist when all major body parts are flexed.

_____

_____

4. Complete Parts D and E of the laboratory report.

LEARN: ACTIVITY

Study the available radiographs of joints by holding the films in front of a light source. Identify the type of joint and the bones incorporated in the joint. Also identify other major visible features.

TABLE 18.1 Angular and Rotational Movements

Movement	Description
Flexion	Decrease of an angle (usually in the sagittal plane)
Extension	Increase of an angle (usually in the sagittal plane)
Hyperextension	Extension beyond anatomical position
Abduction	Movement of a body part away from the midline (usually in the frontal plane)
Adduction	Movement of a body part toward the midline (usually in the frontal plane)
Circumduction	Circular movement (combines flexion, abduction, extension, and adduction)
Rotation	Movement of part around its long axis
Medial (internal)	Inward rotation
Lateral (external)	Outward rotation

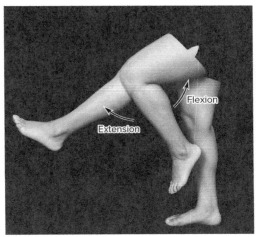

(a) Flexion and extension of the knee joint

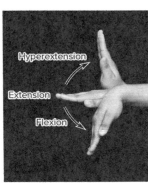

(b) Flexion, extension, and hyper-extension of the hand at the wrist joint

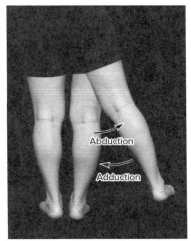

(c) Abduction and adduction of the hip joint

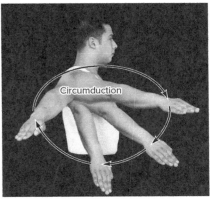

(d) Circumduction of the shoulder joint

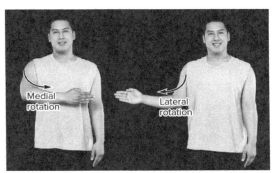

(e) Medial and lateral rotation of the arm at the shoulder joint

FIGURE 18.2   Examples of angular and rotational movements of synovial joints (*a–e*).   **APR**   ((*a, b*) J and J Photography; (*c–e*) J & J Photography)

TABLE 18.2 Special Movements (pertain to specific joints)

Movement	Description
Supination	Rotation of forearm and palm of hand anteriorly or upward
Pronation	Rotation of forearm and palm of hand posteriorly or downward
Elevation	Movement of body part upward
Depression	Movement of body part downward
Dorsiflexion	Movement of ankle joint so dorsum (superior) of foot becomes closer to anterior surface of leg (as standing on heels)
Plantar flexion	Movement of ankle joint so the plantar surface of foot becomes closer to the posterior surface of leg (as standing on toes)
Inversion (called supination in some health professions)	Medial movement of sole of foot at ankle joint
Eversion (called pronation in some health professions)	Lateral movement of sole of foot at ankle joint
Protraction	Anterior movement in the transverse plane
Retraction	Posterior movement in the transverse plane

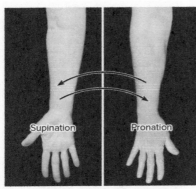

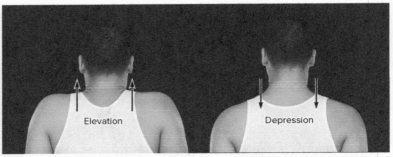

(a) Supination and pronation of the forearm and hand involving movement at the radioulnar joint **APR**

(b) Elevation and depression of the shoulder (scapula) **APR**

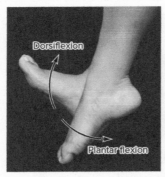

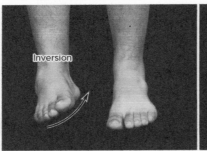

(c) Dorsiflexion and plantar flexion of the foot at the ankle joint **APR**

(d) Inversion and eversion of the right foot at the ankle joint (The left foot is unchanged for comparison.) **APR**

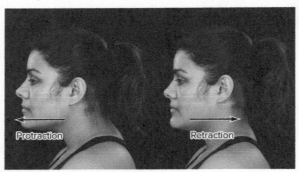

(e) Protraction and retraction of the head

FIGURE 18.3 Special movements of synovial joints (a–e). J and J Photography

Name _____

Date _____

Section _____

The 🅐 corresponds to the indicated Learning Outcome(s) found at the beginning of the laboratory exercise.

Joint Structure and Movements

PART A ASSESSMENTS

Match the terms in column A with the descriptions in column B. Place the letter of your choice in the space provided. 🅐 🅐

Column A	Column B
a. Gomphosis	_____ **1.** Immovable joint between flat bones of the skull united by a thin layer of dense connective tissue
b. Suture	_____ **2.** Fibrocartilage pad fills the slightly movable joint
c. Symphysis	_____ **3.** Temporary joint in which bones are united by bands of hyaline cartilage
d. Synchondrosis	_____ **4.** Slightly movable joint in which bones are united by interosseous membrane
e. Syndesmosis	_____ **5.** Joint formed by union of tooth root in bony socket

PART B ASSESSMENTS

Identify the types of structural and functional joints numbered in figure 18.4. 🅐 🅐

Structural Classification	Functional Classification
1. _____	_____
2. _____	_____
3. _____	_____
4. _____	_____
5. _____	_____
6. _____	_____
7. _____	_____
8. _____	_____
9. _____	_____

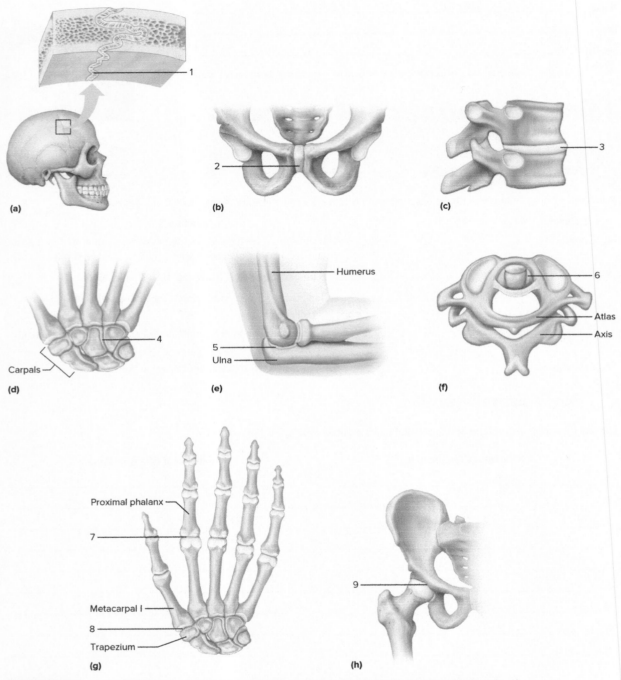

FIGURE 18.4   Identify the types of structural and functional joints numbered in these illustrations (*a–h*).

PART C ASSESSMENTS

Match the types of synovial joints in column A with the examples in column B. Place the letter of your choice in the space provided.

Column A	Column B
a. Ball-and-socket	_____ **1.** Hip joint
	_____ **2.** Metacarpal-phalanx
b. Condylar (ellipsoidal)	_____ **3.** Proximal radius-ulna
	_____ **4.** Humerus-ulna of the elbow joint
c. Hinge	_____ **5.** Phalanx-phalanx
	_____ **6.** Shoulder joint
d. Pivot	_____ **7.** Knee joint
	_____ **8.** Carpal-metacarpal of the thumb
e. Plane (gliding)	_____ **9.** Carpal-carpal
	_____ **10.** Tarsal-tarsal
f. Saddle	

PART D ASSESSMENTS

Complete the missing components of the following table:

Name of Joint	Type of Joint	Bones Included	Types of Movement Possible
Shoulder joint		Humerus, scapula	
Elbow joint			Flexion and extension between humerus and ulna; twisting between radius and humerus; rotation between head of radius and ulna
Hip joint			Movements in all planes and rotation
Knee joint	Hinge (modified), condylar, plane		

PART E ASSESSMENTS

Identify the types of joint movements numbered in figure 18.5.

1. _____ (of head)		**15.** _____ (of vertebral column/trunk)	
2. _____ (of shoulder)		**16.** _____ (of vertebral column/trunk)	
3. _____ (of shoulder)		**17.** _____ (of head and neck)	
4. _____ (of hand at radioulnar joint)		**18.** _____ (of head and neck)	
5. _____ (of hand at radioulnar joint)		**19.** _____ (of arm at shoulder)	
6. _____ (of arm at shoulder)		**20.** _____ (of arm at shoulder)	
7. _____ (of arm at shoulder)		**21.** _____ (of elbow)	
8. _____ (of wrist)		**22.** _____ (of elbow)	
9. _____ (of wrist)		**23.** _____ (of thigh at hip)	
10. _____ (of thigh at hip)		**24.** _____ (of thigh at hip)	
11. _____ (of thigh at hip)		**25.** _____ (of knee)	
12. _____ (of lower limb at hip)		**26.** _____ (of knee)	
13. _____ (of chin/mandible)		**27.** _____ (of foot at ankle)	
14. _____ (of chin/mandible)		**28.** _____ (of foot at ankle)	

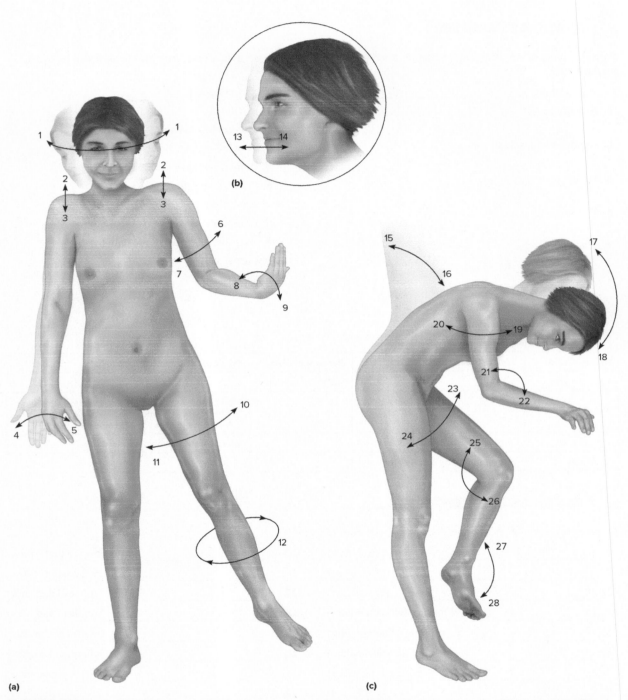

(a)

(b)

(c)

FIGURE 18.5   Identify each of the types of movements numbered and illustrated: (*a*) anterior view; (*b*) lateral view of head; (*c*) lateral view.

PART F ASSESSMENTS

Using the terms provided, label the structures indicated in Figure 18.6.

Terms:
Anterior cruciate ligament
Articular surface of femur
Femur
Fibula
Fibular collateral ligament
Lateral meniscus
Medial meniscus
Tibia
Tibial collateral ligament
Tibial tuberosity
Transverse ligament

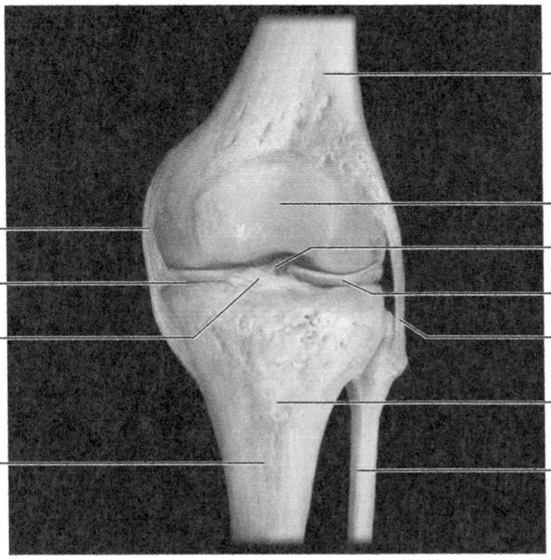

1 _____

2 _____

3 _____

4 _____

5 _____

6 _____

7 _____

8 _____

9 _____

10 _____

11 _____

FIGURE 18.6 All the muscles of the left knee and the patella have been removed. The knee is seen in extension. **APR**
McGraw-Hill Education

Muscular System

LABORATORY EXERCISE

19

Skeletal Muscle Structure and Function

MATERIALS NEEDED

Textbook
Compound light microscope
Prepared microscope slide of skeletal muscle tissue
   (longitudinal section and cross section)
Human muscular model
Model of skeletal muscle fiber
Fresh round beefsteak

⚠ SAFETY

■ Wear disposable gloves when handling the fresh
   beefsteak.
■ Wash your hands before leaving the laboratory.

PURPOSE OF THE EXERCISE

To review the structure of a skeletal muscle.

 ## LEARNING OUTCOMES APR

After completing this exercise, you should be able to

1. Locate the major structures of a skeletal muscle fiber
   on a microscope slide of skeletal muscle tissue and on a
   model.

2. Describe how connective tissue is associated with
   muscle tissue within a skeletal muscle.

3. Distinguish between the origin and insertion of a
   muscle.

4. Describe and demonstrate the general actions of
   agonists, synergists, and antagonists.

A skeletal muscle represents an organ of the muscular system and is composed of several types of tissues. These
tissues include skeletal muscle tissue, nervous tissue, blood,
and various connective tissues.

Each skeletal muscle is encased within and permeated by
connective tissue sheaths. The connective tissue often projects beyond the end of a muscle, providing an attachment to
other muscles or to bones as a tendon. The connective tissues
also extend into the structure of a muscle and separate it into
compartments. The outer layer, called *epimysium,* extends
deeper into the muscle as *perimysium,* containing bundles of
cells (fascicles), and farther inward extends around each muscle cell (fiber) as a thin *endomysium.* The connective tissues
provide support and reinforcement during muscular contractions and allow portions of a muscle to contract somewhat
independently. Some of the collagen fibers are continuous
with the tendon, periosteum, and inward extensions as bone
fibers, allowing a strong structural continuity.

Muscles are named according to location, size, action,
shape, attachments, or number of origins, or the direction
of the fibers. Examples of how muscles are named include
these: gluteus maximus (location and size); adductor
longus (action and size); sternocleidomastoid (attachments);
serratus anterior (shape and location); biceps brachii (two
origins and location); and orbicularis oculi (direction of
fibers and location).

 ## PRACTICE

**PROCEDURE—Skeletal Muscle Structure
and Function**

1. Study section 9.2 titled "Structure of Skeletal Muscle"
   in chapter 9 of the textbook.

2. Examine the microscopic structure of a longitudinal
   section of skeletal muscle by observing a prepared
   microscope slide of this tissue. Use figure 19.1 of skeletal muscle tissue to locate the following features:

   skeletal muscle fiber (cell)
   nuclei
   striations (alternating light and dark)

3. Examine the microscopic structure of a cross section of
   a skeletal muscle by observing a prepared microscope

slide of this tissue. Use figure 19.2 of skeletal muscle tissue to locate the following features:

connective tissue
    perimysium
    endomysium

fascicle

muscle cell
    nucleus
    myofibrils

4. As a review activity, label figure 19.3 and study figure 19.4.
5. Examine the human torso model and locate examples of fascia, tendons, and aponeuroses. An origin tendon is attached to a less movable end of the muscle, whereas the insertion tendon is attached to a more movable end of the muscle. Sheets of connective tissue, called aponeuroses, also serve for some muscle attachments. Locate examples of tendons in your body.
6. Complete Part A of Laboratory Report 19.

 LEARN: ACTIVITY

Examine the fresh round beefsteak. It represents a cross section through the beef thigh muscles. Note the white lines of connective tissue that separate the individual skeletal muscles. Also note how the connective tissue extends into the structure of a muscle and separates it into small compartments of muscle tissue. Locate the epimysium and the perimysium of an individual muscle.

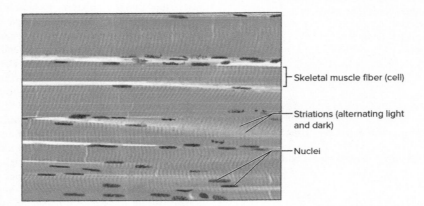

Skeletal muscle fiber (cell)

Striations (alternating light and dark)

Nuclei

FIGURE 19.1 Micrograph of a longitudinal section of skeletal muscle fibers (cells) (400×). **APR** (Al Telser/McGraw-Hill Education)

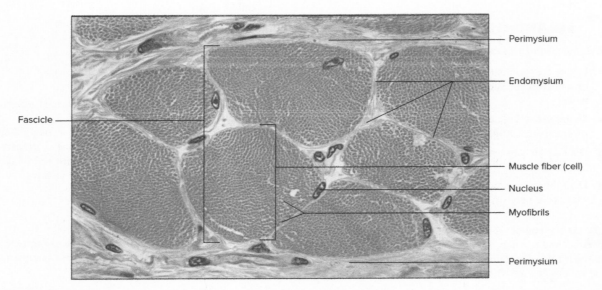

Perimysium

Endomysium

Fascicle

Muscle fiber (cell)

Nucleus

Myofibrils

Perimysium

FIGURE 19.2 Micrograph of a cross section of a fascicle and associated connective tissues (800×). (Ed Reschke/Stone/Getty Images)

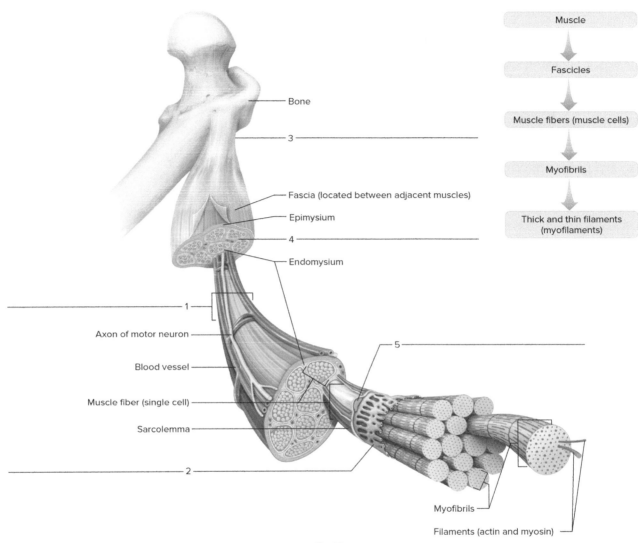

Muscle

Fascicles

Muscle fibers (muscle cells)

Myofibrils

Thick and thin filaments (myofilaments)

Bone

3

Fascia (located between adjacent muscles)

Epimysium

4

Endomysium

1

Axon of motor neuron

Blood vessel

Muscle fiber (single cell)

Sarcolemma

2

5

Myofibrils

Filaments (actin and myosin)

FIGURE 19.3   Label the structures of a skeletal muscle.

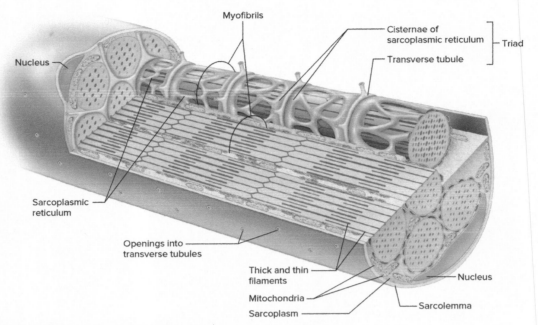

FIGURE 19.4   Structures of a skeletal muscle fiber.

7. Examine the model of the skeletal muscle fiber and locate the following:

sarcolemma

sarcoplasm

myofibril

    thick (myosin) filament

    thin (actin, tropomyosin, and troponin) filament

sarcomere (functional contractile unit)

    A band (dark)

    I band (light)

    H zone

    M line

    Z line (disc)

sarcoplasmic reticulum

    cisternae

transverse (T) tubules

8. Complete Part B of the laboratory report.

9. Review section 9.7 titled "Skeletal Muscle Actions" in chapter 9 of the textbook.

10. Locate the biceps brachii and its origins and insertions in the human torso model and in your body.

11. Make various movements with your upper limb at the shoulder and elbow. For each movement, determine the location of the muscles that function as *agonists* and as *antagonists*. An agonist causes an action, whereas an antagonist works against the action; however, the role of the muscle is dependent on the particular movement involved. Antagonistic muscle pairs pull from opposite sides of the same body region. *Synergistic* muscles often supplement the contraction force of an agonist, or by acting to stabilize nearby joints. **The role of a muscle as an agonist, an antagonist, or a synergist depends upon the movement under consideration, as their roles can change.** These voluntary movements of complex actions are controlled by the nervous system (see table 19.1).

12. Complete Part C of the laboratory report.

TABLE 19.1 Various Roles of Muscles

Functional Category	Description
Agonist*	Muscle responsible for the action (movement)
Antagonist	Muscle responsible for action in the opposite direction of an agonist or for resistance to an agonist
Synergist	Muscle that assists an agonist, often by supplementing the contraction force

*The term *agonist* is often used interchangeably with *prime mover*. When the term prime mover is used, it usually refers to the agonist that provides most of the force of the movement of the action involved.

Name _____

Date _____

Section _____

The ⬟ corresponds to the indicated Learning Outcome(s) found at the beginning of the laboratory exercise.

Skeletal Muscle Structure and Function

♻ PART A ASSESSMENTS

Match the terms in column A with the definitions in column B. Place the letter of your choice in the space provided. ⬟1 ⬟2

Column A		Column B
a. Endomysium	_____	**1.** Membranous channel extending inward from muscle fiber membrane
b. Epimysium	_____	**2.** Cytoplasm of a muscle fiber
c. Fascia	_____	**3.** Connective tissue located between adjacent muscles
d. Fascicle	_____	**4.** Layer of connective tissue that separates a muscle into small bundles called fascicles
e. Myosin	_____	**5.** Cell membrane of a muscle fiber
f. Perimysium	_____	**6.** Layer of connective tissue that surrounds a skeletal muscle
g. Sarcolemma	_____	**7.** Unit of alternating light and dark striations between Z lines
h. Sarcomere	_____	**8.** Layer of connective tissue that surrounds an individual muscle fiber
i. Sarcoplasm	_____	**9.** Cellular organelle in muscle fiber corresponding to the endoplasmic reticulum
j. Sarcoplasmic reticulum	_____	**10.** Cordlike part that attaches a muscle to a bone
k. Tendon	_____	**11.** Protein found within thick filament
l. Transverse (T) tubule	_____	**12.** Small bundle of muscle fibers within a muscle

♻ PART B ASSESSMENTS

Provide the labels for the electron micrograph in figure 19.5. ⬟1

1. _____

2. _____

3. _____

4. _____

Terms:
A band (dark)
I band (light)
Sarcomere
Z line

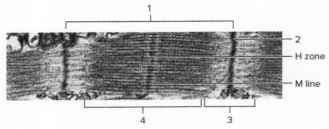

FIGURE 19.5 Identify the bands and lines of the striations of a transmission electron micrograph of relaxed sarcomeres (16,000×), using the terms provided. **APR** (Biology Pics/Science Source)

PART C ASSESSMENTS

Complete the following statements:

1. The _____ of a muscle is typically attached to a more immovable or fixed part. **3**

2. The _____ of a muscle is typically attached to a more movable part on the other side of the joint. **3**

3. When the forearm is extended at the elbow joint, the _____ muscle acts as the agonist. **4**

4. The forearm is flexed at the elbow when the _____ muscle contracts. **4**

5. A muscle responsible for an action is called a(an) _____ . **4**

6. Assisting muscles of an agonist are called as_____ . **4**

7. Antagonists are muscles that resist the actions of _____ and cause movement in the opposite direction. **4**

8. An agonist that provides the most force of a specific action is sometimes referred to as a _____ . **4**

LABORATORY EXERCISE

20

Muscles of the Head and Neck

MATERIALS NEEDED

Textbook
Human torso model with musculature
Human skull
Human skeleton, articulated
Long rubber bands

PURPOSE OF THE EXERCISE

To review the locations, actions, origins, and insertions of the muscles of the head and neck.

LEARNING OUTCOMES APR

After completing this exercise, you should be able to

(1) Locate and identify the muscles of facial expression, the muscles of mastication, and the muscles that move the head.

(2) Describe and demonstrate the action of each of these muscles.

(3) Locate the origin and insertion of each of these muscles in a human skeleton and the musculature of the human torso model.

The skeletal muscles of the face and head include the muscles of facial expression, which lie just beneath the skin; the muscles of mastication, attached to the mandible; and the muscles that move the head, located in the neck.

PRACTICE

PROCEDURE—Muscles of the Head and Neck

1. Review the concept headings titled "Muscles of Facial Expression," "Muscles of Mastication," and "Muscles That Move the Head and Vertebral Column" in section 9.8 in chapter 9 of the textbook.

2. As a review activity, label figures 20.1–20.3.

3. Locate the following muscles in the human torso model and in your body whenever possible:

muscles of facial expression
  epicranius
    frontalis
    occipitalis
  orbicularis oculi
  orbicularis oris
  buccinator
  zygomaticus major
  zygomaticus minor
  platysma

muscles of mastication
  masseter
  temporalis
  medial pterygoid
  lateral pterygoid

muscles that move the head and neck
  sternocleidomastoid
  splenius capitis
  semispinalis capitis
  scalene muscles (scalenes)

4. Demonstrate the action of these muscles in your body.

5. Locate the origins and insertions of these muscles in the human skull and skeleton.

6. Complete Parts A–C of Laboratory Report 20.

LEARN: LAB IN MOTION

A long rubber band can be used to simulate muscle locations, origins, insertions, and actions on the human torso model, the skeleton, or a laboratory partner. Hold one end of the rubber band firmly on the origin location of a muscle; then slightly stretch the rubber band and hold the other end on the insertion site. Allow the insertion end to slowly move toward the origin end to simulate the contraction and action of the muscle. 🅐 🅑 🅒

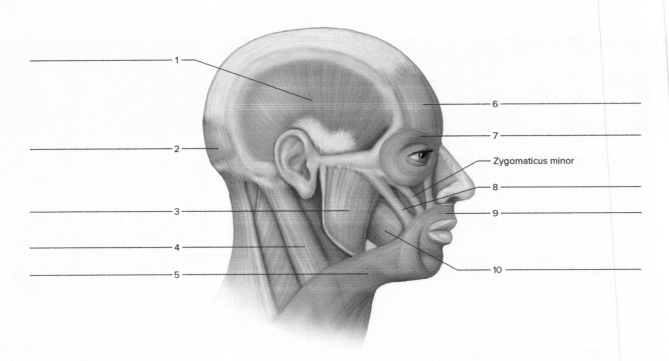

FIGURE 20.1   Label the muscles of expression and mastication.

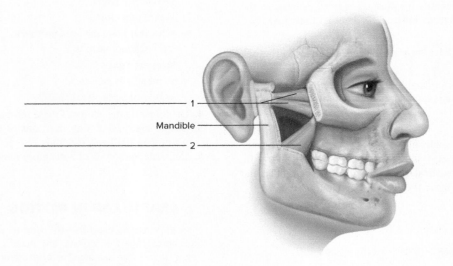

FIGURE 20.2   Label these deep muscles of mastication.

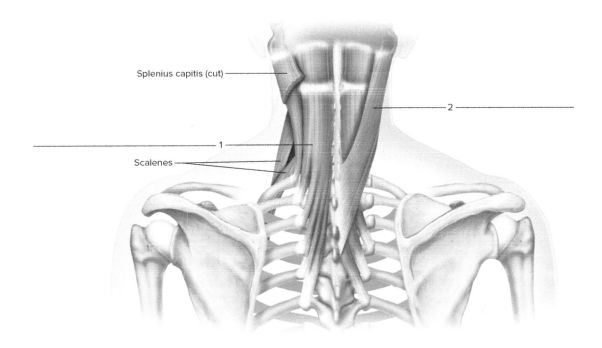

Splenius capitis (cut)

2

1

Scalenes

FIGURE 20.3   Label these deep muscles of the posterior neck (trapezius removed). **1** **APR**

Notes

Name _____

Date _____

Section _____

The Ⓐ corresponds to the indicated Learning Outcome(s) found at the beginning of the laboratory exercise.

Muscles of the Head and Neck

♻ PART A ASSESSMENTS

Complete the following statements:

1. When the _____ contracts, the corner of the mouth is elevated. Ⓐ

2. The _____ acts to compress the wall of the cheeks, as when air is blown out of the mouth. Ⓐ

3. The _____ causes the lips to close and pucker during kissing, whistling, and speaking. Ⓐ

4. The _____ and platysma help to depress the mandible. Ⓐ

5. The _____ acts to elevate and retract the mandible. Ⓐ

6. The _____ pterygoid can elevate the mandible and pull it sideways. Ⓐ

7. The _____ pterygoid can protract the mandible and pull the jaw sideways. Ⓐ

8. The _____ can close the eye, as in blinking. Ⓐ

9. The _____ together pull the head and neck forward and down toward the chest. Ⓐ

10. The _____ can rotate head and laterally flex and extend neck. Ⓐ

11. The muscle used to pout and to express horror by pulling the lower lip downward is the _____ . Ⓐ

12. The muscle used to smile and laugh is the _____ . Ⓐ

♻ PART B ASSESSMENTS

Name the muscle indicated by the following combinations of origin and insertion. Ⓐ

Origin	Insertion	Muscle
1. Occipital bone	Skin around eye	_____
2. Zygomatic bone	Corner of mouth skin and muscle	_____
3. Zygomatic arch	Lateral surface of mandible	_____
4. Sphenoid bone	Anterior surface of mandibular condyle	_____
5. Manubrium of sternum and clavicle	Mastoid process of temporal bone	_____
6. Alveolar processes of mandible and maxilla	Orbicularis oris	_____
7. Fascia in upper chest	Mandible and skin below mouth	_____
8. Temporal bone	Coronoid process of mandible	_____
9. Processes of lower cervical and upper thoracic vertebrae	Occipital bone	_____
10. Transverse processes of cervical vertebrae	Surfaces of ribs 1-2	_____

PART C ASSESSMENTS

 ASSESS

CRITICAL THINKING

Using the terms provided, identify the muscles of facial expression being contracted in each of these photographs by placing the number with the correct term. Ⓐ¹ Ⓐ²

Terms:
Epicranius/frontalis _____
Orbicularis oculi _____
Orbicularis oris _____
Platysma _____
Zygomaticus major _____

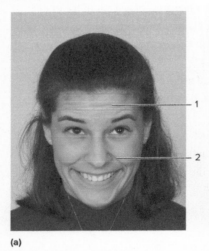

(a)

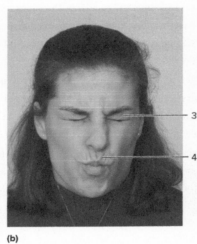

(b)

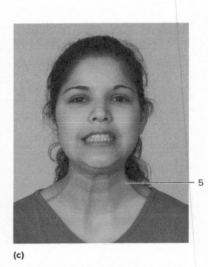

(c)

FIGURE 20.4 Identify the muscles of facial expression being contracted in each of these photographs (a–c), using the terms provided. ((a–c) ©J & J Photography)

PART D ASSESSMENTS

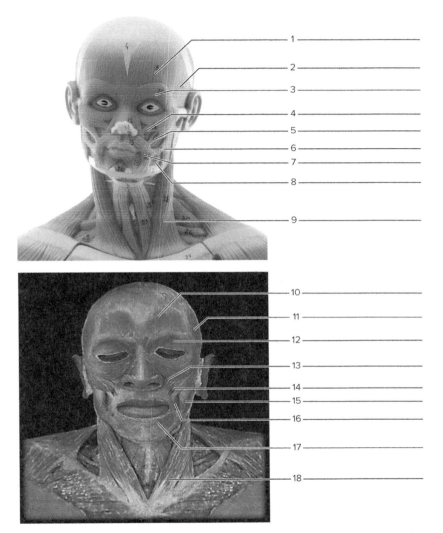

Terms:
Buccinator
Frontalis
Masseter
Orbicularis oculi
Orbicularis oris
Sternocleidomastoid
Temporalis
Zygomaticus major
Zygomaticus minor

FIGURE 20.5 Using the terms provided identify the muscles of the head and neck on the model image and the cadaver image. Each term will be used twice. ((*a*) B59 [1000212] 1/3 Life-Size Muscle Figure, 2-part © 3B Scientific GmbH, Germany, 2017 www.3bscientific.com), (*b*) McGraw-Hill Education/APR)

LABORATORY EXERCISE

21

Muscles of the Chest, Shoulder, and Upper Limb

MATERIALS NEEDED

Textbook
Human torso model
Human skeleton, articulated
Muscular models of the upper limb
Long rubber bands

PURPOSE OF THE EXERCISE

To review the locations, actions, origins, and insertions of the muscles in the chest, shoulder, and upper limb.

 LEARNING OUTCOMES APR

After completing this exercise, you should be able to

1. Locate and identify the muscles of the chest, shoulder, and upper limb.
2. Describe and demonstrate the action of each of these muscles.
3. Locate the origin and insertion of each of these muscles in a human skeleton and on muscular models.

The muscles of the chest and shoulder are responsible for moving the scapula and arm, whereas those within the arm and forearm move joints in the elbow and hand.

 PRACTICE

PROCEDURE—Muscles of the Chest, Shoulder, and Upper Limb

1. Review the concept headings titled "Muscles That Move the Pectoral Girdle," "Muscles That Move the Arm," "Muscles That Move the Forearm," and "Muscles That Move the Hand" in section 9.8 in chapter 9 of the textbook.
2. As a review activity, label figures 21.1–21.4.

3. Locate the following muscles in the human torso model and models of the upper limb. Also locate in your body as many of the muscles as you can.

muscles that move the pectoral girdle
  anterior muscles
    pectoralis minor
    serratus anterior
  posterior muscles
    trapezius
    rhomboid major
    rhomboid minor
    levator scapulae
muscles that move the arm
  origins on axial skeleton
    pectoralis major
    latissimus dorsi
  origins on scapula
    deltoid
    teres major
    coracobrachialis
  rotator cuff (SITS) muscles (origins also on scapula)
    supraspinatus
    infraspinatus
    teres minor
    subscapularis
muscles that move the forearm
  muscle bellies in arm
    biceps brachii
    brachialis
    triceps brachii
  muscle bellies in forearm
    brachioradialis
    supinator
    pronator teres
    pronator quadratus

muscles that move the hand
   anterior flexor muscles
      flexor carpi radialis
      flexor carpi ulnaris
      palmaris longus
      flexor digitorum profundus
      flexor digitorum superficialis

posterior extensor muscles
   extensor carpi radialis longus
   extensor carpi radialis brevis
   extensor carpi ulnaris
   extensor digitorum

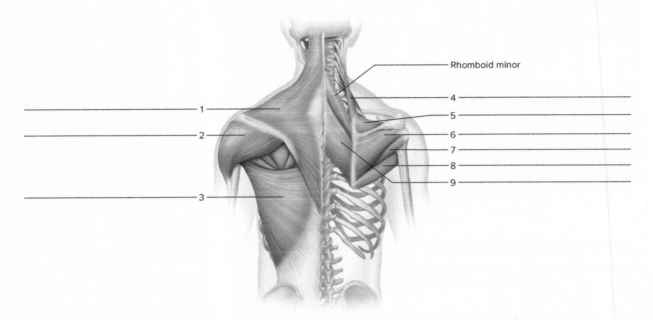

FIGURE 21.1 Label the muscles of the posterior shoulder. Superficial muscles are illustrated on the left side and deep muscles on the right side. 🅐 APR

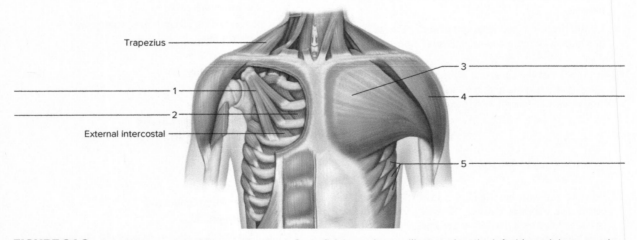

FIGURE 21.2 Label the muscles of the anterior chest. Superficial muscles are illustrated on the left side and deep muscles on the right side. 🅐 APR

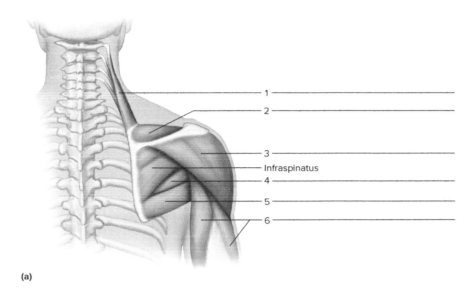

(a)

Infraspinatus

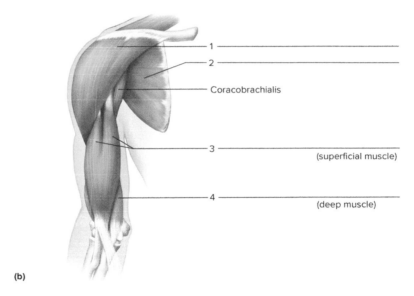

Coracobrachialis

(superficial muscle)

(deep muscle)

(b)

FIGURE 21.3   Label (*a*) the muscles of the posterior shoulder and arm and (*b*) the muscles of the anterior shoulder and arm, with the rib cage removed. 🔺 APR

4. Demonstrate the action of these muscles in your body.
5. Locate the origins and insertions of these muscles in the human skeleton.
6. Complete Parts A, B, and C of Laboratory Report 21.

 LEARN: LAB IN MOTION

A long rubber band can be used to simulate muscle locations, origins, insertions, and actions on muscular models, the skeleton, or a laboratory partner. Hold one end of the rubber band firmly on the origin location of a muscle; then slightly stretch the rubber band and hold the other end on the insertion site. Allow the insertion end to slowly move toward the origin end to simulate the contraction and action of the muscle. 🔺 2 3

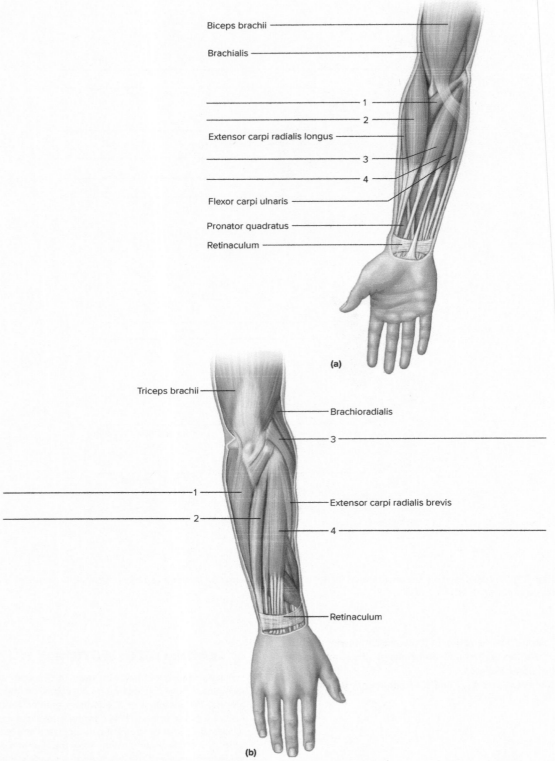

Biceps brachii

Brachialis

1

2

Extensor carpi radialis longus

3

4

Flexor carpi ulnaris

Pronator quadratus

Retinaculum

(a)

Triceps brachii

Brachioradialis

3

1

Extensor carpi radialis brevis

2

4

Retinaculum

(b)

FIGURE 21.4 Label (*a*) the muscles of the anterior forearm and (*b*) the muscles of the posterior forearm. **APR**

Name _____

Date _____

Section _____

The Ⓐ corresponds to the indicated Learning Outcome(s) found at the beginning of the laboratory exercise.

Muscles of the Chest, Shoulder, and Upper Limb

PART A ASSESSMENTS

Match the muscles in column A with the actions in column B. Place the letter of your choice in the space provided. Ⓐ

Column A	Column B
a. Brachialis	_____ **1.** Abducts arm and flexes and extends shoulder
b. Coracobrachialis	_____ **2.** Flexes shoulder, adducts and rotates arm medially
c. Deltoid	_____ **3.** Flexes wrist and adducts hand
d. Extensor carpi ulnaris	_____ **4.** Elevates and retracts scapula
e. Flexor carpi ulnaris	_____ **5.** Depresses and protracts scapula
f. Infraspinatus	_____ **6.** Used to thrust shoulder anteriorly (protraction), as when pushing something
g. Pectoralis major	_____ **7.** Flexes elbow
h. Pectoralis minor	_____ **8.** Flexes shoulder and adducts arm
i. Rhomboid major	_____ **9.** Extends elbow
j. Serratus anterior	_____ **10.** Extends shoulder and adducts and rotates arm medially
k. Teres major	_____ **11.** Extends wrist and adducts hand
l. Triceps brachii	_____ **12.** Rotates arm laterally

PART B ASSESSMENTS

Name the muscle indicated by the following combinations of origin and insertion. [3]

Origin	Insertion	Muscle
1. Spinous processes of upper thoracic vertebrae	Medial border of scapula	_____
2. Anterior surfaces of upper ribs	Medial border of scapula	_____
3. Anterior surfaces of ribs 3–5	Coracoid process of scapula	_____
4. Coracoid process of scapula	Shaft of humerus	_____
5. Lateral border of scapula	Intertubercular sulcus of humerus	_____
6. Anterior surface of scapula	Lesser tubercle of humerus	_____
7. Lateral border of scapula	Greater tubercle of humerus	_____
8. Anterior shaft of humerus	Coronoid process of ulna	_____
9. Medial epicondyle of humerus and coronoid process of ulna	Lateral surface of proximal radius	_____
10. Distal lateral end of humerus	Lateral surface of radius above styloid process	_____
11. Medial epicondyle of humerus	Base of second and third metacarpals	_____
12. Medial epicondyle of humerus	Fascia of palm	_____

 PART C ASSESSMENTS

 ASSESS

CRITICAL THINKING

Identify the muscles indicated in figure 21.5. [1]

1. _____
2. _____
3. _____
4. _____
5. _____
6. _____
7. _____
8. _____
9. _____
10. _____
11. _____

12. _____
13. _____
14. _____
15. _____
16. _____
17. _____
18. _____
19. _____
20. _____
21. _____
22. _____

Terms:
Biceps brachii
Deltoid
External oblique
Pectoralis major
Rectus abdominis
Serratus anterior
Sternocleidomastoid
Trapezius

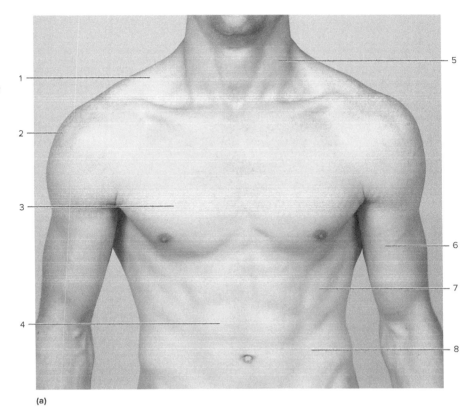

(a)

Terms:
Biceps brachii
Deltoid
Infraspinatus
Latissimus dorsi
Trapezius
Teres major
Triceps brachii

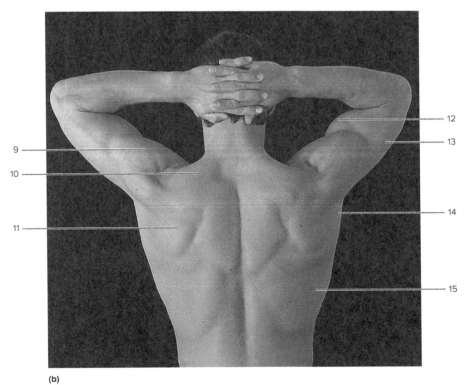

(b)

FIGURE 21.5 Identify the muscles that appear as body surface features in these photographs (a, b, and c), using the terms provided. ((a, c) ©J & J Photography, (b) Dr. Kent M. Van De Graaff)

Terms:
Biceps brachii
Brachioradialis
Deltoid
Pectoralis major
Serratus anterior
Trapezius
Triceps brachii

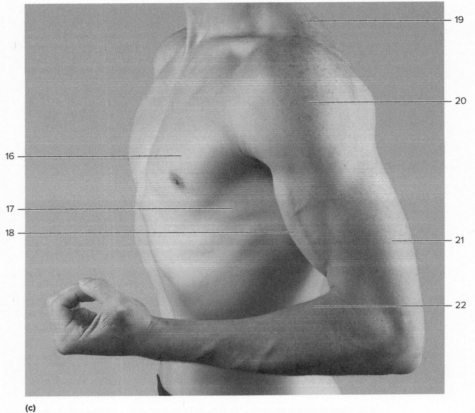

(c)

FIGURE 21.5   *Continued.*

LABORATORY EXERCISE

22

Muscles of the Deep Back, Abdominal Wall, and Pelvic Floor

MATERIALS NEEDED

Textbook
Human torso model with musculature
Human skeleton, articulated
Muscular models of male and female pelves

PURPOSE OF THE EXERCISE

To review the actions, origins, and insertions of the muscles of the deep back, abdominal wall, and pelvic floor.

LEARNING OUTCOMES APR

After completing this exercise, you should be able to

1. Locate and identify the muscles of the deep back, abdominal wall, and pelvic floor.

2. Describe the action of each of these muscles.

3. Locate the origin and insertion of each of these muscles in a human skeleton or on muscular models.

The deep muscles of the back extend the vertebral column. Because the muscles have numerous origins, insertions, and subgroups, the muscles overlap each other. The deep back muscles can not only extend the spine when contracting as a group but also help to maintain posture and normal spine curvatures.

The anterior and lateral walls of the abdomen contain broad, flattened muscles arranged in layers. These muscles connect the rib cage and vertebral column to the pelvic girdle. The abdominal wall muscles compress the abdominal visceral organs, help maintain posture, assist in forceful exhalation, and contribute to trunk flexion and waist rotation.

The muscles of the pelvic floor are arranged in two muscular sheets: (1) a pelvic diaphragm that spans the outlet of the pelvic cavity and (2) a urogenital diaphragm that fills the

space within the pubic arch. The pelvic floor is penetrated by the urethra, vagina, and anus in a female; thus, pelvic floor muscles are important in obstetrics.

PRACTICE

PROCEDURE A—Muscles of the Deep Back and Abdominal Wall

1. Review the concept headings titled "Muscles That Move the Head and Vertebral Column" and "Muscles of the Abdominal Wall" in section 9.8 in chapter 9 of the textbook.

2. As a review activity, label figures 22.1 and 22.2.

3. Locate the following muscles in the human torso model:

   erector spinae group
       iliocostalis (lateral group)
       longissimus (intermediate group)
       spinalis (medial group)
   external oblique
   internal oblique
   transversus abdominis
   rectus abdominis

4. Demonstrate the actions of these muscles in your body.

5. Locate the origin and insertion of each of these muscles in the human skeleton.

6. Complete Part A of Laboratory Report 22.

ASSESS

CRITICAL THINKING

List the muscles from superficial to deep for an appendectomy incision.

_____

_____

_____

_____

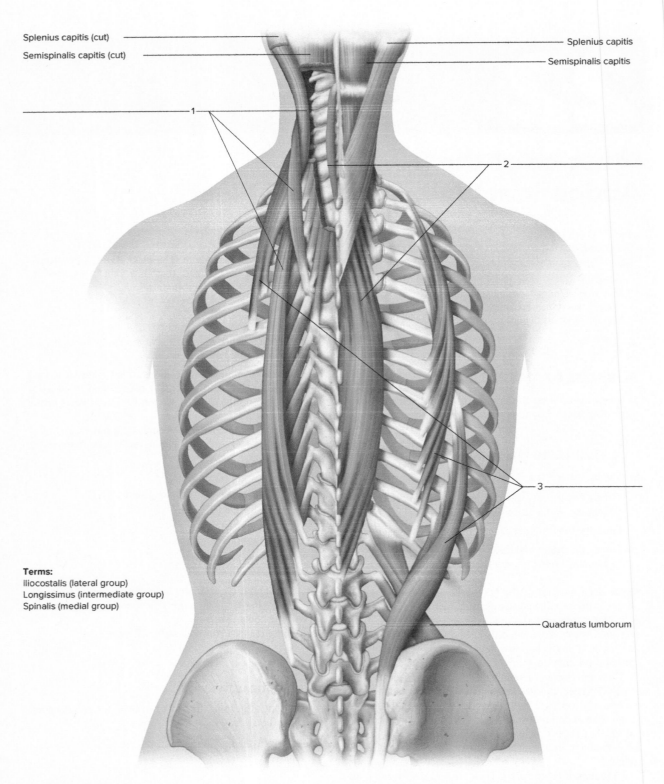

Splenius capitis (cut)

Semispinalis capitis (cut)

Splenius capitis

Semispinalis capitis

1

2

3

Terms:
Iliocostalis (lateral group)
Longissimus (intermediate group)
Spinalis (medial group)

Quadratus lumborum

FIGURE 22.1   Label the three deep back muscle groups of the erector spinae group, using the terms provided. (*Note*: The three erector spinae muscle groups are not illustrated on both sides of the body.) APR

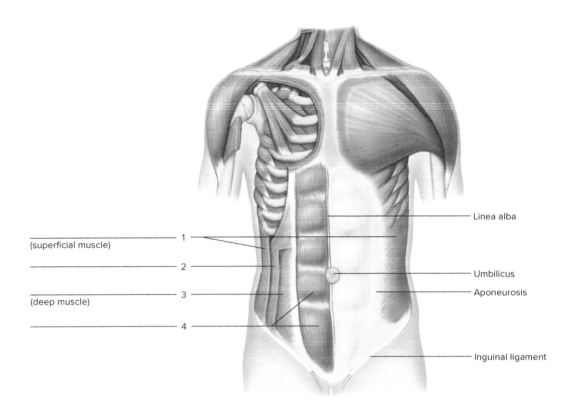

Linea alba

(superficial muscle) — 1

— 2

Umbilicus

(deep muscle) — 3

Aponeurosis

— 4

Inguinal ligament

FIGURE 22.2   Label the muscles of the abdominal wall.    **APR**

PRACTICE

PROCEDURE B—Muscles of the Pelvic Floor

1. Review the concept heading titled "Muscles of the Pelvic Floor and Perineum" in section 9.8 in chapter 9 of the textbook.
2. As a review activity, label figures 22.3 and 22.4.
3. Study figure 22.5 as it relates to the pelvic muscles indicated in figures 22.3 and 22.4.
4. Locate the following muscles in the models of the male and female pelves:

pelvic floor (diaphragm)
   levator ani
   coccygeus

urogenital diaphragm
   deep transversus perinei
   external urethral sphincter

other perineal muscles
   bulbospongiosus
   ischiocavernosus
   external anal sphincter
   superficial transversus perinei

5. Locate the origin and insertion of each of these muscles in the human skeleton.   **3**
6. Complete Part B of the laboratory report.

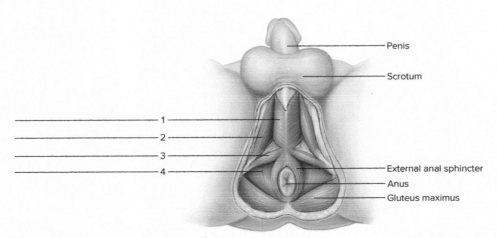

FIGURE 22.3   Label the muscles of the male pelvic floor and perineum. 🅰

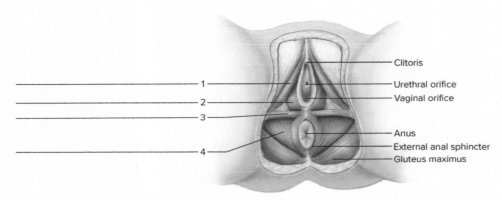

FIGURE 22.4   Label the muscles of the female pelvic floor and perineum. 🅰

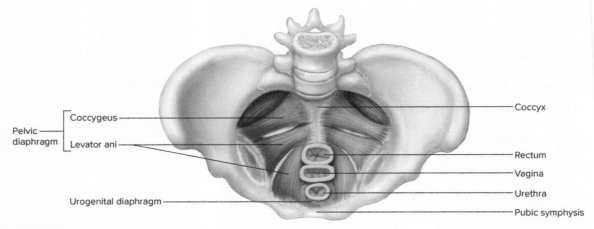

FIGURE 22.5   Internal view of the female pelvis.

Name _____

Date _____

Section _____

The 🄰 corresponds to the indicated Learning Outcome(s) found at the beginning of the laboratory exercise.

Muscles of the Deep Back, Abdominal Wall, and Pelvic Floor

🔁 PART A ASSESSMENTS

Complete the following statements:

1. A band of tough connective tissue in the midline of the anterior abdominal wall called the _____ serves as a muscle attachment. 🄐

2. The _____ muscle spans from the ribs and the xiphoid process of the sternum to the pubic bones. 🄒

3. The _____ forms the third layer (deepest layer) of the abdominal wall muscles. 🄐

4. The action of the external oblique muscle is to compress the abdomen and _____ . 🄑

5. The action of the rectus abdominis is to compress the abdomen and _____ . 🄑

6. The iliocostalis, longissimus, and spinalis muscles together form the _____ . 🄐

🔁 PART B ASSESSMENTS

Complete the following statements:

1. The levator ani and coccygeus together form the _____ . 🄐

2. The action of the external anal sphincter is to _____ . 🄑

3. The action of the coccygeus is to _____
_____ . 🄑

4. The _____ surrounds the base of the penis. 🄐

5. In females, the bulbospongiosus acts to constrict the _____ . 🄑

6. The ischiocavernosus extends from the penis or clitoris of the pubic arch to the _____ . 🄐

7. In the female, the _____ muscles are separated by the urethra, vagina, and rectum. 🄐

8. The action of the deep transversus perinei is to _____ . 🄑

9. The coccygeus extends from the coccyx and sacrum to the _____ . 🄒

10. The _____ assists in closing the urethra. 🄑

LABORATORY EXERCISE

23

Muscles of the Hip and Lower Limb

MATERIALS NEEDED

Textbook
Human torso model with musculature
Human skeleton, articulated
Muscular models of the lower limb
Long rubber bands

PURPOSE OF THE EXERCISE

To review the actions, origins, and insertions of the muscles that move the thigh, leg, and foot.

 LEARNING OUTCOMES APR

After completing this exercise, you should be able to

1. Locate and identify the muscles that move the thigh, leg, and foot.

2. Describe and demonstrate the actions of each of these muscles.

3. Locate the origin and insertion of each of these muscles in a human skeleton and on muscular models.

The muscles that move the thigh are attached to the femur and to some part of the pelvic girdle. Those attached anteriorly primarily act to flex the thigh at the hip, whereas those attached posteriorly act to extend, abduct, or rotate the thigh.

The muscles that move the leg connect the tibia or fibula to the femur or to the pelvic girdle. They function to flex or extend the leg at the knee. Other muscles, located in the leg, act to move the foot.

 PRACTICE

PROCEDURE—Muscles of the Hip and Lower Limb

1. Review the concept headings entitled "Muscles That Move the Thigh," "Muscles That Move the Leg," and "Muscles That Move the Foot" in section 9.8 in chapter 9 of the textbook.

2. As a review activity, label figures 23.1–23.6.
3. Locate the following muscles in the human torso model and in the lower limb models. Also locate as many of them as possible in your body.

muscles that move the thigh
> anterior hip muscles
>> iliopsoas group
>>> psoas major
>>> iliacus
> posterior and lateral hip muscles
>> gluteus maximus
>> gluteus medius
>> gluteus minimus
>> tensor fasciae latae
> medial adductor muscles
>> pectineus
>> adductor longus
>> adductor magnus
>> adductor brevis
>> gracilis

muscles that move the leg
> anterior thigh muscles
>> sartorius
>> quadriceps femoris group
>>> rectus femoris
>>> vastus lateralis
>>> vastus medialis
>>> vastus intermedius
> posterior thigh muscles
>> hamstring group
>>> biceps femoris
>>> semitendinosus
>>> semimembranosus

muscles that move the foot
> anterior leg muscles
>> tibialis anterior
>> fibularis (peroneus) tertius
>> extensor digitorum longus

posterior leg muscles
   gastrocnemius
   soleus
   flexor digitorum longus
   tibialis posterior
  lateral leg muscles
    fibularis (peroneus) longus
    fibularis (peroneus) brevis

4. Demonstrate the action of each of these muscles in your body.
5. Locate the origin and insertion of each of these muscles in the human skeleton.
6. Complete Parts A, B, and C of Laboratory Report 23.

 LEARN: LAB IN MOTION

A long rubber band can be used to simulate muscle locations, origins, insertions, and actions on muscular models, the skeleton, or a laboratory partner. Hold one end of the rubber band firmly on the origin location of a muscle; then slightly stretch the rubber band and hold the other end on the insertion site. Allow the insertion end to slowly move toward the origin end to simulate the contraction and action of the muscle. ⚠ ⚠ ⚠

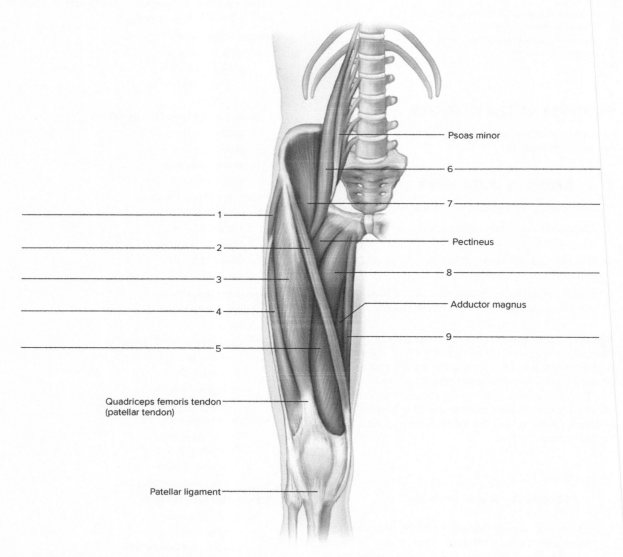

Psoas minor
6
7
Pectineus
8
Adductor magnus
9

1
2
3
4
5

Quadriceps femoris tendon
(patellar tendon)

Patellar ligament

FIGURE 23.1   Label the muscles of the anterior right hip and thigh.   APR

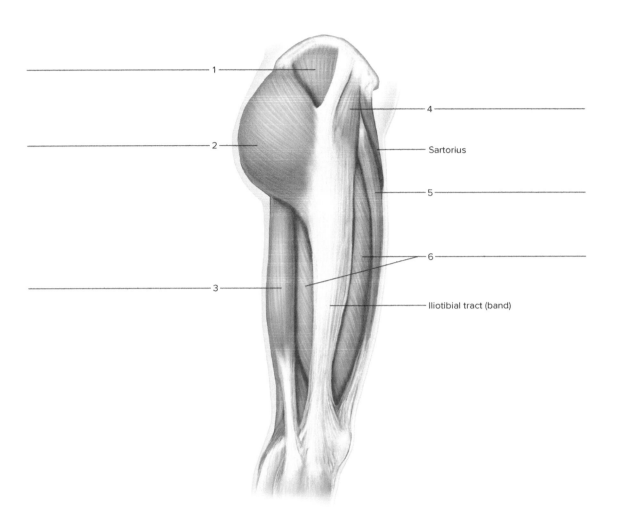

Sartorius

Iliotibial tract (band)

FIGURE 23.2 Label the muscles of the lateral right hip and thigh. 🅐

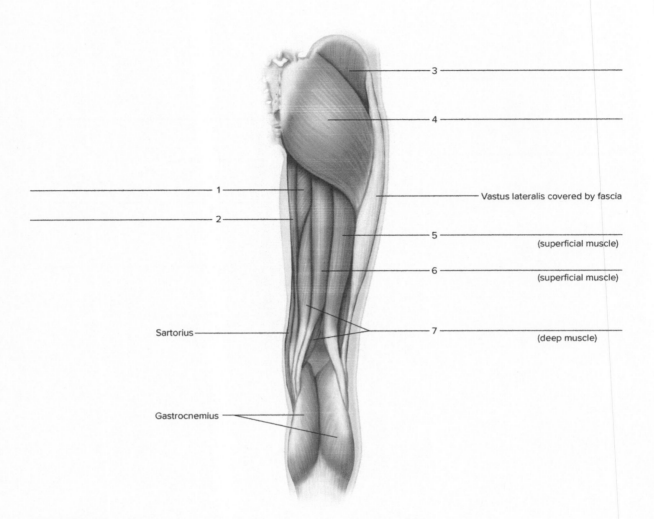

3

4

1

2

Vastus lateralis covered by fascia

5

(superficial muscle)

6

(superficial muscle)

Sartorius

7

(deep muscle)

Gastrocnemius

FIGURE 23.3   Label the muscles of the posterior right hip and thigh.   APR

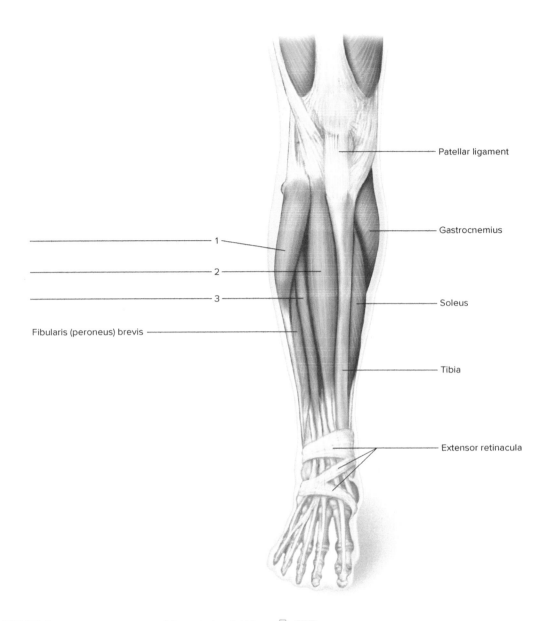

Patellar ligament

Gastrocnemius

1

2

Soleus

3

Fibularis (peroneus) brevis

Tibia

Extensor retinacula

FIGURE 23.4 Label the muscles of the anterior right leg. 1 APR

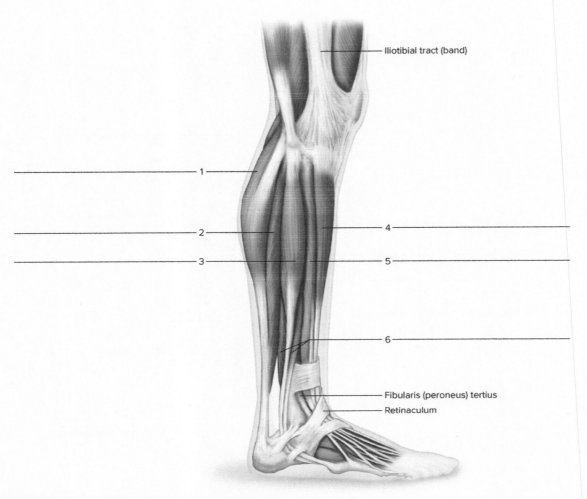

Iliotibial tract (band)

1

2

3

4

5

6

Fibularis (peroneus) tertius
Retinaculum

FIGURE 23.5   Label the muscles of the lateral right leg. 1

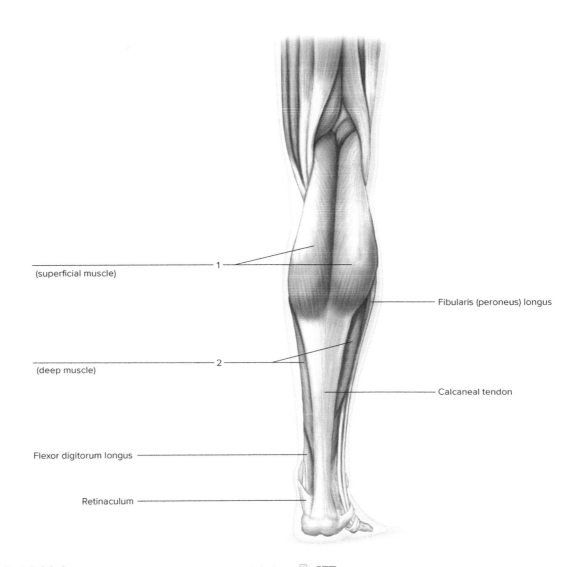

_____ 1
(superficial muscle)

Fibularis (peroneus) longus

_____ 2
(deep muscle)

Calcaneal tendon

Flexor digitorum longus

Retinaculum

FIGURE 23.6   Label the muscles of the posterior right leg.   APR

Notes

Name _____

Date _____

Section _____

The Ⓐ corresponds to the indicated Learning Outcome(s) found at the beginning of the laboratory exercise.

Muscles of the Hip and Lower Limb

🔄 PART A ASSESSMENTS

Match the muscles in column A with the actions in column B. Place the letter of your choice in the space provided. Ⓐ2

Column A	Column B
a. Biceps femoris	_____ 1. Adducts thigh and flexes knee
b. Fibularis (peroneus) longus	_____ 2. Plantar flexion and eversion of foot
c. Fibularis (peroneus) tertius	_____ 3. Flexes hip
d. Gluteus medius	_____ 4. Abducts and rotates thigh laterally; flexes knee and hip
e. Gracilis	_____ 5. Dorsiflexion and eversion of foot
f. Psoas major and iliacus	_____ 6. Abducts thigh and rotates it medially
g. Quadriceps femoris group	_____ 7. Plantar flexion and inversion of foot
h. Sartorius	_____ 8. Flexes knee and laterally rotates leg
i. Tibialis anterior	_____ 9. Extends knee
j. Tibialis posterior	_____ 10. Dorsiflexion and inversion of foot

🔄 PART B ASSESSMENTS

Name the muscle indicated by the following combinations of origin and insertion. Ⓐ3

Origin	Insertion	Muscle
1. Lateral surface of ilium	Greater trochanter of femur	_____
2. Ischial tuberosity	Posterior surface of femur	_____
3. Anterior superior iliac spine	Medial surface of proximal tibia	_____
4. Lateral and medial condyles of femur	Posterior surface of calcaneus	_____
5. Anterior iliac crest	Fascia (iliotibial tract) of the thigh	_____
6. Greater trochanter and posterior surface of femur	Patella and to tibial tuberosity	_____
7. Ischial tuberosity	Medial surface of proximal tibia	_____
8. Medial surface of femur	Patella and to tibial tuberosity	_____
9. Posterior surface of tibia	Distal phalanges of four lateral toes	_____
10. Lateral condyle and lateral surface of tibia	Medial cuneiform and first metatarsal	_____

PART C ASSESSMENTS

ASSESS

CRITICAL THINKING

Identify the muscles indicated in figure 23.7. ⓐ

1. _____

2. _____

3. _____

4. _____

5. _____

6. _____

7. _____

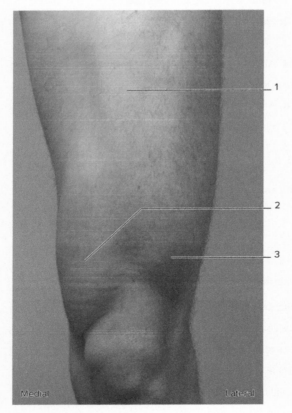

(a) Left thigh, anterior view

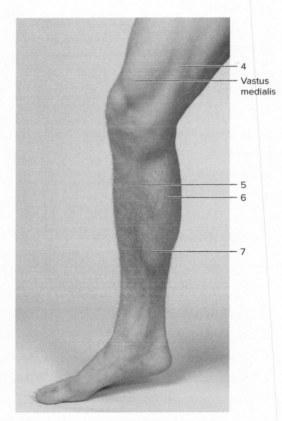

(b) Right lower limb, medial view

Terms:

Gastrocnemius	Tibialis anterior
Rectus femoris	Vastus lateralis
Sartorius	Vastus medialis
Soleus	

FIGURE 23.7   Identify the muscles that appear as lower limb surface features in these photographs (*a* and *b*), using the terms provided. ((*a* and *b*) ©J & J Photography)

PART D ASSESSMENTS

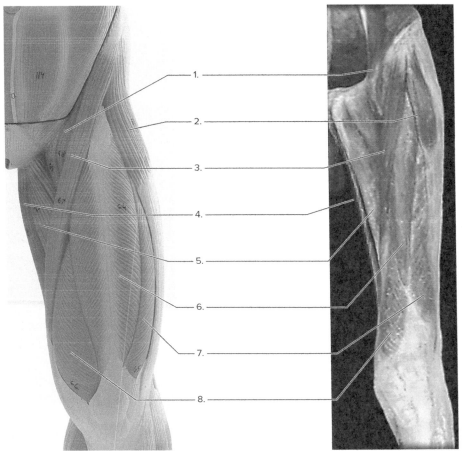

Terms:
Adductor magnus
Gracilis
Iliopsoas
Rectus femoris
Sartorius
Tensor fasciae latae
Vastus lateralis
Vastus mediais

(a)

FIGURE 23.8   (*a*) Using the terms provided, label the indicated muscles on the model and cadaver images. ((*a*) M20 Muscle Leg
© 3B Scientific GmbH, Germany, 2017 www.3bscientific.com; McGraw-Hill Education/APR)

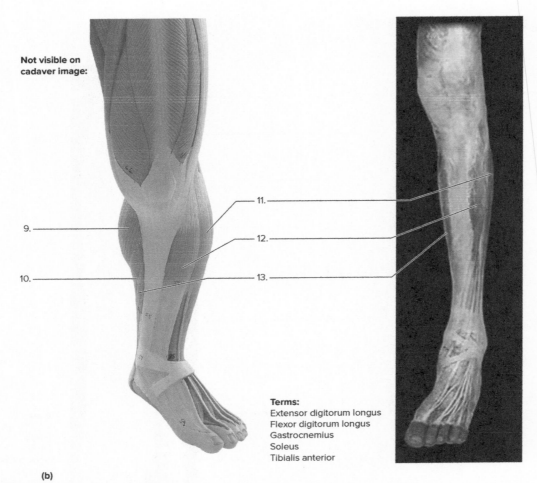

Not visible on cadaver image:

9.

10.

11.

12.

13.

Terms:
Extensor digitorum longus
Flexor digitorum longus
Gastrocnemius
Soleus
Tibialis anterior

(b)

FIGURE 23.8   (*b*) Using the terms provided, label the indicated muscles of the lower limb on the model and cadaver image.   ((*b*) M20 Muscle Leg © 3B Scientific GmbH, Germany, 2017 www.3bscientific.com; McGraw-Hill Education/APR)

Surface Anatomy

LABORATORY EXERCISE

24

Surface Anatomy

MATERIALS NEEDED

Textbook
Small, round stickers
Colored pencils (black and red)

PURPOSE OF THE EXERCISE

To examine the surface features of the human body and the terms used to describe them.

 LEARNING OUTCOMES

After completing this exercise, you should be able to

1. Locate and identify major surface features of the human body.
2. Arrange surface features by body region.
3. Distinguish between surface features that are bony landmarks and soft tissue.

External landmarks, called *surface anatomy*, located on the human body provide an opportunity to examine surface features and to help locate other internal structures. This exercise will focus on bony landmarks and superficial soft tissue structures, primarily related to the skeletal and muscular systems. The technique used for the examination of surface features is called *palpation*, and is accomplished by touching with the hands or fingers. This enables us to determine the location, size, and texture of the underlying anatomical structure. Some of the surface features can be visible without the need of palpation, especially on body builders and very thin individuals. If the individual involved has considerable subcutaneous adipose tissue, additional pressure might be necessary to palpate some of the structures.

These surface features will help us to locate additional surface features and deeper structures during the study of the organ systems. Surface features will help us to determine the pulse locations of superficial arteries and for proper assessments of injection sites; descriptions of pain and injury sites; insertion of examination tubes; and heart, lung, and bowel sounds heard with a stethoscope. For those who choose careers in emergency medical services, nursing, medicine, physical education, physical therapy, chiropractic, and massage therapy, surface anatomy is especially valuable.

 PRACTICE

PROCEDURE—Surface Anatomy

1. Review textbook figures on the body regions and the human organism reference plates in chapter 1, the skeletal system in chapter 7, and the muscular system in chapter 9.
2. Use the textbook figures and the other figures provided to complete figures 24.1 through 24.8. Many features have already been labeled. Features in **boldface** are *bony features*. Other labeled features (not boldface) are *soft tissue features*. Bony features are a part of a bone and the skeletal system. Soft tissue features are composed of tissue other than bone, such as a muscle, tendon, ligament, cartilage, or skin. The missing features are listed by each figure.

Terms:
Hyoid bone
Mandible
Sternocleidomastoid
Trapezius (upper)

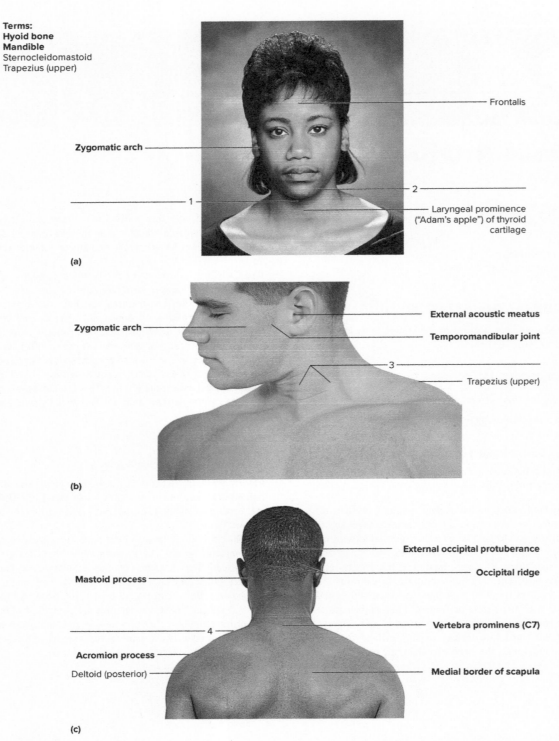

Frontalis

Zygomatic arch

1

2

Laryngeal prominence
("Adam's apple") of thyroid
cartilage

(a)

Zygomatic arch

External acoustic meatus

Temporomandibular joint

3

Trapezius (upper)

(b)

External occipital protuberance

Occipital ridge

Mastoid process

4

Vertebra prominens (C7)

Acromion process

Deltoid (posterior)

Medial border of scapula

(c)

FIGURE 24.1   Label the surface features of (*a*) the anterior head and neck, (*b*) the lateral head and neck, and (*c*) the posterior head and neck, using the terms provided. (**Boldface** indicates bony features; not boldface indicates soft tissue features.) 3 ((*a*) Ken Saladin/Joe DeGrandis,photographer; (*b,c*) ©Eric Wise)

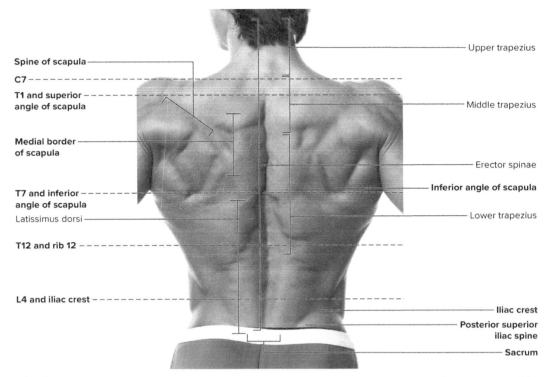

Spine of scapula

C7

T1 and superior angle of scapula

Medial border of scapula

T7 and inferior angle of scapula

Latissimus dorsi

T12 and rib 12

L4 and iliac crest

Upper trapezius

Middle trapezius

Erector spinae

Inferior angle of scapula

Lower trapezius

Iliac crest

Posterior superior iliac spine

Sacrum

FIGURE 24.2   Surface features of the posterior shoulder and torso. (**Boldface** indicates bony features; not boldface indicates soft tissue features.)   (BLACKDAY/Shutterstock)

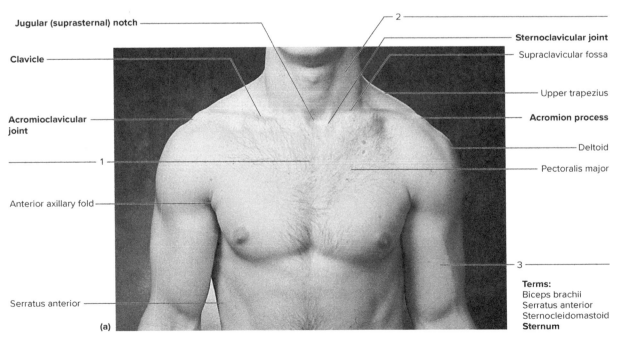

Jugular (suprasternal) notch

Clavicle

Acromioclavicular joint

1

Anterior axillary fold

Serratus anterior

(a)

2

Sternoclavicular joint

Supraclavicular fossa

Upper trapezius

Acromion process

Deltoid

Pectoralis major

3

Terms:
Biceps brachii
Serratus anterior
Sternocleidomastoid
Sternum

FIGURE 24.3   Label the anterior surface features of (*a*) upper torso and (*b*) lower torso, using the terms provided. (**Boldface** indicates bony features; not boldface indicates soft tissue features.)   ((*a*) Ken Saladin/Joe DeGrandis,photographer; (*b*) Juice Images/Alamy Stock Photo)

Terms:
Biceps brachii
Serratus anterior
Sternocleidomastoid
Sternum

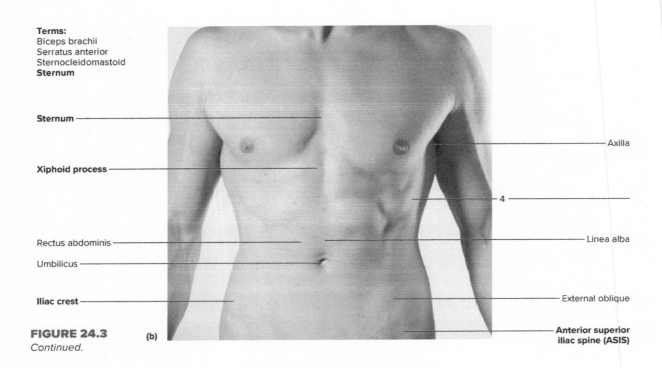

Sternum ——————————————

Axilla

Xiphoid process ————————

4 ——————————

Rectus abdominis ——————

Linea alba

Umbilicus ——————

Iliac crest ————————

External oblique

FIGURE 24.3 (b)
Continued.

**Anterior superior
iliac spine (ASIS)**

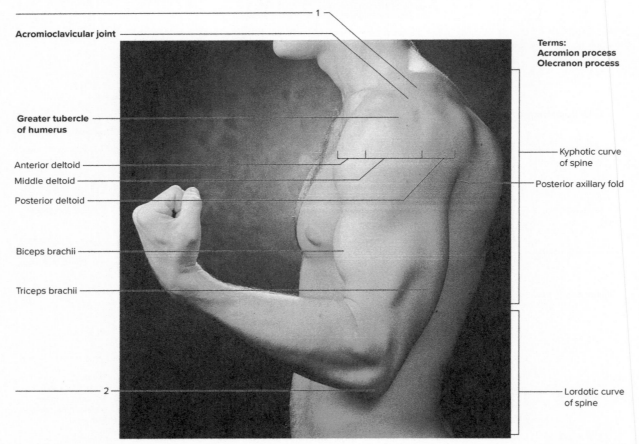

Acromioclavicular joint ————————

1

Terms:
Acromion process
Olecranon process

**Greater tubercle
of humerus** ————————

Kyphotic curve
of spine

Anterior deltoid ————————
Middle deltoid ————————
Posterior deltoid ————————

Posterior axillary fold

Biceps brachii ————————

Triceps brachii ————————

2

Lordotic curve
of spine

FIGURE 24.4   Label the surface features of the lateral shoulder and upper limb, using the terms provided. (**Boldface** indicates bony features; not boldface indicates soft tissue features.) **1** (Ken Saladin/Joe DeGrandis, photographer)

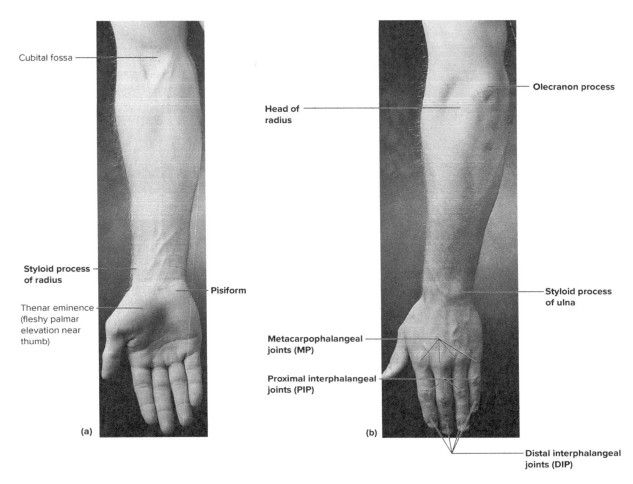

Cubital fossa

Head of radius

Olecranon process

Styloid process of radius

Pisiform

Thenar eminence (fleshy palmar elevation near thumb)

Styloid process of ulna

Metacarpophalangeal joints (MP)

Proximal interphalangeal joints (PIP)

Distal interphalangeal joints (DIP)

(a)

(b)

FIGURE 24.5   Surface features of the upper limb (*a*) anterior view and (*b*) posterior view. (**Boldface** indicates bony features; not boldface indicates soft tissue features.)   ((*a,b*) Ken Saladin/Joe DeGrandis,photographer)

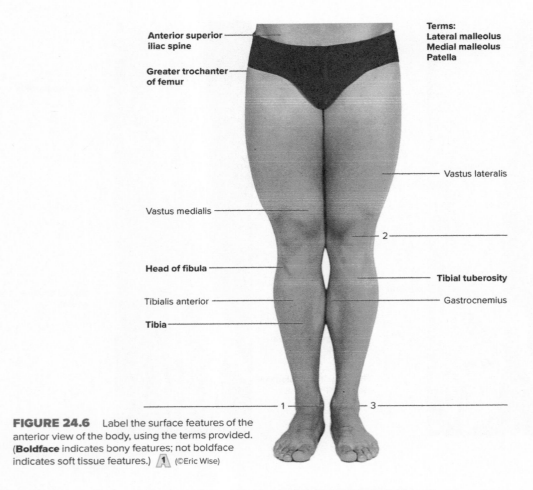

Anterior superior
iliac spine

Greater trochanter
of femur

Terms:
Lateral malleolus
Medial malleolus
Patella

Vastus lateralis

Vastus medialis

2

Head of fibula

Tibial tuberosity

Tibialis anterior

Gastrocnemius

Tibia

1   3

FIGURE 24.6   Label the surface features of the
anterior view of the body, using the terms provided.
(**Boldface** indicates bony features; not boldface
indicates soft tissue features.) _1_ (©Eric Wise)

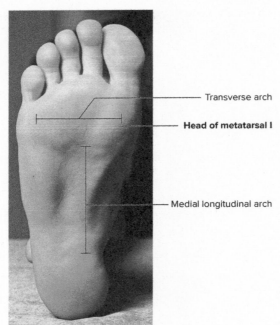

Transverse arch

Head of metatarsal I

Medial longitudinal arch

FIGURE 24.7   Surface features of
the plantar surface of the foot. (**Boldface**
indicates bony features; not boldface
indicates soft tissue features.) (Ken Saladin/
Joe DeGrandis, photographer)

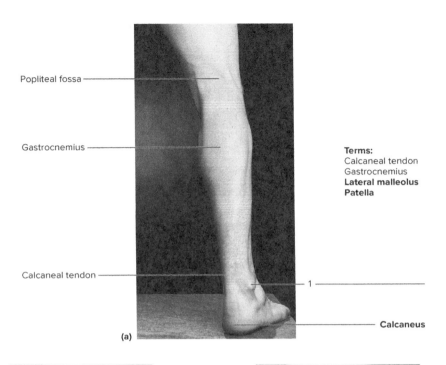

Popliteal fossa

Gastrocnemius

Terms:
Calcaneal tendon
Gastrocnemius
Lateral malleolus
Patella

Calcaneal tendon

1

Calcaneus

(a)

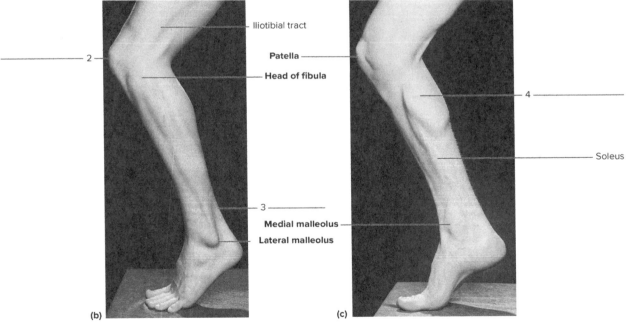

Iliotibial tract

2

Patella

Head of fibula

4

Soleus

3

Medial malleolus

Lateral malleolus

(b)

(c)

FIGURE 24.8   Label the surface features of the (*a*) posterior lower limb, (*b*) lateral lower limb, and (*c*) medial lower limb, using the terms provided. (**Boldface** indicates bony features; not boldface indicates soft tissue features.)   **3**   Joe De Grandis/McGraw-Hill Education

3. Complete Parts A and B of Laboratory Report 24.

4. Use figures 24.1 through 24.8 to determine which items listed in table 24.1 are bony features and which items are soft tissue features. If the feature listed is a bony surface feature, place an "X" in the second column; if the feature listed is a soft tissue feature, place an "X" in the third column. ◭3

5. Complete Part C of the laboratory report.

 LEARN: LAB IN MOTION

Select fifteen surface features presented in the laboratory exercise figures. Using small, round stickers, write the name of a surface feature on each sticker. Working with a laboratory partner, locate the selected surface features on yourself and on your partner. Use the stickers to accurately "label" your partner. ◭1

- Were you able to find all the features on both of you?
  _____

- Was it easier to find the surface features on yourself or your partner? _____

TABLE 24.1 Representative Bony and Soft Tissue Surface Features (*Note*: After indicating your "X" in the appropriate column, use the extra space in columns 2 and 3 to add personal descriptions of the nature of the bony feature or the soft tissue feature palpation.)

Bony and Soft Tissues	Bony Features	Soft Tissue Features
Biceps brachii		
Calcaneal tendon		
Deltoid		
Gastrocnemius		
Greater trochanter		
Head of fibula		
Iliac crest		
Iliotibial tract		
Lateral malleolus		
Mandible		
Mastoid process		
Medial border of scapula		
Rectus abdominis		
Sacrum		
Serratus anterior		
Sternocleidomastoid		
Sternum		
Thenar eminence		
Tibialis anterior		
Zygomatic arch		

Name _____

Date _____

Section _____

The corresponds to the indicated Learning Outcome(s) found at the beginning of the laboratory exercise.

Surface Anatomy

PART A ASSESSMENTS

Match the body regions in column A with the surface features in column B. Place the letter of your choice in the space provided. **1** **2**

Column A	Column B
a. Head	_____ **1.** Umbilicus
	_____ **2.** Medial malleolus
	_____ **3.** Iliac crest
	_____ **4.** Transverse arch
b. Trunk (torso, including shoulder, and hip)	_____ **5.** Spine of scapula
	_____ **6.** External occipital protuberance
	_____ **7.** Metacarpophalangeal joints
	_____ **8.** Sternum
	_____ **9.** Olecranon process
c. Upper limb	_____ **10.** Zygomatic arch
	_____ **11.** Mastoid process
	_____ **12.** Thenar eminence
	_____ **13.** Popliteal fossa
d. Lower limb	_____ **14.** Cubital fossa

PART B ASSESSMENTS

Label figure 24.9 with the surface features provided.

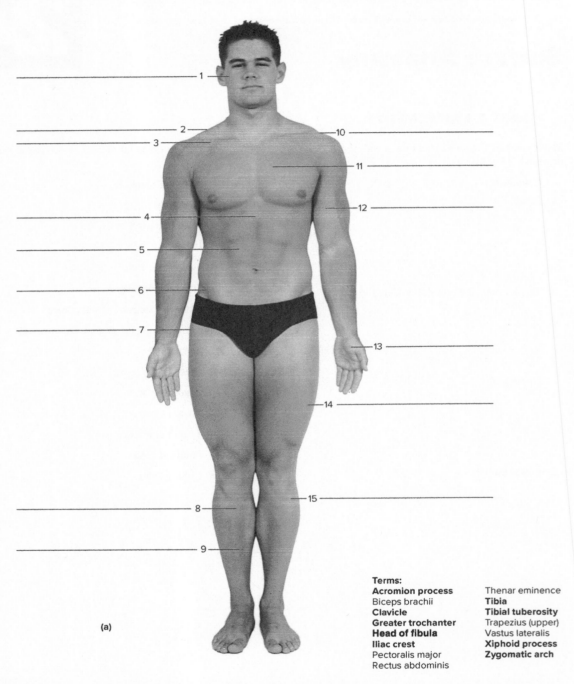

(a)

Terms:
Acromion process   Thenar eminence
Biceps brachii   **Tibia**
Clavicle   **Tibial tuberosity**
Greater trochanter   Trapezius (upper)
Head of fibula   Vastus lateralis
Iliac crest   **Xiphoid process**
Pectoralis major   **Zygomatic arch**
Rectus abdominis

FIGURE 24.9   Label the surface features of the (a) anterior view of the body and (b) posterior view of the body, using the terms provided. (**Boldface** indicates bony features; not boldface indicates soft tissue features.) ⚠ ((a, b) ©Eric Wise)

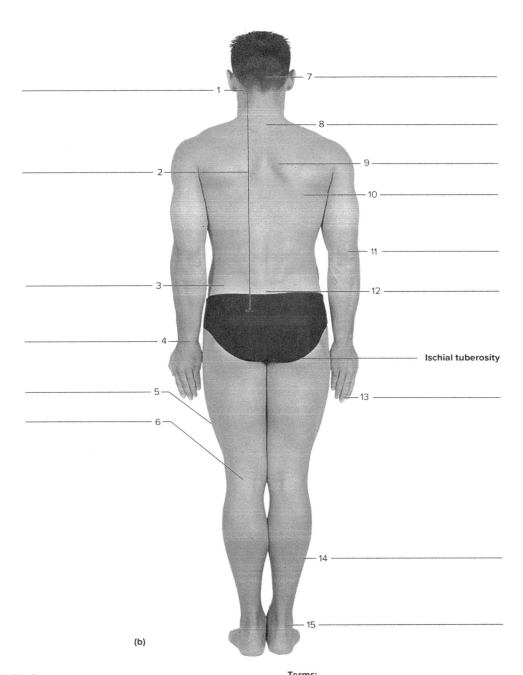

(b)

FIGURE 24.9   *Continued.*

Ischial tuberosity

Terms:
C7
Calcaneal tendon
Distal interphalangeal joint (DIP)
Erector spinae
External occipital protuberance
Iliac crest
Iliotibial tract
Inferior angle of scapula

Mastoid process
Medial border of scapula
Olecranon process
Popliteal fossa
Sacrum
Soleus
Styloid process of ulna

PART C ASSESSMENTS

Using table 24.1, indicate the locations of the surface features with an "X" on figure 24.10. Use a black "X" for the bony features and a red "X" for the soft tissue features.

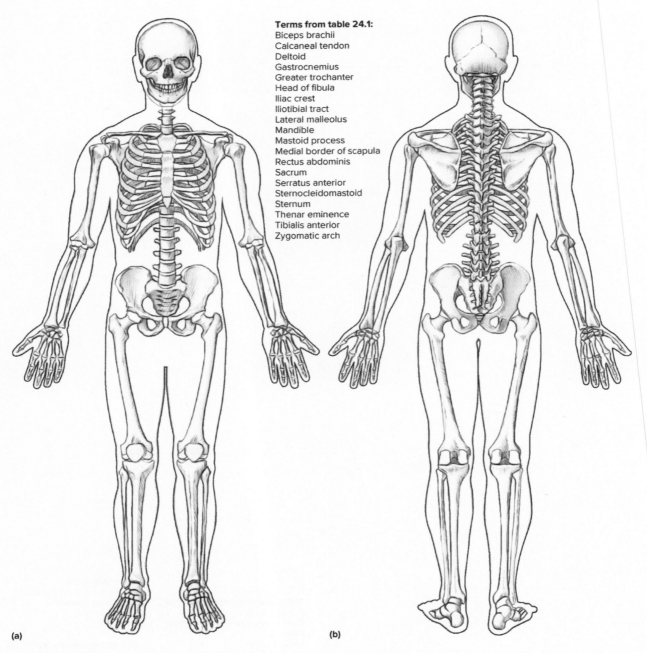

Terms from table 24.1:
Biceps brachii
Calcaneal tendon
Deltoid
Gastrocnemius
Greater trochanter
Head of fibula
Iliac crest
Iliotibial tract
Lateral malleolus
Mandible
Mastoid process
Medial border of scapula
Rectus abdominis
Sacrum
Serratus anterior
Sternocleidomastoid
Sternum
Thenar eminence
Tibialis anterior
Zygomatic arch

(a)   (b)

FIGURE 24.10 Indicate bony surface features with a black "X" and soft tissue surface features with a red "X," using the results from table 24.1, on the (*a*) anterior and (*b*) posterior diagrams of the body. 🔺

Nervous System

25

Nervous Tissue and Nerves

MATERIALS NEEDED

Textbook
Compound light microscope
Prepared microscope slides of the following:
  Spinal cord (smear)
  Posterior (dorsal) root ganglion (section)
  Neuroglia (astrocytes)
  Peripheral nerve (cross section and longitudinal
    section)
Neuron model
Prepared microscope slide of Purkinje cells from
  cerebellum

PURPOSE OF THE EXERCISE

To review the characteristics of nervous tissue and to observe neurons, neuroglia, and various features of a nerve.

LEARNING OUTCOMES APR

After completing this exercise, you should be able to

1. Describe the location and characteristics of nervous tissue.
2. Distinguish structural and functional characteristics of neurons and neuroglia.
3. Identify and sketch the major structures of a neuron and a nerve.

Nervous tissue, which occurs in the brain, spinal cord, and nerves, contains neurons and neuroglia. The neurons are the basic structural and functional units of the nervous system involved in making decisions, detecting stimuli, and conducting impulses. The neuroglia perform various supportive and protective functions for neurons.

PRACTICE

**PROCEDURE—Nervous Tissue
and Nerves** APR

1. Review section 10.2 titled "Nervous Tissue Cells: Neurons and Neuroglia" in chapter 10 of the textbook.
2. As a review activity, label figures 25.1 and 25.2.
3. Complete Parts A and B of Laboratory Report 25.
4. Obtain a prepared microscope slide of a spinal cord smear. Using low-power magnification, search the slide and locate the relatively large, deeply stained cell bodies of motor neurons (multipolar neurons).
5. Observe a single, multipolar motor neuron, using high-power magnification, and note the following features:

    cell body
      nucleus
      nucleolus
      chromatophilic substance (Nissl bodies)
      neurofibrils (threadlike structures extending into the
        nerve fibers)
    dendrites
    axon (nerve fiber)

    Compare the slide to the neuron model and to figure 25.3. You also may note small, darkly stained nuclei of neuroglia around the motor neuron.
6. Sketch and label a motor (efferent) neuron in the space provided in Part C of the laboratory report.
7. Obtain a prepared microscope slide of a posterior (dorsal) root ganglion. Search the slide and locate a cluster of sensory neuron cell bodies. You also may note bundles of nerve fibers passing among groups of neuron cell bodies (fig. 25.4).
8. Sketch and label a sensory (afferent) neuron cell body in the space provided in Part C of the laboratory report.

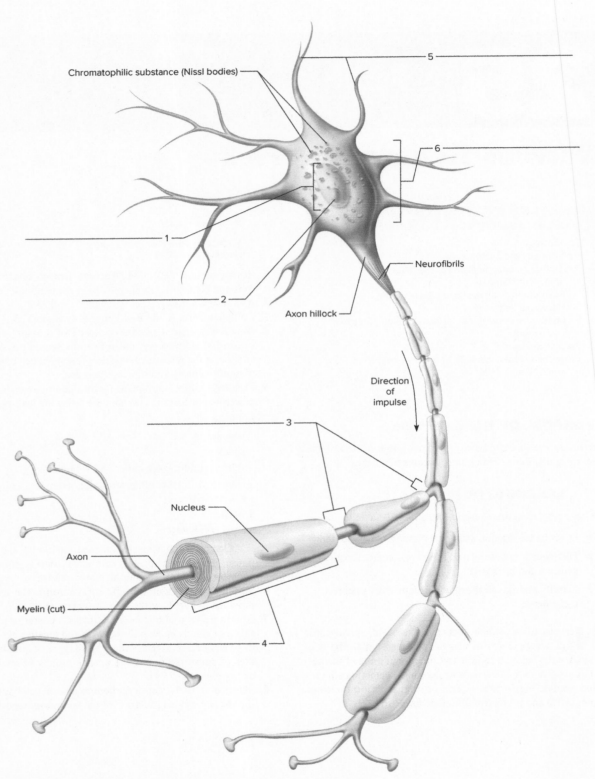

Chromatophilic substance (Nissl bodies)

5

6

1

2

Neurofibrils

Axon hillock

Direction
of
impulse

3

Nucleus

Axon

Myelin (cut)

4

FIGURE 25.1   Label this diagram of a multipolar motor neuron. 🄰 APR

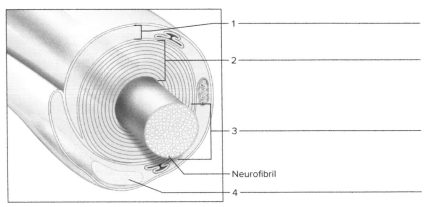

FIGURE 25.2   Label the features of the myelinated axon (nerve fiber). ⚠️ **APR**

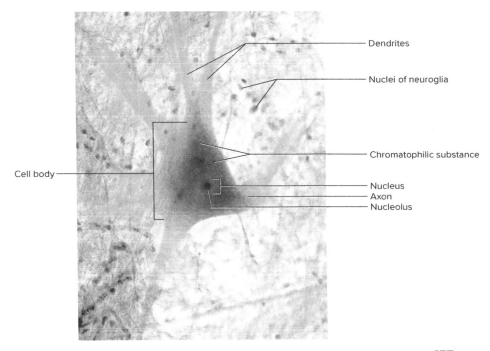

FIGURE 25.3   Micrograph of a multipolar neuron and neuroglia from a spinal cord smear (600×). **APR** (Alvin Telser/McGraw-Hill Education)

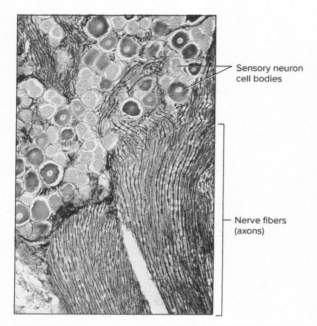

FIGURE 25.4   Micrograph of a posterior (dorsal) root ganglion containing cell bodies of sensory neurons (100×). **APR** (Ed Reschke)

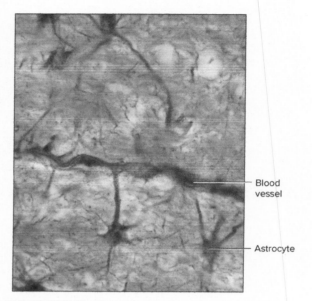

FIGURE 25.5   Micrograph of astrocytes (1,000×). (Ed Reschke)

9. Obtain a prepared microscope slide of neuroglia. Search the slide and locate some darkly-stained astrocytes with numerous long, slender processes (fig. 25.5).
10. Sketch neuroglia in the space provided in Part C of the laboratory report.
11. Obtain a prepared microscope slide of a nerve. Locate the cross section of the nerve and note the many round nerve fibers inside. Also note the dense layer of connective tissue (perineurium) that encircles a fascicle of nerve fibers and holds them together in a bundle. The individual nerve fibers are surrounded by a layer of more delicate connective tissue (endoneurium) (fig. 25.6).
12. Using high-power magnification, observe a single nerve fiber and note the following features:

axon

myelin sheath around the axon of Schwann cell (most of the myelin may have been dissolved and lost during the slide preparation)

neurilemma of Schwann cell

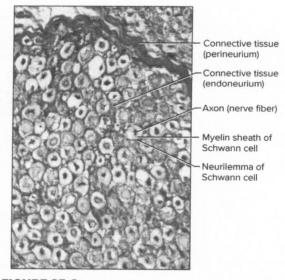

FIGURE 25.6   Cross section of a bundle of axons within a nerve (400×). (Ed Reschke)

13. Sketch and label a nerve fiber with Schwann cell (cross section) in the space provided in Part D of the laboratory report.

14. Locate the longitudinal section of the nerve on the slide (fig. 25.7). Note the following:

axon

myelin sheath of Schwann cell

neurilemma of Schwann cell

node of Ranvier

15. Sketch and label a nerve fiber with Schwann cell (longitudinal section) in the space provided in Part D of the laboratory report.

 LEARN: ACTIVITY

Obtain a prepared microscope slide of Purkinje cells (fig. 25.8). To locate these neurons, search the slide for large, flask-shaped cell bodies. Each cell body has one or two large, thick dendrites that give rise to extensive branching networks of dendrites. These large cells are located in a particular region of the brain (cerebellar cortex).

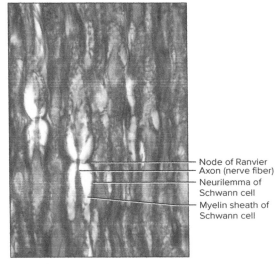

Node of Ranvier
Axon (nerve fiber)
Neurilemma of Schwann cell
Myelin sheath of Schwann cell

FIGURE 25.7   Longitudinal section of a nerve (650×).

APR (Ed Reschke)

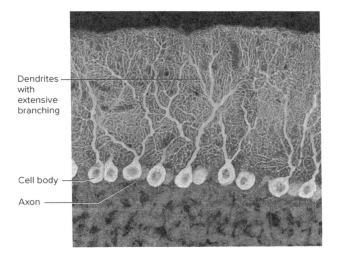

Dendrites with extensive branching

Cell body

Axon

FIGURE 25.8   Large multipolar Purkinje cell from cerebellum of the brain (400×). (Thomas Deerinck, NCMIR/Science Source)

Notes

Name _____

Date _____

Section _____

The ⬡ corresponds to the indicated Learning Outcome(s) found at the beginning of the laboratory exercise.

Nervous Tissue and Nerves

♻ PART A ASSESSMENTS

Match the terms in column A with the descriptions in column B. Place the letter of your choice in the space provided. ⬡ ⬡

Column A	Column B
a. Astrocyte	_____ **1.** Sheath of Schwann cell containing cytoplasm and nucleus that encloses myelin
b. Axon	_____ **2.** Corresponds to rough endoplasmic reticulum in other cells
c. Chromatophilic substance (Nissl bodies)	_____ **3.** Network of fine threads extending into nerve fiber
	_____ **4.** Substance of Schwann cell composed of lipoprotein that insulates axons and increases impulse speed
d. Collateral	
e. Dendrite	_____ **5.** Neuron process with many branches that conducts an impulse toward the cell body
f. Myelin sheath	_____ **6.** Branch of an axon
g. Neurilemma	_____ **7.** Star-shaped neuroglia between neurons and blood vessels
h. Neurofibrils	_____ **8.** Nerve fiber arising from a slight elevation of the cell body that conducts an impulse away from the cell body

♻ PART B ASSESSMENTS

Match the terms in column A with the descriptions in column B. Place the letter of your choice in the space provided. ⬡ ⬡

Column A	Column B
a. Effector	_____ **1.** Transmits impulse from sensory to motor neuron within central nervous system
b. Ependymal cell	_____ **2.** Transmits impulse out of the brain or spinal cord to effectors (muscles and glands)
c. Ganglion	_____ **3.** Transmits impulse into brain or spinal cord from receptors
d. Interneuron	_____ **4.** Myelin-forming neuroglia in brain and spinal cord
e. Microglia	_____ **5.** Phagocytic neuroglia
	_____ **6.** Structure capable of responding to motor impulse
f. Motor (efferent) neuron	_____ **7.** Specialized mass of neuron cell bodies outside the brain or spinal cord
g. Oligodendrocyte	
h. Sensory (afferent) neuron	_____ **8.** Cells that cover the inside spaces of the brain ventricles and help regulate cerebrospinal fluid

PART C ASSESSMENTS

In the space that follows, sketch the indicated cells. Label any of the cellular structures observed, and indicate the magnification of each sketch. **2** **3**

Motor neuron (_____×)

Sensory neuron cell body (_____×)

Neuroglia (_____×)

PART D ASSESSMENTS

In the space that follows, sketch the indicated view of a nerve fiber (axon). Label any structures observed and indicate the magnification. **2** **3**

Nerve fiber cross section with Schwann cell (_____×)

Nerve fiber longitudinal section with Schwann cell (_____×)

26

Brain and Cranial Nerves

MATERIALS NEEDED

Textbook
Dissectible model of the human brain
Preserved human brain
Anatomical charts of the human brain

PURPOSE OF THE EXERCISE

To review the structural and functional characteristics of the human brain and cranial nerves.

 ## LEARNING OUTCOMES APR

After completing this exercise, you should be able to

1. Identify the major external and internal structures in the human brain.

2. Locate the major functional regions of the brain.

3. Identify the twelve pairs of cranial nerves.

4. Differentiate the functions of the cranial nerves.

The brain, the largest and most complex part of the nervous system, contains nerve centers associated with sensory functions and is responsible for sensations and perceptions. It issues motor commands to skeletal muscles and carries on higher mental activities. It also functions to coordinate muscular movements, and it contains centers and neural pathways necessary for the regulation of internal organs.

Twelve pairs of cranial nerves arise from the inferior surface of the brain and are designated by number and name. The cranial nerves are part of the PNS; most arise from the brainstem region of the brain. Although most of these nerves conduct both sensory and motor impulses, some contain only sensory fibers associated with special sense organs. Others are primarily composed of motor fibers and are involved with the activities of muscles and glands.

 ## PRACTICE

PROCEDURE A—Human Brain

1. Review section 11.4 titled "The Brain" in chapter 11 of the textbook.

2. As a review activity, label figures 26.1 and 26.2.

3. Complete Part A of Laboratory Report 26.

4. Observe the anatomical charts, dissectible model, and preserved specimen of the human brain. Locate each of the following features:

cerebrum

cerebral hemispheres

corpus callosum

gyri (convolutions)

sulci

    central sulcus

    lateral sulcus

fissures

    longitudinal fissure

    transverse fissure

lobes

    frontal lobe

    parietal lobe

    temporal lobe

    occipital lobe

    insula (insular lobe)

cerebral cortex

basal nuclei (basal ganglia, a widely used clinical term)

    caudate nucleus

    putamen

    globus pallidus

ventricles

   lateral ventricles

   third ventricle

   fourth ventricle

   choroid plexuses

   cerebral aqueduct

diencephalon

   thalamus

   hypothalamus

   optic chiasma

   mammillary bodies

   pineal gland

cerebellum

   lateral (right and left) hemispheres

   falx cerebelli

   vermis

   cerebellar cortex

   arbor vitae

brainstem

   midbrain (mesencephalon)

     cerebral aqueduct

     cerebral peduncles

     corpora quadrigemina

   pons

   medulla oblongata

5. Locate the labeled areas in figure 26.3 that represent the following functional regions of the cerebrum:

motor area for voluntary muscle control

motor speech area (Broca's area)

cutaneous sensory area

auditory area

visual area

sensory speech area (Wernicke's area)

6. Complete Part B of the laboratory report.

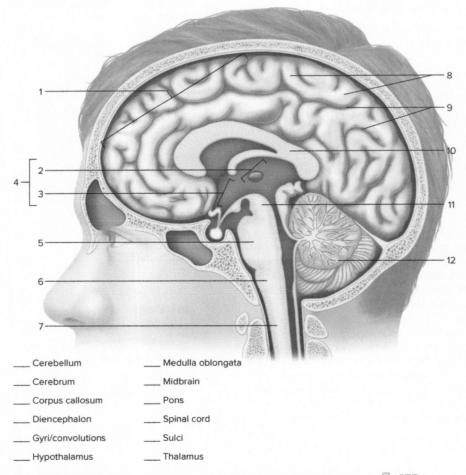

   ___ Cerebellum   ___ Medulla oblongata

   ___ Cerebrum   ___ Midbrain

   ___ Corpus callosum   ___ Pons

   ___ Diencephalon   ___ Spinal cord

   ___ Gyri/convolutions ___ Sulci

   ___ Hypothalamus   ___ Thalamus

FIGURE 26.1   Label this diagram by placing the correct numbers in the spaces provided. **A** **APR**

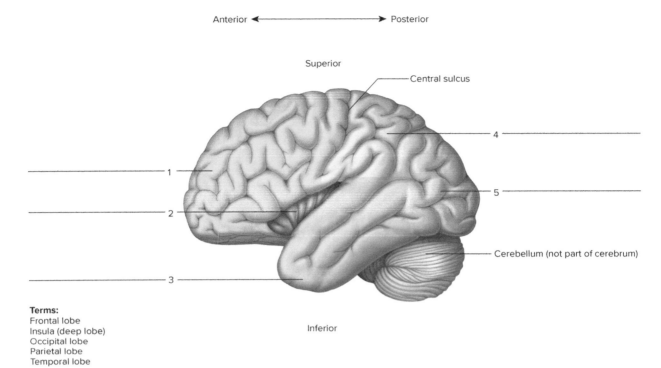

Anterior ←————————→ Posterior

Superior

Central sulcus

4

1

5

2

3

Cerebellum (not part of cerebrum)

Inferior

Terms:
Frontal lobe
Insula (deep lobe)
Occipital lobe
Parietal lobe
Temporal lobe

FIGURE 26.2   Label the five lobes of the left cerebral hemisphere, using the terms provided. Lobe 3 is retracted to expose the deep lobe 2 in this figure. 🄰 APR

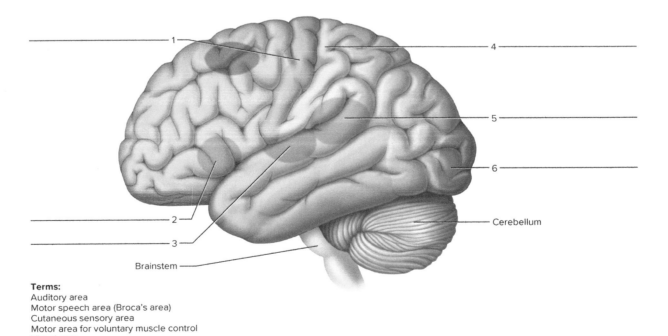

1

4

5

6

2

Cerebellum

3

Brainstem

Terms:
Auditory area
Motor speech area (Broca's area)
Cutaneous sensory area
Motor area for voluntary muscle control
Visual area
Sensory speech area (Wernicke's area)

FIGURE 26.3   Label the functional areas of the cerebrum, using the terms provided. (*Note*: These areas are not visible as distinct parts of the brain.) 🄰

PRACTICE

PROCEDURE B—Cranial Nerves

1. Review the concept heading titled "Cranial Nerves" in section 11.6 in chapter 11 of the textbook.

2. As a review activity, label figure 26.4. The cranial nerves arise from the base of the brain and can be viewed from the inferior surface.

3. Observe the model and preserved specimen of the human brain, and locate as many of the following cranial nerves as possible as you differentiate some of their associated functions:

olfactory nerves (I)—smell

optic nerves (II)—vision

oculomotor nerves (III)—pupil constriction and opening of eyelid

trochlear nerves (IV)—stimulation of superior oblique eye muscles

trigeminal nerves (V)—sensation from face and teeth and mastication movements

abducens nerves (VI)—lateral eye movements

facial nerves (VII)—salivation, tear secretions, and taste

vestibulocochlear nerves (VIII)—hearing and balance (equilibrium)

glossopharyngeal nerves (IX)—control of swallowing movements

vagus nerves (X)—regulation of many visceral organs, including the heart rate

accessory nerves (XI)—control of neck and shoulder muscles

hypoglossal nerves (XII)—control of tongue movements

The following mnemonic device will help you learn the twelve pairs of cranial nerves in the proper order:

Old Opie **oc**casionally **t**ries **trig**onometry, **a**nd **f**eels **v**ery **glo**omy, **vag**ue, **a**nd **hypo**active.[1]

4. Complete Part C of the laboratory report.

---

1. From *HAPS-EDucator,* Winter 2002. An official publication of the Human Anatomy & Physiology Society (HAPS).

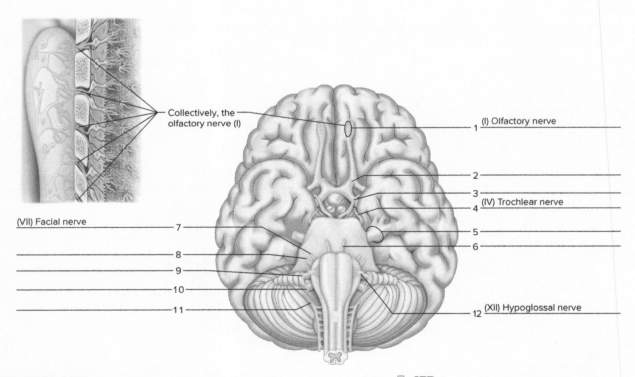

Collectively, the olfactory nerve (I)

(I) Olfactory nerve — 1

2

3

(IV) Trochlear nerve — 4

(VII) Facial nerve — 7

5

6

8

9

10

11

12 — (XII) Hypoglossal nerve

FIGURE 26.4   Provide the names of the cranial nerves in this inferior view. **3 APR**

Name _____

Date _____

Section _____

The ⟁ corresponds to the indicated Learning Outcome(s) found at the beginning of the laboratory exercise.

Brain and Cranial Nerves

⟳ PART A ASSESSMENTS

Match the terms in column A with the descriptions in column B. Place the letter of your choice in the space provided. ⟁

Column A	Column B
a. Central sulcus	_____ 1. Structure formed by the crossing-over of the optic nerves
b. Cerebral cortex	_____ 2. Part of diencephalon that forms lower walls and floor of third ventricle
c. Corpus callosum	_____ 3. Cone-shaped gland attached to upper posterior portion of diencephalon
d. Falx cerebelli	_____ 4. Connects cerebral hemispheres
e. Gyrus (convolution)	_____ 5. Ridge on surface of cerebrum
f. Hypothalamus	_____ 6. Separates frontal and parietal lobes
g. Insula	_____ 7. Part of brainstem between diencephalon and pons
h. Medulla oblongata	_____ 8. Rounded bulge on underside of brainstem
i. Midbrain	_____ 9. Part of brainstem continuous with the spinal cord
j. Optic chiasma	_____ 10. Layer of dura mater that separates cerebellar hemispheres
k. Pineal gland	_____ 11. Cerebral lobe located deep within lateral sulcus
l. Pons	_____ 12. Thin layer of gray matter on surface of cerebrum

⟳ PART B ASSESSMENTS

Identify the features indicated in the sagittal (median) section of the right half of the human brain in figure 26.5. ⟁

1. _____   6. _____

2. _____   7. _____

3. _____   8. _____

4. _____   9. _____

5. _____   10. _____

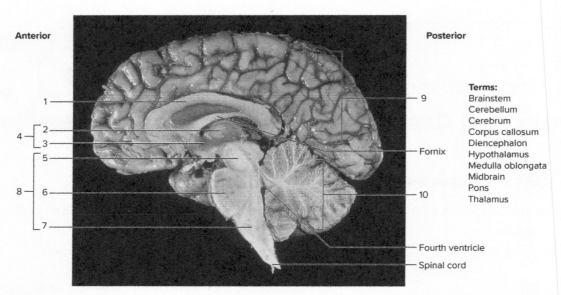

Anterior   Posterior

Terms:
Brainstem
Cerebellum
Cerebrum
Corpus callosum
Diencephalon
Hypothalamus
Medulla oblongata
Midbrain
Pons
Thalamus

Fornix

Fourth ventricle

Spinal cord

FIGURE 26.5   Identify the features on this sagittal (median) section of the right half of the human brain, using the terms provided. **APR** (Martin M. Rotker/Science Source)

 PART C ASSESSMENTS

Match the cranial nerves in column A with the associated functions in column B. Place the letter of your choice in the space provided. **A**

Column A	Column B
a. Abducens	_____ **1.** Regulates heart rate
b. Accessory	_____ **2.** Equilibrium and hearing
c. Facial	_____ **3.** Stimulates superior oblique muscle of eye
d. Glossopharyngeal	_____ **4.** Sensory impulses from teeth and face
e. Hypoglossal	_____ **5.** Pupil constriction and eyelid opening
f. Oculomotor	_____ **6.** Smell
g. Olfactory	_____ **7.** Controls neck and shoulder movements
h. Optic	_____ **8.** Controls tongue movements
i. Trigeminal	_____ **9.** Vision
j. Trochlear	_____ **10.** Stimulates lateral rectus muscle of eye
k. Vagus	_____ **11.** Taste, salivation, and secretion of tears
l. Vestibulocochlear	_____ **12.** Swallowing

PART D ASSESSMENTS

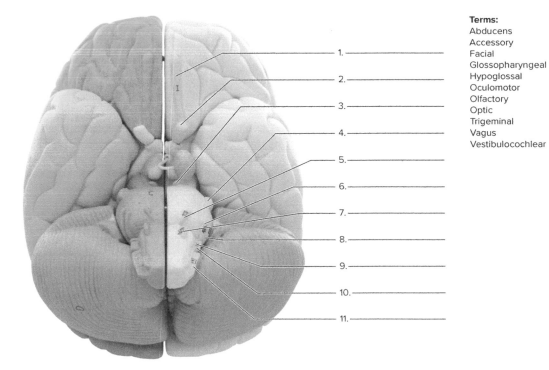

Terms:
Abducens
Accessory
Facial
Glossopharyngeal
Hypoglossal
Oculomotor
Olfactory
Optic
Trigeminal
Vagus
Vestibulocochlear

1. _____

2. _____

3. _____

4. _____

5. _____

6. _____

7. _____

8. _____

9. _____

10. _____

11. _____

FIGURE 26.6   Using the terms provided identify the visible cranial nerves on the model image. (Trochlear nerve not visible). (C22 Neuro-Anatomical Brain © 3B Scientific GmbH, Germany. 2017 www.3bscientific.com)

LABORATORY EXERCISE

27

Dissection of the Sheep Brain

MATERIALS NEEDED

Dissectible model of human brain
Preserved sheep brain
Dissecting tray
Dissection instruments
Long knife
Disposable gloves
Frontal sections of sheep brains

 SAFETY

- Wear disposable gloves and safety goggles when handling the sheep brains.
- Save or dispose of the brains as instructed.
- Wash your hands before leaving the laboratory.

PURPOSE OF THE EXERCISE

To observe the major features of the sheep brain and to compare these features with those of the human brain.

 LEARNING OUTCOMES APR

After completing this exercise, you should be able to

(1) Examine the major structures of the sheep brain.

(2) Summarize several differences and similarities between the sheep brain and the human brain.

(3) Locate the most developed cranial nerves of the sheep brain.

Mammalian brains have many features in common. Human brains may not be available, so sheep brains often are dissected as an aid to understanding mammalian brain structure. However, the adaptations of the sheep differ from the adaptations of the human, so comparisons of their structural features may not be precise. The sheep is a quadruped; therefore, the spinal cord is horizontal, unlike the vertical orientation in a bipedal human. Preserved sheep brains have a different appearance and are firmer than those that are removed directly from the cranial cavity; this is caused by the preservatives used.

 PRACTICE

PROCEDURE—Dissection of the Sheep Brain

1. Obtain a preserved sheep brain and rinse it thoroughly in water to remove as much of the preserving fluid as possible.
2. Examine the surface of the brain for the presence of meninges. (The outermost layers of these membranes may have been mostly lost during removal of the brain from the cranial cavity.) Locate the following:

   dura mater—the thick, opaque outer layer
   arachnoid mater—the weblike, semitransparent middle layer spanning the fissures of the brain
   pia mater—the thin, delicate, innermost layer that adheres to the gyri and sulci of the surface of the brain

3. Remove any remaining dura mater by pulling it gently from the surface of the brain.
4. Position the brain with its ventral surface down in the dissecting tray. Study figure 27.1, and locate the following structures on the specimen:

   cerebral hemispheres
   gyri (convolutions)
   sulci
   longitudinal fissure
   frontal lobe
   parietal lobe
   temporal lobe
   occipital lobe
   cerebellum
   medulla oblongata
   spinal cord

5. Gently separate the cerebral hemispheres along the longitudinal fissure and expose the transverse band of white fibers within the fissure that connects the hemispheres. This band is the *corpus callosum*.
6. Bend the cerebellum and medulla oblongata slightly downward and away from the cerebrum (fig. 27.2). This will expose the *pineal gland* in the upper midline and the *corpora quadrigemina*, which consists of four

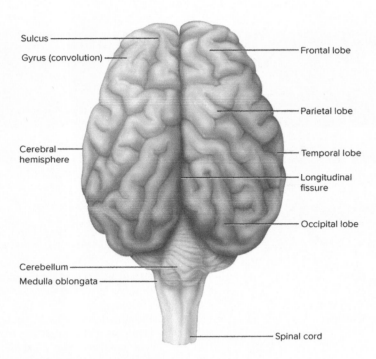

Sulcus

Gyrus (convolution)

Cerebral hemisphere

Cerebellum

Medulla oblongata

Frontal lobe

Parietal lobe

Temporal lobe

Longitudinal fissure

Occipital lobe

Spinal cord

FIGURE 27.1   Dorsal surface of the sheep brain. APR

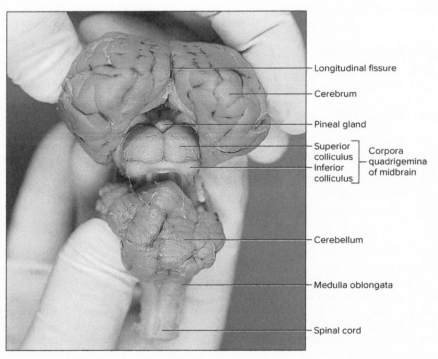

Longitudinal fissure

Cerebrum

Pineal gland

Superior colliculus

Inferior colliculus

Corpora quadrigemina of midbrain

Cerebellum

Medulla oblongata

Spinal cord

FIGURE 27.2   Gently bend the cerebellum and medulla oblongata away from the cerebrum to expose the pineal gland of the diencephalon and the corpora quadrigemina of the midbrain. (© J and J Photography)

rounded structures, called *colliculi*, associated with the midbrain.

7. Position the brain with its ventral surface upward. Study figures 27.3 and 27.4, and locate the following structures on the specimen:

longitudinal fissure
olfactory bulbs
optic nerves

optic chiasma
optic tract
mammillary body (single in sheep)
infundibulum (pituitary stalk)
midbrain
pons

8. Although some of the cranial nerves may be missing or are quite small and difficult to find, locate as many of

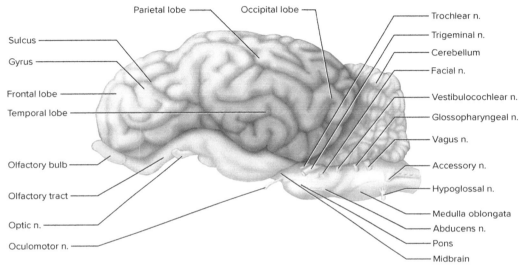

FIGURE 27.3   Lateral surface of the sheep brain. (The "n." stands for nerve.)

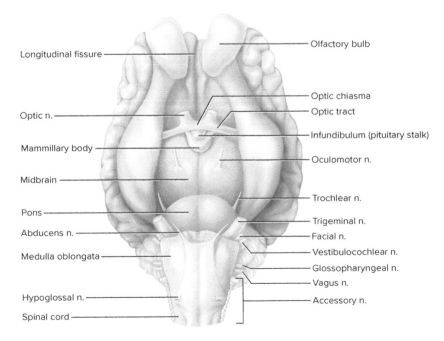

FIGURE 27.4   Ventral surface of the sheep brain. APR

the following as possible, using figures 27.3 and 27.4 as references:

oculomotor nerves
trochlear nerves
trigeminal nerves
abducens nerves
facial nerves
vestibulocochlear nerves
glossopharyngeal nerves
vagus nerves
accessory nerves
hypoglossal nerves

9. Using a long, sharp knife, cut the sheep brain along the midline to produce a median section. Study figures 27.2 and 27.5, and locate the following structures on the specimen:

cerebrum
olfactory bulb
corpus callosum
cerebellum
    white matter
    gray matter
lateral ventricle (one in each cerebral hemisphere)
third ventricle (within diencephalon)
fourth ventricle (between brainstem and cerebellum)

diencephalon
    optic chiasma
    infundibulum
    pituitary gland (this structure may be missing)
    mammillary body (single in sheep)
    thalamus
    hypothalamus
    pineal gland
midbrain
    corpora quadrigemina
        superior colliculi
        inferior colliculi
pons
medulla oblongata

 LEARN: ACTIVITY

Observe a frontal section from a sheep brain. Note the longitudinal fissure, gray matter, white matter, corpus callosum, lateral ventricles, third ventricle, and thalamus.

10. Dispose of the sheep brain as directed by the laboratory instructor.
11. Complete Parts A, B, and C of Laboratory Report 27.

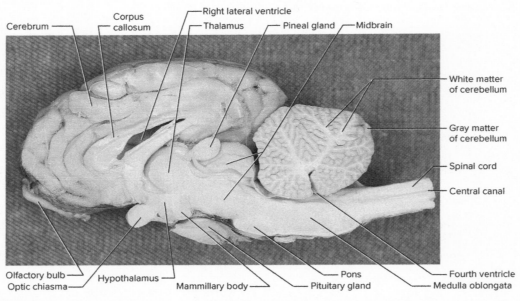

FIGURE 27.5   Sagittal (median) section of the right half of the sheep brain dissection.   **APR** (©J & J Photography)

Name _____

Date _____

Section _____

The  corresponds to the indicated Learning Outcome(s) found at the beginning of the laboratory exercise.

LABORATORY
REPORT

Dissection of the Sheep Brain

♻ PART A ASSESSMENTS

Answer the following questions:

1. Describe the location of any meninges observed to be associated with the sheep brain. **1** _____

   _____

   _____

2. Compare the relative sizes of the sheep and human cerebral hemispheres. **2** _____

   _____

   _____

3. How do the gyri and sulci of the sheep cerebrum compare with the human cerebrum in numbers? **2** _____

   _____

   _____

4. What is the significance of the differences you noted in your answers for questions 2 and 3? **2** _____

   _____

   _____

5. What difference did you note in the structures of the sheep cerebellum and the human cerebellum? **2** _____

   _____

   _____

6. How do the sizes of the olfactory bulbs of the sheep brain compare with those of the human brain? **2** _____

   _____

   _____

7. Based on their relative sizes, which of the cranial nerves seems to be most highly developed in the sheep brain? **3** _____

   _____

   _____

8. What is the significance of the observations you noted in your answers for questions 6 and 7? **2** _____

   _____

   _____

PART B ASSESSMENTS

ASSESS

CRITICAL THINKING

Prepare a list of at least six features to illustrate ways in which the brains of sheep and humans are similar. 2

1. _____

2. _____

3. _____

4. _____

5. _____

6. _____

Interpret the significance of these similarities. 2 _____

_____

_____

_____

PART C ASSESSMENTS

Identify the features indicated in the sagittal (median) section of the sheep brain in figure 27.6. 1

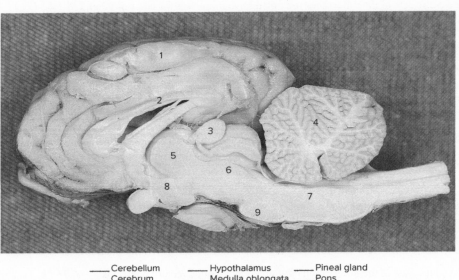

_____ Cerebellum   _____ Hypothalamus   _____ Pineal gland
_____ Cerebrum   _____ Medulla oblongata _____ Pons
_____ Corpus callosum _____ Midbrain   _____ Thalamus

FIGURE 27.6   Label the features of this sagittal (median) section of the sheep brain by placing the correct numbers in the spaces provided. (©J & J Photography)

28

Spinal Cord and Meninges

MATERIALS NEEDED

Textbook
Compound light microscope
Prepared microscope slide of a spinal cord cross section with spinal nerve roots
Spinal cord model with meninges
Preserved spinal cord with meninges intact

PURPOSE OF THE EXERCISE

To review the characteristics of the spinal cord and the meninges and to observe the major features of these structures.

LEARNING OUTCOMESS APR

After completing this exercise, you should be able to

1. Identify the major features and functions of the spinal cord.

2. Locate the ascending and descending tracts of the spinal cord.

3. Arrange the layers of the meninges and describe the structure of each.

The spinal cord is a column of nerve fibers that extends down through the vertebral canal. Together with the brain, it makes up the central nervous system.

Neurons within the spinal cord provide a two-way communication system between the brain and body parts outside the central nervous system. The cord also contains the processing centers for spinal reflexes.

Three membranes, or meninges, surround the entire CNS. The meninges consist of layers of membranes located between the bones of the skull and vertebral column and the soft tissues of the central nervous system. They include the dura mater, the arachnoid mater, and the pia mater.

PRACTICE

PROCEDURE A—Structure of the Spinal Cord

1. Review section 11.5 titled "The Spinal Cord" in chapter 11 of the textbook.

2. As a review activity, label figures 28.1 and 28.2.

3. Complete Part A of Laboratory Report 28.

4. Study figure 28.3 and complete Part B of the laboratory report.

5. Obtain a prepared microscope slide of a spinal cord cross section. Use the low power of the microscope to locate the following features:

posterior median sulcus
anterior median fissure
central canal
gray matter
    gray commissure
    posterior (dorsal) horn
    lateral horn
    anterior (ventral) horn
white matter
    posterior (dorsal) funiculus (column)
    lateral funiculus (column)
    anterior (ventral) funiculus (column)
roots of spinal nerve
    posterior (dorsal) roots
    posterior (dorsal) root ganglia
    anterior (ventral) roots

6. Observe the model of the spinal cord, and locate the features listed in step 5.

7. Complete Part C of the laboratory report.

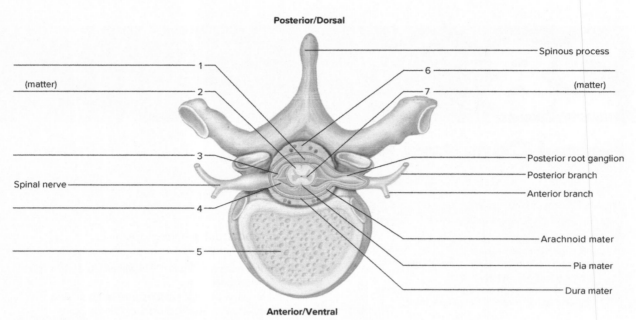

Posterior/Dorsal

Spinous process

1

6

(matter)

7

(matter)

3

Spinal nerve

Posterior root ganglion

Posterior branch

Anterior branch

4

Arachnoid mater

5

Pia mater

Dura mater

Anterior/Ventral

FIGURE 28.1   Label the features of the spinal cord and the surrounding structures.

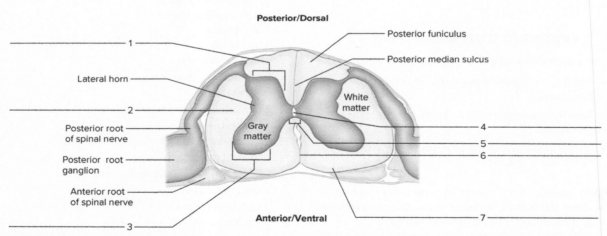

Posterior/Dorsal

1

Posterior funiculus

Posterior median sulcus

Lateral horn

White
matter

2

Gray
matter

4

Posterior root
of spinal nerve

5

6

Posterior root
ganglion

Anterior root
of spinal nerve

3

7

Anterior/Ventral

FIGURE 28.2   Label this cross section of the spinal cord, including the features of the white and gray matter.

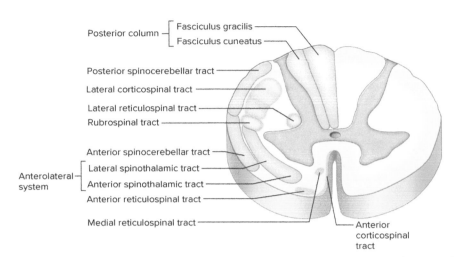

Posterior column — [Fasciculus gracilis —
                     Fasciculus cuneatus —

Posterior spinocerebellar tract —

Lateral corticospinal tract —

Lateral reticulospinal tract —
Rubrospinal tract —

Anterior spinocerebellar tract —

Anterolateral system — [Lateral spinothalamic tract —
                         Anterior spinothalamic tract —
Anterior reticulospinal tract —

Medial reticulospinal tract —

Anterior corticospinal tract

FIGURE 28.3   Major ascending (sensory) and descending (motor) tracts within a cross section of the spinal cord. Ascending tracts are in pink, descending tracts in light brown, and are shown only on one side. This pattern varies with the level of the spinal cord. This pattern is representative of the midcervical region. (*Note:* These tracts are not visible as individually stained structures on microscope slides.)

 PRACTICE

PROCEDURE B—Meninges

1. Review section 11.2 titled "Meninges" in chapter 11 of the textbook.
2. Study figure 28.4, showing the meninges associated with the brain and spinal cord.
3. Observe the spinal cord model with meninges and locate the following features:
   dura mater
   arachnoid mater
   subarachnoid space
   pia mater
4. As a review activity, label figure 28.5.
5. Complete Part D of the laboratory report.

 LEARN: ACTIVITY

Observe the preserved section of spinal cord with the meninges intact. Note the appearance of the central gray matter and the surrounding white matter. Note the heavy covering of dura mater, firmly attached to the cord on each side by a set of ligaments (denticulate ligaments) originating in the pia mater. The intermediate layer of meninges, the arachnoid mater, is devoid of blood vessels, but in a live human being, the space beneath this layer contains cerebrospinal fluid. The pia mater, closely attached to the surface of the spinal cord, contains many blood vessels.

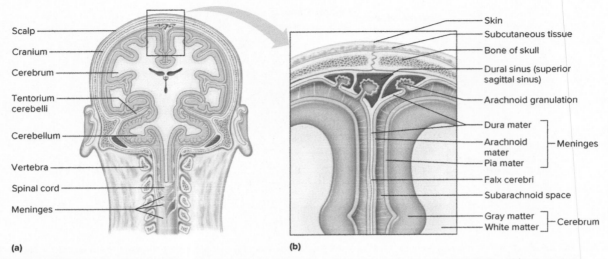

Scalp

Cranium

Cerebrum

Tentorium cerebelli

Cerebellum

Vertebra

Spinal cord

Meninges

(a)

Skin

Subcutaneous tissue

Bone of skull

Dural sinus (superior sagittal sinus)

Arachnoid granulation

Dura mater

Arachnoid mater

Pia mater

Meninges

Falx cerebri

Subarachnoid space

Gray matter

White matter

Cerebrum

(b)

FIGURE 28.4   Meninges. (*a*) Membranes called meninges enclose the brain and spinal cord. (*b*) The meninges include three layers: dura mater, arachnoid mater, and pia mater. **APR**

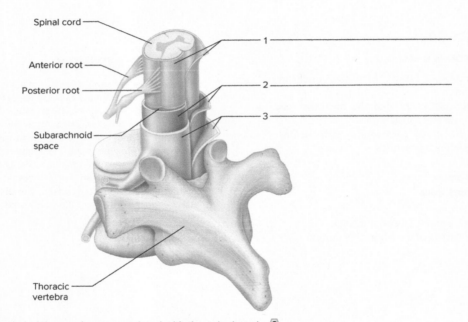

Spinal cord

Anterior root

Posterior root

Subarachnoid space

1

2

3

Thoracic vertebra

FIGURE 28.5   Label the meninges associated with the spinal cord. **A**

Name _____

Date _____

Section _____

The 🔺 corresponds to the indicated Learning Outcome(s) found at the beginning of the laboratory exercise.

LABORATORY
REPORT

28

Spinal Cord and Meninges

♻ PART A ASSESSMENTS

Complete the following statements:

1. The spinal cord gives rise to 31 pairs of _____ . 🔺

2. The bulge in the spinal cord that gives off nerves to the upper limbs is called the _____ . 🔺

3. The bulge in the spinal cord that gives off nerves to the lower limbs is called the _____ . 🔺

4. The _____ is a groove that extends the length of the spinal cord posteriorly. 🔺

5. In a spinal cord cross section, the posterior _____ of the gray matter resemble the upper wings of a butterfly. 🔺

6. The cell bodies of motor neurons are found in the _____ horns of the spinal cord. 🔺

7. The _____ connects the gray matter on the left and right sides of the spinal cord. 🔺

8. The _____ in the gray commissure of the spinal cord contains cerebrospinal fluid and is continuous with the ventricles of the brain. 🔺

9. The white matter of the spinal cord is divided into anterior, lateral, and posterior _____ . 🔺

10. The longitudinal bundles of nerve fibers within the spinal cord constitute major nerve pathways called _____ . 🔺

♻ PART B ASSESSMENTS

Match the nerve tracts in column A with the descriptions in column B. Place the letter of your choice in the space provided. 🔺

Column A	Column B
a. Corticospinal	_____ **1.** Ascending tract to the brain to interpret touch, pressure, and body movements
b. Fasciculus gracilis	_____ **2.** Descending tract whose fibers conduct motor impulses to sweat glands and muscles to control tone
c. Lateral spinothalamic	_____ **3.** Descending tract whose fibers conduct motor impulses to skeletal muscles
d. Posterior spinocerebellar	_____ **4.** Ascending tract to the cerebellum necessary for coordination of skeletal muscles
e. Reticulospinal	_____ **5.** Ascending tract to the brain to give rise to sensations of temperature and pain

PART C ASSESSMENTS

Identify the features indicated in the spinal cord cross section of figure 28.6. 🅐

Posterior/Dorsal

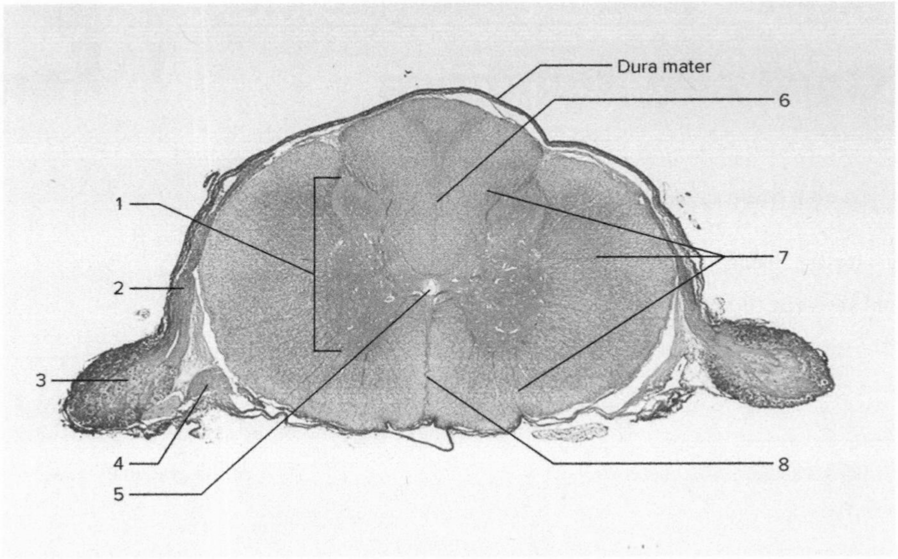

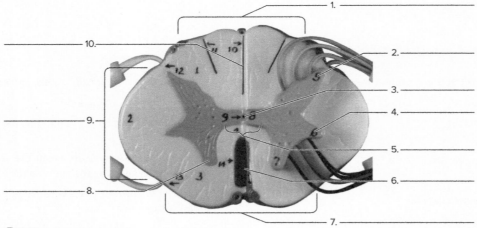

Anterior/Ventral

_____ Anterior median fissure	_____ Gray matter	_____ Posterior root of spinal nerve
_____ Anterior root of spinal nerve	_____ Posterior median sulcus	_____ White matter
_____ Central canal	_____ Posterior root ganglion	

FIGURE 28.6   Micrograph of a spinal cord cross section with spinal nerve roots (7.5×). Label the features by placing the correct numbers in the spaces provided. (Ed Reschke)

Terms:
Anterior/ventral funiculus/column
Anterior/ventral horn
Anterior median fissure
Central canal
Gray commissure
Lateral funiculus/column
Lateral horn
Posterior/dorsal funiculus/column
Posterior/dorsal horn
Posterior median sulcus

FIGURE 28.7   Using the terms provided, identify the structures indicated on the cross section model of the spinal cord image. (2017 Denoyer-Geppert Science Company denoyer.com)

PART D ASSESSMENTS

Match the terms in column A with the descriptions in column B. Place the letter of your choice in the space provided.

Column A		Column B
a. Arachnoid mater	_____	**1.** Band of pia mater that anchors dura mater to cord
b. Denticulate ligament	_____	**2.** Channel through which venous blood flows
c. Dural sinus	_____	**3.** Outermost layer of meninges
d. Dura mater	_____	**4.** Follows irregular contours of spinal cord surface
e. Epidural space	_____	**5.** Contains cerebrospinal fluid
f. Pia mater	_____	**6.** Thin, weblike middle membrane
g. Subarachnoid space	_____	**7.** Separates dura mater from bone of vertebra

29

Reflex Arc and Reflexes

MATERIALS NEEDED

Textbook
Rubber percussion hammer

PURPOSE OF THE EXERCISE

To review the characteristics of reflex arcs and reflex behavior and to demonstrate some of the reflexes that occur in the human body.

 ## LEARNING OUTCOMES APR

After completing this exercise, you should be able to

1. Describe the components of a reflex arc.
2. Demonstrate and record stretch reflexes that occur in humans.
3. Analyze the components and patterns of stretch reflexes.

A reflex arc represents the simplest type of neural pathway found in the nervous system. This pathway begins with a receptor at the dendrite end of a sensory (afferent) neuron. The sensory neuron leads into the central nervous system and may communicate with one or more interneurons. Some of these interneurons, in turn, communicate with motor (efferent) neurons, whose axons (nerve fibers) lead outward to effectors.

Thus, when a sensory receptor is stimulated by a change occurring inside or outside the body, impulses may pass through a reflex arc, and, as a result, effectors may respond. Such an automatic, subconscious response is called a *reflex*.

A *stretch reflex* involves a single synapse (monosynaptic) between a sensory and a motor neuron within the gray matter of the spinal cord. Examples of stretch reflexes include the patellar, calcaneal, biceps, triceps, and plantar reflexes. Other, more complex *withdrawal reflexes* involve interneurons in combination with sensory and motor neurons; thus, they are polysynaptic. Examples of withdrawal reflexes include responses to touching hot objects or stepping on sharp objects.

Reflexes demonstrated in this laboratory exercise are stretch reflexes. When a muscle is stretched by a tap over its tendon, stretch receptors (proprioceptors) called *muscle spindles* are stretched within the muscle, which initiates an impulse over a reflex arc. A sensory neuron conducts an impulse from the muscle spindle into the gray matter of the spinal cord, where it synapses with a motor neuron, which conducts the impulse to the effector muscle. The stretched muscle responds by contracting to resist or reverse further stretching. These stretch reflexes are important to maintaining proper posture, balance, and movements. Observations of many of these reflexes in clinical tests on patients may indicate damage to a level of the spinal cord or peripheral nerves of the particular reflex arc.

 ## PRACTICE

PROCEDURE—Reflex Arc and Reflexes

1. Review the concept heading entitled "Reflex Arcs" in section 11.5 in chapter 11 of the textbook and the concept heading entitled "Proprioception" in section 12.2 in chapter 12 of the textbook.
2. Study figure 29.1.
3. Complete Part A of Laboratory Report 29.
4. Work with a laboratory partner to demonstrate each of the reflexes listed (see fig. 29.2). *It is important that the muscles involved in the reflexes be totally relaxed to observe proper responses.* If a person is trying too hard to experience the reflex or is trying to suppress the reflex, assign a multitasking activity while the stimulus with the rubber hammer occurs. For example, assign a physical task with the upper limbs along with a complex mental activity during the patellar (knee-jerk) demonstration. After each demonstration, record your observations in the table provided in Part B of the laboratory report.
   a. *Patellar (knee-jerk) reflex.* Have your laboratory partner sit on a table (or sturdy chair) with legs relaxed and hanging freely over the edge without touching the floor. Gently strike your partner's patellar ligament (just below the patella) with the blunt side of a rubber percussion hammer (fig. 29.2a). The reflex involves the femoral nerve and the spinal cord. The normal response is a moderate extension of the leg at the knee joint.

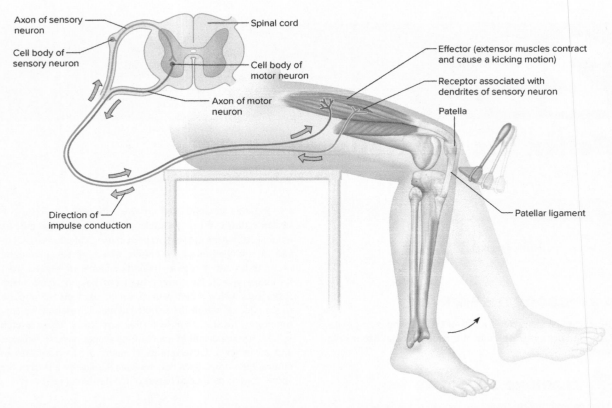

FIGURE 29.1 Diagram of a stretch (monosynaptic) reflex arc. The patellar reflex represents a specific example of a stretch reflex. **APR**

b. *Calcaneal (ankle-jerk) reflex.* Have your partner kneel on a chair with back toward you and with feet slightly dorsiflexed over the edge and relaxed. Gently strike the calcaneal tendon (just above its insertion on the calcaneus) with the blunt side of the rubber hammer (fig. 29.2*b*). The reflex involves the tibial nerve and the spinal cord. The normal response is plantar flexion of the foot.

c. *Biceps (biceps-jerk) reflex.* Have your partner place a bare arm bent about 90° at the elbow on the table. Press your thumb on the inside of the elbow over the tendon of the biceps brachii, and gently strike your thumb with the rubber hammer (fig. 29.2*c*). The reflex involves the musculocutaneous nerve and the spinal cord. Watch the biceps brachii for a response. The response might be a slight twitch of the muscle or flexion of the forearm at the elbow joint.

d. *Triceps (triceps-jerk) reflex.* Have your partner lie supine with an upper limb bent about 90° across the abdomen. Gently strike the tendon of the triceps brachii near its insertion just proximal to the olecranon process at the tip of the elbow (fig. 29.2*d*). The reflex involves the radial nerve and the spinal cord. Watch the triceps brachii for a response. The response might be a slight twitch of the muscle or extension of the forearm at the elbow joint.

e. *Plantar reflex.* Have your partner remove a shoe and sock and lie supine with the lateral surface of the foot resting on the table. Draw the metal tip of the rubber hammer, applying firm pressure, over the sole from the heel to the base of the large toe (fig. 29.2*e*). The normal response is flexion (curling) of the toes and plantar flexion of the foot. If the toes spread apart and dorsiflexion of the great toe occurs, the reflex is the abnormal *Babinski reflex* response (normal in infants until the nerve fibers have complete myelinization). If the Babinski reflex occurs later in life, it may indicate damage to the corticospinal tract of the CNS.

5. Complete Part B of the laboratory report.

(a) Patellar (knee-jerk) reflex

(b) Calcaneal (ankle-jerk) reflex

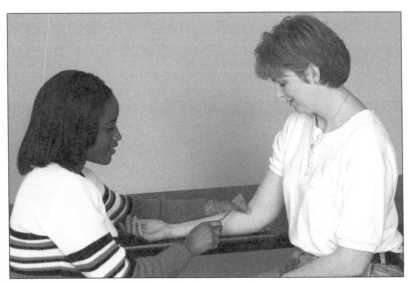

(c) Biceps reflex

FIGURE 29.2   Demonstrate each of the following reflexes: (*a*) patellar reflex; (*b*) calcaneal reflex; (*c*) biceps reflex; (*d*) triceps reflex; and (*e*) plantar reflex.   ((*a–e*) © J & J Photography)

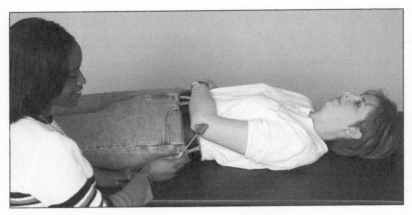

(d) Triceps reflex

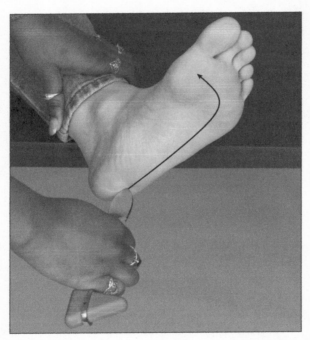

(e) Plantar reflex

FIGURE 29.2 *Continued.*

Name _____

Date _____

Section _____

The [A] corresponds to the indicated Learning Outcome(s) found at the beginning of the laboratory exercise.

Reflex Arc and Reflexes

♻ PART A ASSESSMENTS

Complete the following statements:

1. Neural _____ are routes followed by nerve impulses as they pass through the nervous system. [A]

2. Interneurons in a withdrawal reflex are located in the _____ . [A]

3. _____ are automatic subconscious responses to external or internal stimuli. [A]

4. Effectors of a reflex arc are glands and _____ . [A]

5. A patellar reflex employs only _____ and motor neurons. [A]

6. The effector muscle of the patellar reflex is the _____ . [A]

7. The sensory stretch receptors (muscle spindles) of the patellar reflex are located in the _____
muscle. [A]

8. The patellar reflex helps the body to maintain upright _____ . [A]

9. The sensory receptors of a withdrawal reflex are located in the _____ . [A]

10. _____ muscles in the limbs are the effectors of a withdrawal reflex. [A]

11. The normal plantar reflex results in _____ of toes. [A]

12. Stroking the sole of the foot in infants results in dorsiflexion and toes that spread apart, called the
_____ reflex. [A]

PART B ASSESSMENTS

1. Complete the following table: 2

Reflex Tested	Response Observed	Degree of Response (hypoactive, normal, or hyperactive)	Effector Muscle Involved
Patellar (knee-jerk)			
Calcaneal (ankle-jerk)			
Biceps			
Triceps			
Plantar			Gastrocnemius, soleus, and flexor digitorum longus

2. List the major events that occur in the patellar (knee-jerk) reflex, from the striking of the patellar ligament to the resulting response. 3 _____

_____

_____

_____

_____

 ASSESS

CRITICAL THINKING

What characteristics do the reflexes you demonstrated have in common? 3

_____

_____

_____

General and Special Senses

30

Receptors and General Senses

MATERIALS NEEDED

Textbook
Marking pen (washable)
Millimeter ruler
Bristle or sharp pencil
Forceps (fine points) or measuring calipers
Blunt metal probes
Three beakers (250 mL)
Warm tap water or 45°C (113°F) water bath
Cold water (ice water)
Thermometer
Prepared microscope slides of tactile (Meissner's)
    and lamellated (Pacinian) corpuscles
Compound light microscope

PURPOSE OF THE EXERCISE

To review the characteristics of sensory receptors and general senses and to investigate some of the general senses associated with the skin.

 LEARNING OUTCOMES

After completing this exercise, you should be able to

1. Associate types of sensory receptors with general senses throughout the body.

2. Determine and record the distribution of touch, warm, and cold receptors in various regions of the skin.

3. Measure the two-point threshold of various regions of the skin.

Sensory receptors are sensitive to changes that occur within the body and its surroundings. Each type of receptor is particularly sensitive to a distinct kind of environmental change and is much less sensitive to other forms of stimulation. When receptors are stimulated, they initiate nerve impulses that travel into the central nervous system. The raw form in which these receptors send information to the brain is called *sensation*. The way our brains interpret this information is called *perception*.

The sensory receptors found widely distributed throughout skin, muscles, joints, and visceral organs are associated with **general senses**. These senses include touch, pressure, temperature, pain, and the senses of muscle movement and body position.

Sensory receptors that are more specialized and confined to the head are associated with **special senses**. Laboratory Exercises 31 through 35 describe the special senses.

 PRACTICE

PROCEDURE A—Receptors

1. Review section 12.2 titled "Receptors, Sensation, and Perception" and section 12.3 titled "General Senses" in chapter 12 of the textbook.
2. Complete Part A of Laboratory Report 30.

 LEARN: ACTIVITY

Observe the tactile (Meissner's) corpuscle with the microscope set up by the laboratory instructor. This type of receptor is abundant in the superficial dermis in outer regions of the body, such as in the fingertips, soles, lips, and external genital organs. It is responsible for the sensation of light touch (see fig. 30.1 and chapter 6 of the textbook).

Observe the lamellated (Pacinian) corpuscle in the second demonstration microscope. This corpuscle is composed of many layers of connective tissue cells and has a nerve fiber in its central core. Lamellated corpuscles are numerous in the hands, feet, joints, and external genital organs. They are responsible for the sense of deep pressure (fig. 30.2). How are tactile and lamellated corpuscles similar?

_____

_____

How are they different?

_____

_____

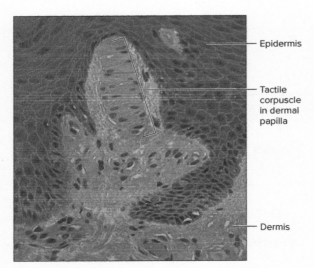

FIGURE 30.1 Tactile (Meissner's) corpuscles, such as this one, are responsible for the sensation of light touch (250×). (©Ed Reschke)

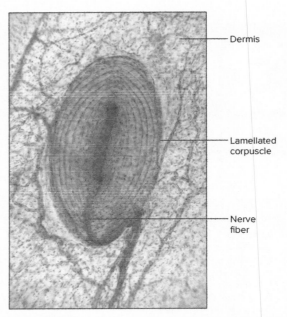

FIGURE 30.2 Lamellated (Pacinian) corpuscles, such as this one, are responsible for the sensation of deep pressure (100×). (©Ed Reschke)

 **PRACTICE**

PROCEDURE B—Sense of Touch

1. Investigate the distribution of touch receptors in your laboratory partner's skin. To do this, follow these steps:
   a. Use a marking pen and a millimeter ruler to prepare a square with 2.5 cm on each side on the skin on your partner's inner wrist, near the palm.
   b. Divide the square into smaller squares with 0.5 cm on a side, producing a small grid.
   c. Ask your partner to rest the marked wrist on the tabletop and to keep his/her eyes closed throughout the remainder of the experiment.
   d. Press the end of a bristle on the skin in some part of the grid, using just enough pressure to cause the bristle to bend. A sharp pencil can be used as an alternate device.
   e. Ask your partner to report whenever the touch of the bristle is felt. Record the results in Part B of the laboratory report.
   f. Continue this procedure until you have tested twenty-five different locations on the grid. Move randomly through the grid to help prevent anticipation of the next stimulation site.
2. Test two other areas of exposed skin in the same manner, and record the results in Part B of the laboratory report.
3. Answer the questions in Part B of the laboratory report.

 **PRACTICE**

PROCEDURE C—Two-Point Threshold

1. Test your partner's ability to recognize the difference between one and two points of skin being stimulated simultaneously. To do this, follow these steps:
   a. Have your partner place a hand with the palm up on the table and close his/her eyes.
   b. Hold the tips of a forceps tightly together and gently touch the skin on your partner's index finger.
   c. Ask your partner to report if it feels like one or two points are touching the finger.
   d. Allow the tips of the forceps or the points of the measuring calipers to spread so they are 1 mm apart, press both points against the skin simultaneously, and ask your partner to report as before.
   e. Repeat this procedure, allowing the tips of the forceps to spread more each time until your partner can feel both tips being pressed against the skin. The minimum distance between the tips of the forceps when both can be felt is called the *two-point threshold*. As soon as you are able to distinguish two points, two separate receptors are being stimulated instead of only one receptor. Areas of skin with a small threshold are more sensitive than areas with large two-point thresholds.
   f. Record the two-point threshold for the skin of a fingertip in Part C of the laboratory report.

2. Repeat this procedure to determine the two-point threshold of the palm, the back of the hand, the back of the neck, the forearm, and the leg. Record the results in Part C of the laboratory report.

3. Answer the questions in Part C of the laboratory report.

 PRACTICE

PROCEDURE D—Sense of Temperature

1. Investigate the distribution of *warm receptors* in your partner's skin. To do this, follow these steps:

   a. Mark a square with 2.5 cm sides on your partner's palm.

   b. Prepare a grid by dividing the square into smaller squares, 0.5 cm on a side.

   c. Have your partner rest the marked palm on the table and close his/her eyes.

   d. Heat a blunt metal probe by placing it in a beaker of warm (hot) water (about 40–45°C/104–113°F) for a minute or so. *(Be sure the probe does not get so hot that it burns the skin.)* Use a thermometer to monitor the appropriate warm water from the tap or the water bath.

   e. Wipe the probe dry and touch it to the skin on some part of the grid.

   f. Ask your partner to report if the probe feels warm. Then record the results in Part D of the laboratory report.

   g. Keep the probe warm, and repeat the procedure until you have randomly tested several different locations on the grid.

2. Investigate the distribution of *cold receptors* by repeating the procedure. Use a blunt metal probe that has been cooled by placing it in ice water for a minute or so. Record the results in Part D of the laboratory report.

3. Answer the questions in Part D of the laboratory report.

 LEARN: LAB IN MOTION

Prepare three beakers of water of different temperatures. One beaker should contain warm water (about 40°C/104°F), one should be room temperature (about 22°C/72°F), and one should contain cold water (about 10°C/50°F). Place the index finger of one hand in the warm water and, at the same time, place the index finger of the other hand in the cold water for about 2 minutes. Then, simultaneously move both index fingers into the water at room temperature. What temperature do you sense with each finger? How do you explain the resulting perceptions?

_____

_____

_____

_____

Notes

Name _____

Date _____

Section _____

The Ⓐ corresponds to the indicated Learning Outcome(s) found at the beginning of the laboratory exercise.

Receptors and General Senses

🔁 PART A ASSESSMENTS

Complete the following statements:

1. _____ are receptors that are sensitive to changes in the concentrations of chemicals. Ⓐ

2. Whenever tissues are damaged, _____ receptors are likely to be stimulated. Ⓐ

3. Receptors that are sensitive to temperature changes are called as _____ . Ⓐ

4. _____ are sensitive to changes in the intensity of light energy. Ⓐ

5. A sensation may seem to fade away when receptors are continuously stimulated as a result of _____ .
   Ⓐ

6. Tactile (Meissner's) corpuscles are responsible for the sense of light _____ . Ⓐ

7. Lamellated (Pacinian) corpuscles are responsible for the sense of deep _____ . Ⓐ

8. Warm receptors are most sensitive to temperatures between _____ . Ⓐ

9. Cold receptors are most sensitive to temperatures between _____ . Ⓐ

10. Widely distributed sensory receptors throughout the body are associated with _____ senses, in contrast to special senses. Ⓐ

🔁 PART B ASSESSMENTS

Skin of wrist

1. Record a "+" to indicate where the bristle was felt and a "0" to indicate where it was not felt. Ⓐ

Region tested _____   Region tested _____

2. Show the distribution of touch receptors in two other regions of skin. ▲2
3. Answer the following questions:

    a. How do you describe the pattern of distribution for touch receptors in the regions of the skin you tested? ▲2

    _____

    _____

    b. How does the concentration of touch receptors seem to vary from region to region? ▲2 _____

    _____

PART C ASSESSMENTS

1. Record the two-point threshold, in millimeters, for skin in each of the following regions: ▲3

    Fingertip _____

    Palm _____

    Back of hand _____

    Back of neck _____

    Forearm _____

    Leg _____

2. Answer the following questions:

    a. What region of the skin tested has the greatest ability to discriminate two points? ▲3 _____

    b. What region of the skin has the least sensitivity to this test? ▲3 _____

    c. What is the significance of the observations in questions *a* and *b*? ▲3 _____

    _____

    _____

PART D ASSESSMENTS

Skin of palm

1. Record a "+" to indicate where warm was felt and a "0" to indicate where it was not felt. ▲2

Skin of palm

2. Record a "+" to indicate where cold was felt and a "0" to indicate where it was not felt. ▲2

3. Answer the following questions:

    a. How do temperature receptors appear to be distributed in the skin of the palm? ▲2 _____

    _____

    b. Compare the distribution and concentration of warm and cold receptors in the skin of the palm. ▲2 _____

    _____

    _____

LABORATORY EXERCISE

31

Smell and Taste

PURPOSE OF THE EXERCISE

To review the structures of the organs of smell and taste and to investigate the abilities of smell and taste receptors to discriminate various chemical substances.

LEARNING OUTCOMES APR

After completing this exercise, you should be able to

1. Compare the characteristics of the smell (olfactory) receptors and taste (gustatory) receptors.
2. Explain how the senses of smell and taste function.
3. Record recognized odors and the time needed for olfactory sensory adaptation to occur.
4. Locate the distribution of taste receptors on the surface of the tongue and mouth cavity.

The senses of smell (olfaction) and taste (gustation) are dependent upon chemoreceptors that are stimulated by various chemicals dissolved in liquids. When these receptors are stimulated by appropriate chemicals, action potentials travel to the brain for interpretation. The receptors of smell are found in the olfactory organs, which are located in the superior parts of the nasal cavity and in a portion of the nasal septum. The receptors of taste occur in the taste buds, which are sensory organs primarily found on the surface of the tongue, but receptor cells are also distributed in other areas of the mouth cavity and pharynx. Taste sensations are grouped into five recognized categories: sweet, sour, salty, bitter, and umami. The sweet sensation is produced from sugars, the sour sensation from acids, the salty sensation from inorganic salts, the bitter sensation from alkaloids and spoiled foods, and the umami sensation from aspartic and glutamic acids or a derivative such as monosodium glutamate.

The senses of smell and taste function closely together, because substances that are tasted often are smelled at the same moment, and they play important roles in the selection of foods. The taste and aroma of foods are also influenced by appearance, texture, temperature, and the person's mood. Chemicals are considered odorless and tasteless if receptor sites for them are absent.

PRACTICE

PROCEDURE A—Sense of Smell (Olfaction)

1. Review the concept heading titled "Sense of Smell: Olfaction" in section 12.4 in chapter 12 of the textbook.
2. As a review activity, label figure 31.1.
3. Complete Part A of Laboratory Report 31.
4. Test your laboratory partner's ability to recognize the odors of the bottled substances available in the laboratory. To do this, follow these steps:
   a. Have your partner keep his/her eyes closed.
   b. Remove the stopper from one of the bottles and hold it about 4 cm under your partner's nostrils for about 2 seconds.
   c. Ask your partner to identify the odor, and then replace the stopper.
   d. Record your partner's response in Part B of the laboratory report.
   e. Repeat steps *b–d* for each of the bottled substances.
5. Repeat the preceding procedure, using the same set of bottled substances, but present them to your partner in a different sequence. Record the results in Part B of the laboratory report.

6. Wait 10 minutes and then determine the time it takes for your partner to experience olfactory sensory adaptation. To do this, follow these steps:
   a. Ask your partner to breathe in through the nostrils and exhale through the mouth.
   b. Remove the stopper from one of the bottles, and hold it about 4 cm under your partner's nostrils.
   c. Keep track of the time that passes until your partner is no longer able to detect the odor of the substance.
   d. Record the result in Part B of the laboratory report.
   e. Wait 5 minutes and repeat this procedure, using a different bottled substance.
   f. Test a third substance in the same manner.
   g. Record the results as before.
7. Complete Part B of the laboratory report.

 LEARN: ACTIVITY

Observe the olfactory epithelium in the demonstration microscope. The olfactory receptor cells are spindle-shaped, bipolar neurons with spherical nuclei. They also have six to eight cilia (olfactory hairs) at their apical ends. The supporting cells are pseudostratified columnar epithelial cells. However, in this region the tissue lacks goblet cells (see fig. 31.2 and chapter 5 of the textbook).

___ Cilia

___ Columnar epithelial cells

___ Cribriform plate of ethmoid bone

___ Nasal cavity

___ Nerve fibers within olfactory bulb

___ Olfactory nerves

___ Olfactory receptor cells

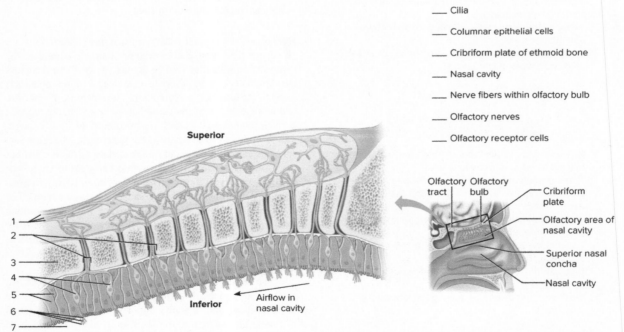

FIGURE 31.1   Label this diagram of the olfactory organ by placing the correct numbers in the spaces provided. **APR**

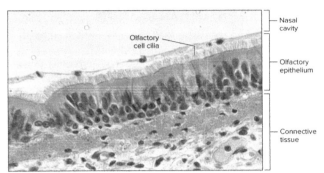

FIGURE 31.2 Micrograph of olfactory epithelium (250×). Olfactory receptors have cilia (olfactory hairs) at their apical ends. **APR** (Biophoto Associates/Science Source)

 PRACTICE

PROCEDURE B—Sense of Taste (Gustation)

1. Review the concept heading titled "Sense of Taste: Gustation" in section 12.4 in chapter 12 of the textbook.
2. As a review activity, label figure 31.3.
3. Complete Part C of the laboratory report.
4. Map the distribution of the receptors for the primary taste sensations on your partner's tongue. To do this, follow these steps:
   a. Ask your partner to rinse his/her mouth with water and then partially dry the surface of the tongue with a paper towel.
   b. Moisten a clean cotton swab with 5% sucrose solution, and touch several regions of your partner's tongue with the swab.
   c. Each time you touch the tongue, ask your partner to report what sensation is experienced.
   d. Test the tip, sides, and back of the tongue in this manner.

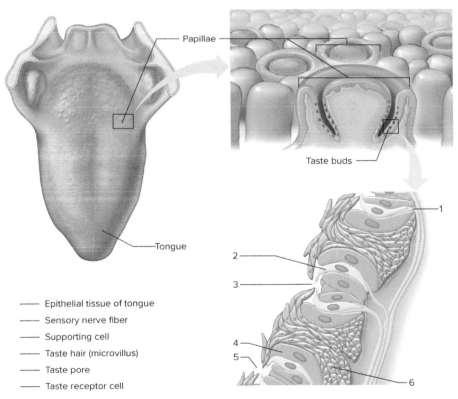

_____ Epithelial tissue of tongue

_____ Sensory nerve fiber

_____ Supporting cell

_____ Taste hair (microvillus)

_____ Taste pore

_____ Taste receptor cell

FIGURE 31.3 Label this diagram by placing the correct numbers in the spaces provided. **APR**

e. Test some other representative areas inside the mouth cavity such as the cheek, gums, and roof of the mouth (hard and soft palate) for a sweet sensation.

f. Record your partner's responses in Part D of the laboratory report.

g. Have your partner rinse his/her mouth and dry the tongue again, and repeat the preceding procedure, using each of the other four test solutions—NaCl, acetic acid, quinine or Epsom salt, and MSG solution. Be sure to use a fresh swab for each test substance and dispose of used swabs and paper towels as directed.

5. Complete Part D of the laboratory report.

 LEARN: ACTIVITY

Observe the oval-shaped taste bud in the demonstration microscope. Note the surrounding epithelial cells. The taste pore, an opening into the taste bud, may be filled with taste hairs (microvilli). Within the taste bud, there are supporting cells and thinner taste-receptor cells, which often have lightly stained nuclei (fig. 31.4).

 LEARN: LAB IN MOTION

Test your laboratory partner's ability to recognize the tastes of apple, potato, carrot, and onion. (A package of mixed flavors of LifeSavers is a good alternative.) To do this, follow these steps:

1. Have your partner close his/her eyes and hold the nostrils shut.
2. Place a small piece of one of the test substances on your partner's tongue.
3. Ask your partner to identify the substance without chewing or swallowing it.
4. Repeat the procedure for each of the other substances.

How do you explain the results of this experiment?

_____

_____

_____

_____

_____

_____

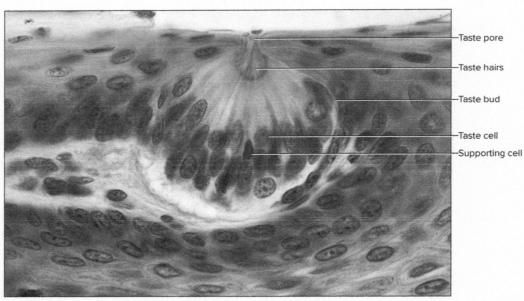

FIGURE 31.4   Micrograph of a taste bud (400×). Taste receptors are found in taste buds. **APR** (©Ed Reschke)

Name _____

Date _____

Section _____

The  corresponds to the indicated Learning Outcome(s) found at the beginning of the laboratory exercise.

Smell and Taste

♻ PART A ASSESSMENTS

Complete the following statements:

1. Olfactory, or smell, receptors are _____ neurons surrounded by columnar epithelial cells. ⓐ1

2. The distal ends of the olfactory neurons are covered with hairlike _____ . ⓐ1

3. Before gaseous substances can stimulate the olfactory receptors, they must be dissolved in _____ that surrounds the cilia. ⓐ1

4. The axons of olfactory receptors pass through small openings in the _____ of the ethmoid bone. ⓐ2

5. Olfactory bulbs lie on either side of the _____ of the ethmoid bone. ⓐ2

6. The sensory impulses pass from the olfactory bulbs through the _____ to the interpreting centers of the brain. ⓐ2

7. The olfactory interpreting centers are located deep within the temporal lobes and at the base of the _____ lobes of the cerebrum. ⓐ2

8. Olfactory sensations usually fade rapidly as a result of _____ . ⓐ1

9. Olfactory receptor neurons are the only parts of the nervous system that come in contact with the _____ directly. ⓐ1

10. A chemical would be considered _____ if a person lacked a particular receptor site on the cilia of the olfactory neurons. ⓐ1

PART B ASSESSMENTS

1. Record the results (as "+" if recognized; as "0" if unrecognized) from the tests of odor recognition in the following table: ▲3

Substance Tested	Odor Reported	
	First Trial	Second Trial

2. Record the results of the olfactory sensory adaptation time in the following table: ▲3

Substance Tested	Adaptation Time in Seconds

3. Complete the following:

   a. How do you describe your partner's ability to recognize the odors of the substances you tested? 2 _____

   _____

   _____

   b. Compare your experimental results with those of others in the class. Did you find any evidence to indicate that individuals may vary in their ability to recognize odors? Explain your answer. 2 _____

   _____

   _____

ASSESS

CRITICAL THINKING

Does the time it takes for sensory adaptation to occur seem to vary with the substances tested? Explain your answer.

_____

_____

_____

PART C ASSESSMENTS

Complete the following statements:

1. Taste receptor cells are modified _____ cells. 1

2. The opening to a taste bud is called a(an) _____. 1

3. The _____ of a taste cell are its sensitive part. 1

4. Before the taste of a substance can be detected, the substance must be dissolved in _____. 1

5. Substances that stimulate taste cells bind with _____ sites on the surfaces of taste hairs. 1

6. Sour receptors are mainly stimulated by _____. 1

7. Salt receptors are mainly stimulated by ionized inorganic _____. 1

8. Alkaloids usually have a(an) _____ taste. 1

PART D ASSESSMENTS

1. *Taste receptor distribution.* Record a "+" to indicate where a taste sensation seemed to originate and a "0" if no sensation occurred when the spot was stimulated.

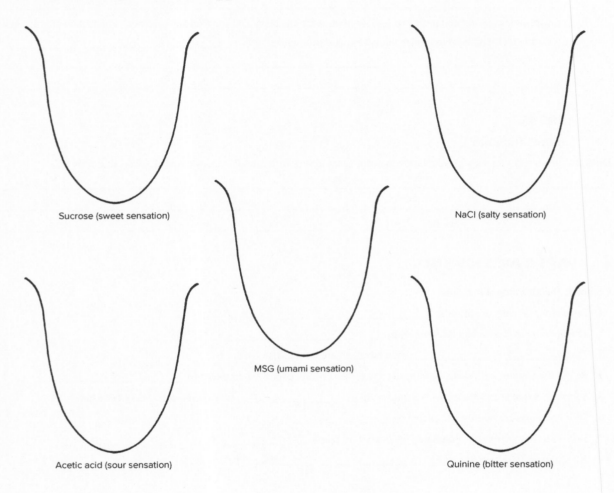

Sucrose (sweet sensation)

NaCl (salty sensation)

MSG (umami sensation)

Acetic acid (sour sensation)

Quinine (bitter sensation)

2. Complete the following:

   a. Describe how each type of taste receptor is distributed on the surface of your partner's tongue. _____

   _____

   _____

   _____

   b. Describe other locations inside the mouth where any sensations of sweet, salty, sour, bitter, or umami were located.

   _____

   _____

   c. How does your taste distribution map on the tongue compare to those of other students in the class? _____

   _____

   _____

LABORATORY EXERCISE

32

Ear and Hearing

MATERIALS NEEDED

Textbook
Dissectible ear model
Watch that ticks
Tuning fork (128 or 256 cps)
Rubber hammer
Cotton
Meterstick
Compound light microscope
Prepared microscope slide of cochlea (section)
Audiometer

PURPOSE OF THE EXERCISE

To review the structural and functional characteristics of the ear and to conduct some ordinary hearing tests.

 ## LEARNING OUTCOMES APR

After completing this exercise, you should be able to

1. Locate the major structures of the ear.
2. Describe the functions of the structures of the ear.
3. Trace the pathway of sound vibrations from the tympanic membrane to the hearing receptors.
4. Conduct four ordinary hearing tests and summarize the results.

The ear is composed of outer (external), middle, and inner (internal) parts. The outer structures gather sound waves and direct them inward to the tympanic membrane. The parts of the middle ear, in turn, transmit vibrations from the tympanic membrane (eardrum) to the inner ear, where the hearing receptors are located. As they are stimulated, these receptors initiate impulses that are conducted over the vestibulocochlear nerve into the auditory cortex of the brain, where the impulses are interpreted and the sensations of hearing are created.

Although the ear is considered here as a hearing organ, it also has important functions for equilibrium (balance). The inner ear semicircular canals, utricle, and saccule have equilibrium functions. Equilibrium will be studied in Laboratory Exercise 33.

 ## PRACTICE

PROCEDURE A—Structure and Function of the Ear APR

1. Review the concept heading titled "Sense of Hearing" in section 12.4 in chapter 12 of the textbook.
2. As a review activity, label figures 32.1 and 32.2, and study figure 32.3.
3. Examine the dissectible model of the ear and locate the following features:

   outer (external) ear
      auricle (pinna)
      external acoustic meatus
      tympanic membrane (eardrum)
   middle ear
      tympanic cavity
      auditory ossicles (fig. 32.4)
         malleus
         incus
         stapes
      oval window
      tensor tympani
      stapedius
   auditory tube (pharyngotympanic tube; eustachian tube)
   inner (internal) ear
      bony (osseous) labyrinth
      membranous labyrinth
      cochlea
      semicircular canals and ducts
      vestibule
         utricle
         saccule
   vestibulocochlear nerve
      vestibular nerve (balance branch)
      cochlear nerve (hearing branch)
4. Complete Parts A and B of Laboratory Report 32.

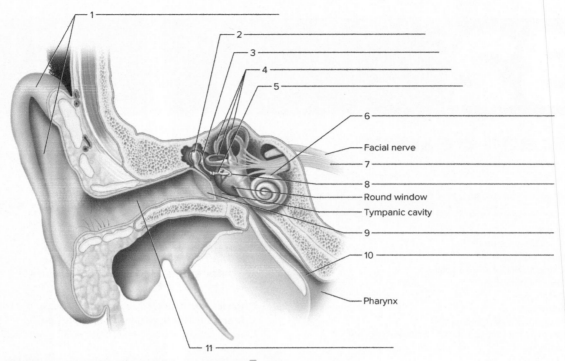

FIGURE 32.1   Label the major structures of the ear. 🄰 APR

 LEARN: ACTIVITY

Observe the section of the cochlea in the demonstration microscope. Locate one of the turns of the cochlea, and *using figures 32.3 and 32.5 as a guide, identify the scala vestibuli, cochlear duct (scala media), scala tympani, vestibular membrane, basilar membrane, and spiral organ.*

 PRACTICE

PROCEDURE B—Hearing Tests

Perform the following tests in a quiet room, using your laboratory partner as the test subject.

1. *Auditory acuity test.* To conduct this test, follow these steps:
   a. Have the test subject sit with eyes closed.
   b. Pack one of the subject's ears with cotton.
   c. Hold a ticking watch close to the open ear, and slowly move it straight out and away from the ear.
   d. Have the subject indicate when the sound of the ticking can no longer be heard.
   e. Use a meterstick to measure the distance in centimeters from the ear to the position of the watch.
   f. Repeat this procedure to test the acuity of the other ear.
   g. Record the test results in Part C of the laboratory report.

2. *Sound localization test.* To conduct this test, follow these steps:
   a. Have the subject sit with eyes closed.
   b. Hold the ticking watch somewhere within the audible range of the subject's ears, and ask the subject to point to the watch.
   c. Move the watch to another position and repeat the request. In this manner, determine how accurately the subject can locate the watch when it is in each of the following positions: in front of the head, behind the head, above the head, on the right side of the head, and on the left side of the head.
   d. Record the test results in Part C of the laboratory report.

3. *Rinne test.* This test is done to assess possible conduction deafness by comparing bone and air conduction. To conduct this test, follow these steps:
   a. Obtain a tuning fork and strike it with a rubber hammer or on the heel of your hand, causing it to vibrate.
   b. Place the end of the fork's handle against the subject's mastoid process behind one ear. Have the prongs of the fork pointed downward and away from the ear, and be sure nothing is touching them (fig. 32.6a). The sound sensation is that of bone conduction. If no sound is experienced, nerve deafness exists.
   c. Ask the subject to indicate when the sound is no longer heard.

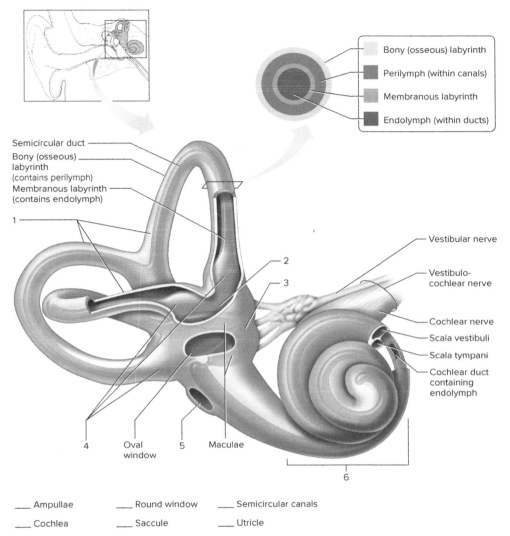

Bony (osseous) labyrinth

Perilymph (within canals)

Membranous labyrinth

Endolymph (within ducts)

Semicircular duct

Bony (osseous) labyrinth (contains perilymph)

Membranous labyrinth (contains endolymph)

1

2

3

Vestibular nerve

Vestibulo-cochlear nerve

Cochlear nerve

Scala vestibuli

Scala tympani

Cochlear duct containing endolymph

4   Oval window   5   Maculae

6

___ Ampullae   ___ Round window   ___ Semicircular canals

___ Cochlea   ___ Saccule   ___ Utricle

FIGURE 32.2   Label the structures of the inner ear by placing the correct numbers in the spaces provided.

d. Then quickly remove the fork from the mastoid process and position it in the air close to the opening of the nearby external acoustic meatus (fig. 32.6b).

If hearing is normal, the sound (from air conduction) will be heard again; if there is conductive impairment, the sound will not be heard. Conductive impairment involves outer or middle ear defects. Hearing aids can improve hearing for conductive deafness because bone conduction transmits the sound into the inner ear. Surgery could possibly correct this type of defect.

e. Record the test results in Part C of the laboratory report.

4. *Weber test.* This test is used to distinguish possible conduction or sensory deafness. To conduct this test, follow these steps:

a. Strike the tuning fork with the rubber hammer.

b. Place the handle of the fork against the subject's forehead in the midline (fig. 32.7).

c. Ask the subject to indicate if the sound is louder in one ear than in the other or if it is equally loud in both ears.

If hearing is normal, the sound will be equally loud in both ears. If there is conductive impairment, the sound will appear louder in the affected ear. If some degree of sensory (nerve) deafness exists, the sound will be louder in the normal ear. The impairment involves the spiral organ or the cochlear nerve. Hearing aids will not improve sensory deafness.

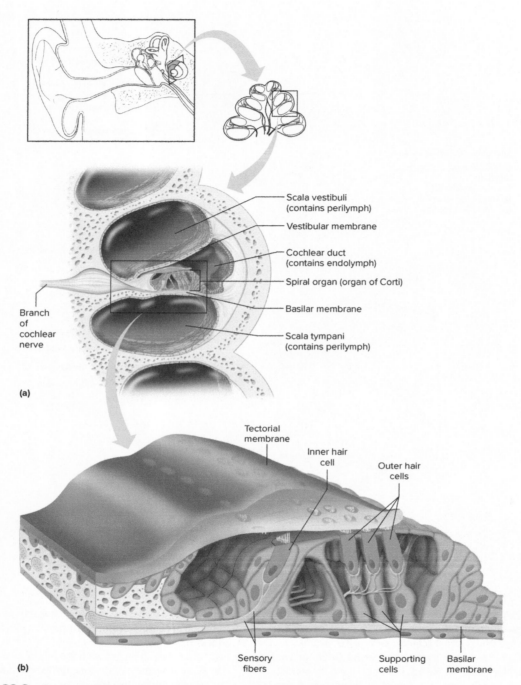

FIGURE 32.3 Cochlea. (*a*) Cross section of the cochlea. (*b*) Spiral organ and the tectorial membrane.

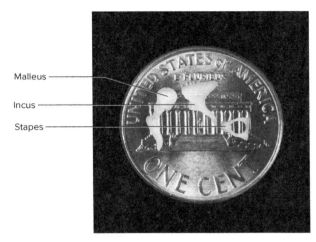

FIGURE 32.4   Middle ear bones (auditory ossicles) superimposed on a penny to show their relative size. The arrangement of the malleus, incus, and stapes in the enlarged photograph resembles the articulations in the middle ear. (©J & J Photography)

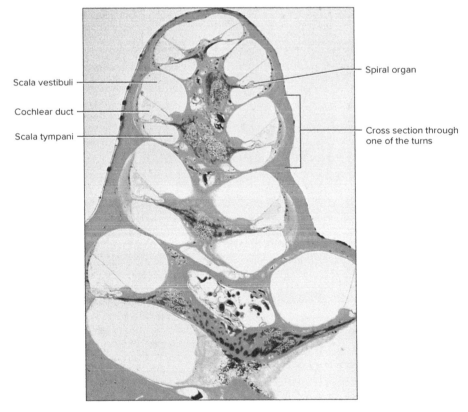

FIGURE 32.5   A section through the cochlea (22×). **APR** (Biophoto Associates/Science Source)

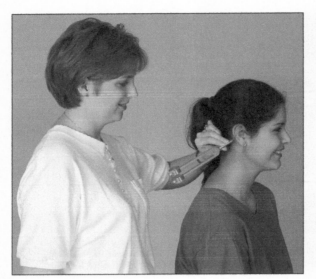

(a)

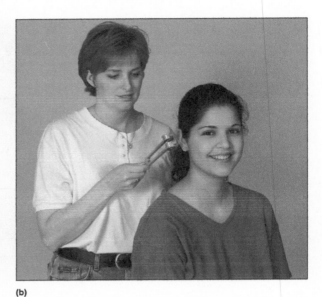

(b)

FIGURE 32.6   Rinne test: (a) first placement of vibrating tuning fork until sound is no longer heard; (b) second placement of tuning fork to assess air conduction. (J and J Photography)

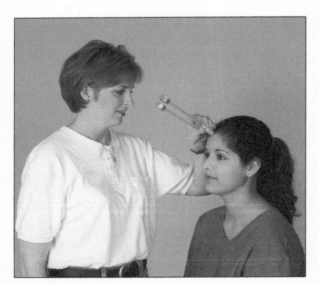

FIGURE 32.7   Weber test. (J and J Photography)

d. Have the subject experience the effects of conductive impairment by packing one ear with cotton and repeating the Weber test. Usually, the sound appears louder in the plugged (impaired) ear because extraneous sounds from the room are blocked out.

e. Record the test results in Part C of the laboratory report.

5. Complete Part C of the laboratory report.

ASSESS

CRITICAL THINKING

Ear structures from the outer ear into the inner ear are progressively smaller. Using results obtained from the hearing tests, explain the advantage of this arrangement.

_____

_____

_____

_____

LEARN: ACTIVITY

Ask the laboratory instructor to demonstrate the use of the audiometer. This instrument produces sound vibrations of known frequencies transmitted to one or both ears of a test subject through earphones. The audiometer can be used to determine the threshold of hearing for different sound frequencies, and, in the case of hearing impairment, it can be used to determine the percentage of hearing loss for each frequency.

Name _____

Date _____

Section _____

LABORATORY
REPORT

The △ corresponds to the indicated Learning Outcome(s) found at the beginning of the laboratory exercise.

Ear and Hearing

32

PART A ASSESSMENTS

Match the terms in column A with the descriptions in column B. Place the letter of your choice in the space provided. △1 △2

Column A	Column B
a. Auditory tube	_____ **1.** Auditory ossicle attached to tympanic membrane
b. Bony (osseous) labyrinth	_____ **2.** Air-filled space containing auditory ossicles within middle ear
c. Ceruminous gland	_____ **3.** Contacts hairs of hearing receptors
d. External acoustic meatus	_____ **4.** Leads from oval window to apex of cochlea and contains perilymph
e. Malleus	_____ **5.** S-shaped tube leading to tympanic membrane
f. Membranous labyrinth	_____ **6.** Wax-secreting structure
g. Scala tympani	_____ **7.** Cone-shaped, semitransparent membrane attached to malleus
h. Scala vestibuli	_____ **8.** Auditory ossicle attached to oval window
i. Stapes	_____ **9.** Tube containing endolymph within bony (osseous) labyrinth
j. Tectorial membrane	_____ **10.** Bony canal of inner ear in temporal bone
k. Tympanic cavity	_____ **11.** Connects middle ear and pharynx
l. Tympanic membrane (eardrum)	_____ **12.** Extends from apex of cochlea to round window and contains perilymph

PART B ASSESSMENTS

Label the structures indicated in the micrograph of the spiral organ in figure 32.8. △1

Terms:
Basilar membrane
Cochlear duct
Hair cells
Scala tympani
Tectorial membrane

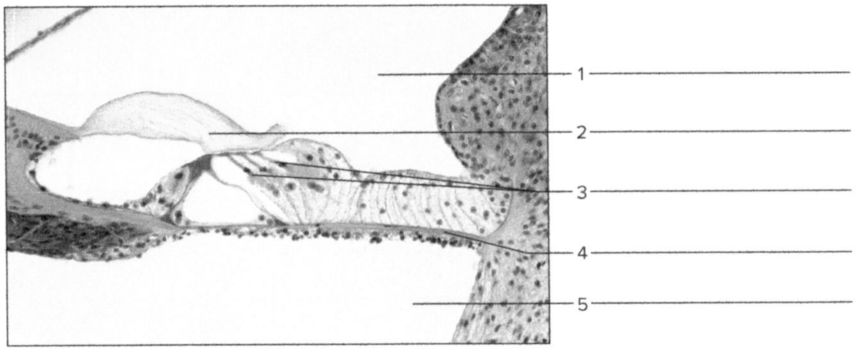

FIGURE 32.8   Label the structures associated with this spiral organ region of a cochlea, using the terms provided (400×). **APR** (Biophoto Associates/Science Source)

PART C ASSESSMENTS

1. Results of auditory acuity test: A

Ear Tested	Audible Distance (cm)
Right	
Left	

2. Results of sound localization test: A

Actual Location	Reported Location
Front of the head	
Behind the head	
Above the head	
Right side of the head	
Left side of the head	

3. Results of experiments using tuning forks: A

Test	Left Ear (normal or impaired)	Right Ear (normal or impaired)
Rinne		
Weber		

4. Summarize the results of the hearing tests you conducted on your laboratory partner. A

_____

_____

_____

_____

PART D ASSESSMENTS

Terms:
Auditory tube
Auricle
Cochlea
External
acoustic
meatus
Incus
Malleus
Semicircular
canals
Stapes
Tympanic
membrane
Vestibule

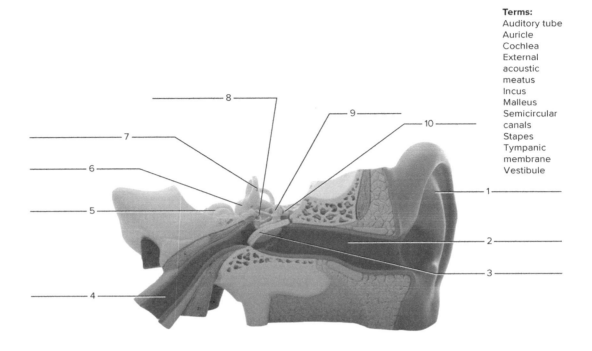

FIGURE 32.9 Using the terms provided label the indicated structures of the outer, middle, and inner ear on the ear model. (Model # E10 [1000250] ©3B Scientific GmbH, Germany, 2015 www.3bscientific.com/Photo by Christine Eckel, Ph.D.)

33

Ear and Equilibrium

MATERIALS NEEDED

Textbook
Swivel chair
Bright light
Compound light microscope
Prepared microscope slide of semicircular duct
(cross section through ampulla)

⚠ SAFETY

- Do not pick subjects who have frequent motion sickness.
- Have four people surround the subject in the swivel chair, in case the person falls from vertigo or loss of balance.
- Stop your experiment if the subject becomes nauseated.

PURPOSE OF THE EXERCISE

To review the structure and function of the organs of equilibrium and to conduct some tests of equilibrium.

LEARNING OUTCOMES

After completing this exercise, you should be able to

1. Locate the organs of static and dynamic equilibrium and describe their functions.

2. Explain how vision contributes in the maintenance of equilibrium.

3. Conduct and record the results of the Romberg and Bárány tests of equilibrium.

The sense of equilibrium involves two sets of sensory organs. One set helps to maintain the stability of the head and body when they are motionless or during linear acceleration and produces a sense of static (gravitational) equilibrium. The other set is concerned with balancing the head and body when

they are moved suddenly or during angular acceleration and produces a sense of dynamic (rotational) equilibrium.

The organs associated with the sense of static equilibrium are located within the utricle and saccule of the vestibules of the inner ears. Those associated with the sense of dynamic equilibrium are found within the ampullae of the three semicircular ducts of the inner ear.

PRACTICE

PROCEDURE A—Organs of Equilibrium

1. Review the concept heading titled "Sense of Equilibrium" in section 12.4 in chapter 12 of the textbook.
2. Complete Part A of Laboratory Report 33.

LEARN: ACTIVITY

Observe the cross section of the semicircular duct through the ampulla in the demonstration microscope. Note the crista ampullaris (figs. 33.1 and 33.2) projecting into the lumen of the membranous labyrinth, which in a living person is filled with endolymph. The space between the membranous and osseous (bony) labyrinths is filled with perilymph.

PRACTICE

PROCEDURE B—Tests of Equilibrium

Perform the following tests, using a person as a test subject who is not easily disturbed by dizziness or rotational movement. Also have some other students standing close by to help prevent the test subject from falling during the tests. *The tests should be stopped immediately if the test subject begins to feel uncomfortable or nauseated.*

1. *Vision and equilibrium test.* To demonstrate the importance of vision in the maintenance of equilibrium, follow these steps:
   a. Have the test subject stand erect on one foot for 1 minute with eyes open.

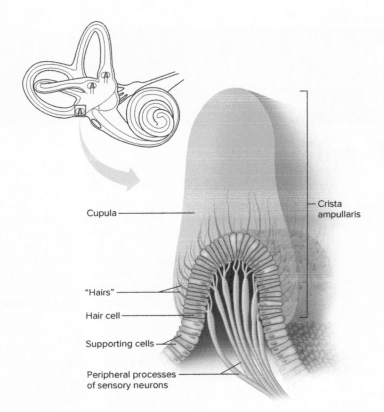

Cupula

Crista
ampullaris

"Hairs"

Hair cell

Supporting cells

Peripheral processes
of sensory neurons

FIGURE 33.1   The crista ampullaris is located within the ampulla of each semicircular duct.

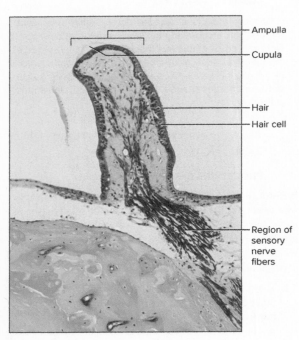

Ampulla

Cupula

Hair

Hair cell

Region of
sensory
nerve
fibers

FIGURE 33.2   A micrograph of a crista ampullaris
(400×). (Biophoto Associates/Science Source)

b. Observe the subject's degree of unsteadiness.

c. Repeat the procedure with the subject's eyes closed.
Be prepared to prevent the subject from falling.

d. Answer the questions related to the vision and equilibrium test in Part B of the laboratory report.

2. *Romberg test.* The purpose of this test is to evaluate how
the organs of static equilibrium in the vestibule enable
one to maintain balance. To conduct this test, follow
these steps:

   a. Position the test subject close to a chalkboard with
   the back toward the board.

   b. Place a bright light in front of the subject so that a
   shadow of the body is cast on the board.

   c. Have the subject stand erect with feet close together and
   eyes staring straight ahead for a period of 3 minutes.

   d. During the test, make marks on the chalkboard along
   the edge of the shadow of the subject's shoulders to
   indicate the range of side-to-side swaying.

   e. Measure the maximum sway in centimeters and
   record the results in Part B of the laboratory report.

   f. Repeat the procedure with the subject's eyes closed.

   g. Position the subject so one side is toward the chalkboard.

   h. Repeat the procedure with the eyes open.

   i. Repeat the procedure with the eyes closed.

The Romberg test is used to evaluate a person's ability to integrate sensory information from proprioceptors and receptors within the organs of static equilibrium and to relay appropriate motor impulses to postural muscles. A person who shows little unsteadiness when standing with feet together and eyes open, but who becomes unsteady when the eyes are closed, has a positive Romberg test.

3. *Bárány test.* The purpose of this test is to evaluate the effects of rotational acceleration on the semicircular ducts and dynamic equilibrium. To conduct this test, follow these steps:

   a. Have the test subject sit on a swivel chair with eyes open and focused on a distant object, the head tilted forward about 30° and the hands gripped firmly to the seat. Position four people around the chair for safety. *Be prepared to prevent the subject and the chair from tipping over.*

   b. Rotate the chair ten rotations within 20 seconds.

   c. Abruptly stop the movement of the chair. The subject will still have the sensation of continuous movement and might experience some dizziness (vertigo).

   d. Have the subject look forward and immediately note the nature of the eye movements and their direction. (Such reflex eye twitching movements are called *nystagmus.*) Also note the time it takes for the nystagmus to cease. Nystagmus will continue until the cupula is returned to an original position.

   e. Record your observations in Part B of the laboratory report.

   f. Allow the subject several minutes of rest; then repeat the procedure with the subject's head tilted nearly 90° onto one shoulder.

g. After another rest period, repeat the procedure with the subject's head bent forward so that the chin is resting on the chest.

In this test, when the head is tilted about 30°, the lateral semicircular ducts receive maximal stimulation, and the nystagmus is normally from side to side. When the head is tilted at 90°, the superior ducts are stimulated, and the nystagmus is up and down. When the head is bent forward with the chin on the chest, the posterior ducts are stimulated, and the nystagmus is rotary. A membranous semicircular duct occupies the space within the osseous (bony) semicircular canal within a temporal bone. Observe a semicircular canal located in a temporal bone (fig. 33.3).

4. Complete Part B of the laboratory report.

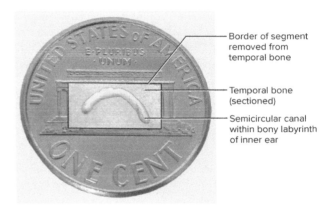

FIGURE 33.3   Semicircular canal superimposed on a penny to show its relative size. A semicircular duct of the same shape would occupy the semicircular canal within the temporal bone. (©J & J Photography)

Notes

Name _____

Date _____

Section _____

The Ⓐ corresponds to the indicated Learning Outcome(s) found at the beginning of the laboratory exercise.

Ear and Equilibrium

🔁 PART A ASSESSMENTS

Complete the following statements:

1. The organs of static equilibrium are located within two expanded chambers within the vestibule called the
   _____ and the saccule. Ⓐ

2. All of the balance organs are located within the _____ bone of the skull. Ⓐ

3. The receptor cells of these organs are found in the wall of the membranous labyrinth in a structure called the
   _____ . Ⓐ

4. Otoliths are small grains composed of _____ . Ⓐ

5. Sensory impulses travel from the organs of equilibrium to the brain on the _____ nerve. Ⓐ

6. The sensory organ of a semicircular duct lies within a swelling of the duct called the _____ . Ⓐ

7. The sensory organ within the ampulla of a semicircular duct is called a(an) _____ . Ⓐ

8. The _____ of this sensory organ consists of a dome-shaped, gelatinous mass. Ⓐ

9. When the head is moved, the fluid inside the membranous portion of a semicircular duct tends to remain stationary
   because of the _____ of the fluid. Ⓐ

10. Parts of the _____ of the brain interpret impulses from the equilibrium receptors. Ⓐ

🔁 PART B ASSESSMENTS

1. Vision and equilibrium test results:
   a. When the eyes are open, what sensory organs provide information needed to maintain equilibrium? ②

   _____

   _____

   b. When the eyes are closed, what sensory organs provide such information? ② _____

   _____

   _____

2. Romberg test results:

   a. Record the test results in the following table:

Conditions	Maximal Movement (cm)
Back toward board, eyes open	
Back toward board, eyes closed	
Side toward board, eyes open	
Side toward board, eyes closed	

   b. Did the test subject's unsteadiness increase when the eyes were closed? _____ What is the significance of this observation? **2** _____

_____

_____

   c. Why would you expect a person with impairment of the organs of equilibrium to become more unsteady when the eyes are closed? **2** _____

_____

_____

3. Bárány test results:

   a. Record the test results in the following table: **3**

Position of Head	Description of Eye Movements	Time for Movement to Cease
Tilted 30 degrees forward		
Tilted 90 degrees onto shoulder		
Tilted forward, chin on chest		

   b. Summarize the results of this test. **3** _____

_____

_____

 ASSESS

CRITICAL THINKING

What additional sensory information would you expect persons with impairment of organs of equilibrium to use to supplement their relative lack of some sensory information?

_____

_____

_____

34

Eye Structure

MATERIALS NEEDED

Textbook
Dissectible eye model
Compound light microscope
Prepared microscope slide of a mammalian eye
    (sagittal section)
Sheep or beef eye (fresh or preserved)
Dissecting tray
Dissecting instruments—forceps, sharp scissors, and
    dissecting needle
Ophthalmoscope

SAFETY

- Do not allow light to reach the macula lutea region during the eye exam for longer than 1 second at a time.
- Wear disposable gloves and safety goggles when working on the eye dissection.
- Dispose of tissue remnants and gloves as instructed.
- Wash the dissecting tray and instruments as instructed.
- Wash your laboratory table.
- Wash your hands before leaving the laboratory.

PURPOSE OF THE EXERCISE

To review the structure and function of the eye and to dissect a mammalian eye.

LEARNING OUTCOMES APR

After completing this exercise, you should be able to

1. Locate the major structures of an eye.
2. Describe the functions of the structures of an eye.
3. Trace the structures through which light passes as it travels from the cornea to the retina.
4. Dissect a mammalian eye and locate its major features.
5. Recognize the anatomical terms pertaining to the eye structure.

The eye contains photoreceptors, modified neurons located on its inner wall. Other parts of the eye provide protective functions or make it possible to move the eyeball. Still other structures focus light entering the eye so that a sharp image is projected onto the receptor cells. The impulses generated when the receptors are stimulated are conducted along the optic nerves to the brain, which interprets the impulses and creates the sensation of sight.

PRACTICE

PROCEDURE A—Structure and Function of the Eye APR

1. Review the concept headings titled "Visual Accessory Organs," "Structure of the Eye," and "Photoreceptors" in section 12.4 in chapter 12 of the textbook.
2. As a review activity, label figures 34.1–34.3.
3. Complete Part A of Laboratory Report 34.
4. Examine the dissectible model of the eye and locate the following features:

eyelid
conjunctiva
orbicularis oculi
levator palpebrae superioris
lacrimal apparatus
    lacrimal gland
    canaliculi
    lacrimal sac
    nasolacrimal duct
extrinsic muscles
    superior rectus
    inferior rectus
    medial rectus
    lateral rectus
    superior oblique
    inferior oblique
trochlea (pulley)
outer (fibrous) tunic (layer)
    sclera
    cornea

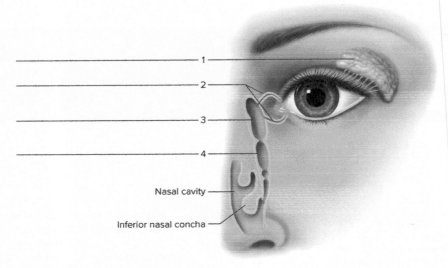

FIGURE 34.1   Label the structures of the lacrimal apparatus.   A

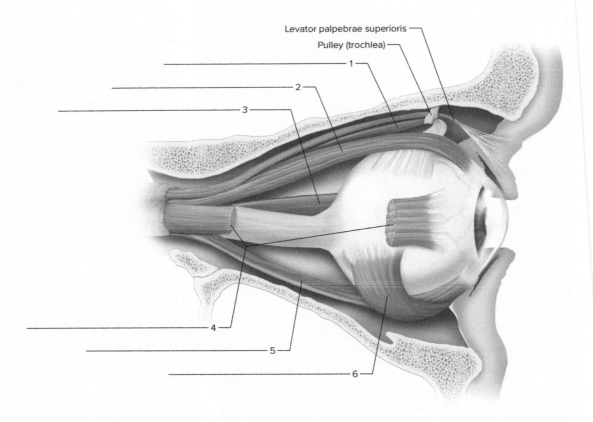

FIGURE 34.2   Label the extrinsic muscles of the right eye (lateral view).   A APR

Anterior

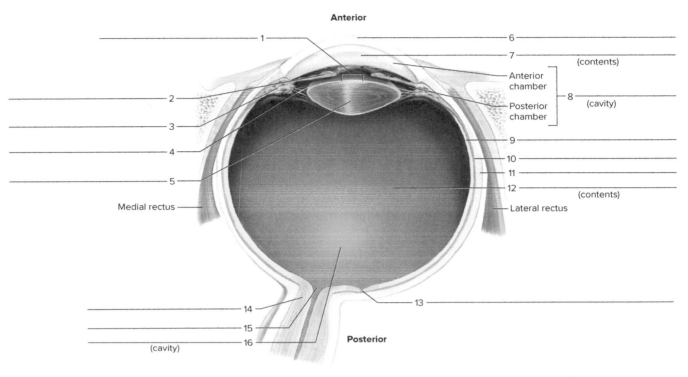

FIGURE 34.3   Label the structures indicated in this transverse section of the right eye (superior view). /1\ **APR**

middle (vascular) tunic (layer)
   choroid coat
   ciliary body
      ciliary processes
      ciliary muscle
      suspensory ligaments
   iris
   pupil (opening in center of iris)
inner (nervous) tunic (layer)
   retina
      macula lutea
      fovea centralis
      optic disc
   optic nerve
lens
anterior cavity
   anterior chamber
   posterior chamber
   aqueous humor
posterior cavity
   vitreous humor

5. Reexamine the labeled figures 34.1–34.3 and study the functions and nerve innervations of the muscles of the eyelids and eyes in table 34.1.
6. Obtain a microscope slide of a mammalian eye section, and locate as many of the features in step 4 as possible.
7. Using low-power magnification, observe the posterior portion of the eye wall, and locate the choroid coat and retina (fig. 34.4).
8. Using high-power magnification, examine the retina portion of the eye wall in more detail, and note its layered structure (fig. 34.4). Locate the following:

layer of nerve fiber (innermost layer of the retina)
layer of ganglion cells
layer of bipolar neurons
layer of photoreceptors (rods and cones)
pigmented epithelium (outermost layer of the retina)

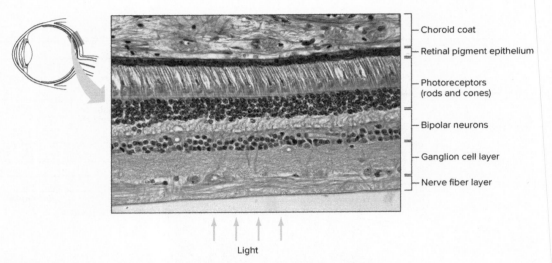

Light

FIGURE 34.4   The cells of the retina are arranged in distinct layers (75×).   **APR** (©Ed Reschke)

TABLE 34.1 Muscles Associated with the Eyelids and Eyes

Skeletal Muscles	Innervation	Function
Muscles of the eyelids		
Orbicularis oculi	Facial nerve (VII)	Closes eye
Levator palpebrae superioris	Oculomotor nerve (III)	Opens eye
Extrinsic muscles of the eyes		
Superior rectus	Oculomotor nerve (III)	Rotates eye upward and toward midline
Inferior rectus	Oculomotor nerve (III)	Rotates eye downward and toward midline
Medial rectus	Oculomotor nerve (III)	Rotates eye toward midline
Lateral rectus	Abducens nerve (VI)	Rotates eye away from midline
Superior oblique	Trochlear nerve (IV)	Rotates eye downward and away from midline
Inferior oblique	Oculomotor nerve (III)	Rotates eye upward and away from midline
Smooth Muscles	**Innervation**	**Function**
Ciliary muscle	Oculomotor nerve (III) parasympathetic fibers	Relax suspensory ligaments
Iris, circular muscle set	Oculomotor nerve (III) parasympathetic fibers	Constrict pupil
Iris, radial muscle set	Sympathetic fibers	Dilate pupil

LEARN: LAB IN MOTION

Use an ophthalmoscope to examine the interior of your laboratory partner's eye. This instrument consists of a set of lenses held in a rotating disc, a light source, and some mirrors that reflect the light into the test subject's eye (fig. 34.5).

The examination should be conducted in a dimly lit room. Have your partner seated and staring straight ahead at eye level. Move the rotating disc of the ophthalmoscope so that the *O* appears in the lens selection window. Hold the instrument in your right hand with the end of your index finger on the rotating disc. Direct the light at a slight angle from a distance of about 15 cm into the subject's right eye (fig. 34.6). The light beam should pass along the inner edge of the pupil. Look through the instrument, and you should see a reddish, circular area—the interior of the eye. Rotate the disc of lenses to higher values until sharp focus is achieved.

Move the ophthalmoscope to within about 5 cm of the eye being examined, *being very careful that the instrument does not touch the eye,* and again rotate the lenses to sharpen the focus (fig. 34.6). Locate the optic disc and the blood vessels that pass through it. Also locate the macula lutea by having your partner stare directly into the light of the instrument (fig. 34.7). *Do not allow light to reach the macula lutea region for longer than 1 second at a time.*

Examine the subject's iris by viewing it from the side and by using a lens with a +15 or +20 value.

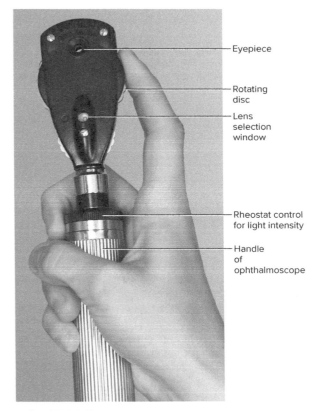

— Eyepiece

— Rotating disc

— Lens selection window

— Rheostat control for light intensity

— Handle of ophthalmoscope

FIGURE 34.5 The ophthalmoscope is used to examine the interior of the eye. (J and J Photography)

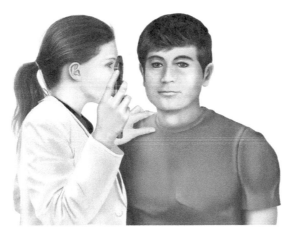

FIGURE 34.6 Rotate the disc of lenses until sharp focus is achieved and then move the ophthalmoscope to within 5 cm of the eye to examine the optic disc of the right eye.

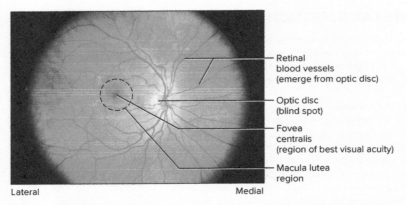

Retinal
blood vessels
(emerge from optic disc)

Optic disc
(blind spot)

Fovea
centralis
(region of best visual acuity)

Macula lutea
region

Lateral   Medial

FIGURE 34.7   Retina of the left eye as viewed through the pupil using an ophthalmoscope. Only small portions of the retina are able to be viewed at a time. **APR** (Science Source)

 PRACTICE

PROCEDURE B—Eye Dissection

1. Obtain a mammalian eye, place it in a dissecting tray, and dissect it as follows:
   a. Trim away the fat and other connective tissues but leave the stubs of the *extrinsic muscles* and the *optic nerve.* This nerve projects outward from the posterior region of the eyeball.
   b. The *conjunctiva* lines the inside of the eyelid and is reflected over the anterior surface of the eye, except for the cornea. Lift some of this thin membrane away from the eye with forceps and examine it.

   c. Locate and observe the *cornea, sclera,* and *iris.* Also note the *pupil* and its shape. The cornea from a fresh eye is transparent; when preserved, it becomes opaque.
   d. Use sharp scissors to make a frontal section of the eye. To do this, cut through the wall about 1 cm from the margin of the cornea and continue all the way around the eyeball. Try not to damage the internal structures of the eye (fig. 34.8).
   e. Gently separate the eyeball into anterior and posterior portions. Usually, the jellylike vitreous humor will remain in the posterior portion, and the lens may adhere to it. Place the parts in the dissecting tray with their contents facing upward.

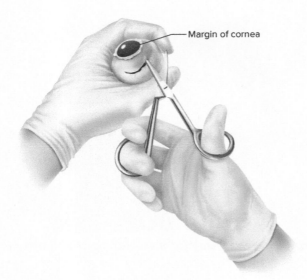

Margin of cornea

FIGURE 34.8   Prepare a frontal section of the eye.

f. Examine the anterior portion of the eye, and locate the *ciliary body*, which appears as a dark, circular structure. Also note the *iris* and the *lens* if it remained in the anterior portion. The lens is normally attached to the ciliary body by many *suspensory ligaments*, which appear as delicate, transparent threads that detach as a lens is removed.

g. Use a dissecting needle to gently remove the lens, and examine it. If the lens is still transparent, hold it up and look through it at something in the distance and note that the lens inverts the image. The lens of a preserved eye is usually too opaque for this experience. If the lens of the human eye becomes opaque, the defect is called a cataract (fig. 34.9).

h. Examine the posterior portion of the eye. Note the *vitreous humor*. This jellylike mass helps to hold the lens in place anteriorly and to hold the *retina* against the choroid coat.

i. Carefully remove the vitreous humor and examine the retina. This layer will appear as a thin, nearly colorless to cream-colored membrane that detaches easily from the choroid coat. Compare the structures identified to figure 34.10.

j. Locate the *optic disc*—the point where the retina is attached to the posterior wall of the eyeball and where the optic nerve originates. There are no receptor cells in the optic disc, so this region is also called the "blind spot."

k. Note the iridescent area of the choroid coat beneath the retina. This colored surface in ungulates (mammals having hoofs) is called the *tapetum fibrosum*. It reflects light back through the retina, an action thought to aid the night vision of some animals. The tapetum fibrosum is lacking in the human eye.

2. Discard the tissues of the eye as directed by the laboratory instructor.

3. Complete Parts B and C of the laboratory report.

 ASSESS

CRITICAL THINKING

A strong blow to the head might cause the retina to detach. From observations made during the eye dissection, explain why this can happen.

_____

_____

_____

_____

_____

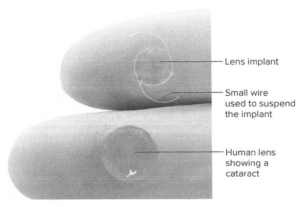

FIGURE 34.9   Human lens showing a cataract and one type of lens implant (intraocular lens replacement) on tips of fingers to show their relative size. A normal lens is transparent. (J and J Photography)

- Lens implant
- Small wire used to suspend the implant
- Human lens showing a cataract

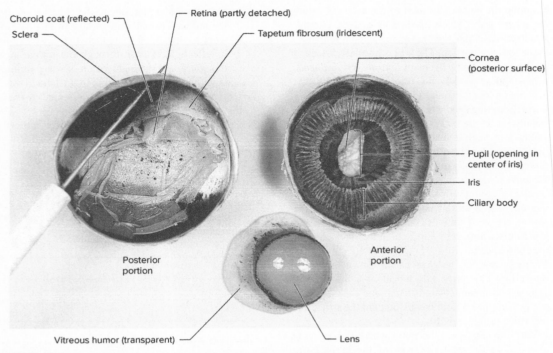

Choroid coat (reflected)
Sclera
Retina (partly detached)
Tapetum fibrosum (iridescent)
Cornea (posterior surface)
Pupil (opening in center of iris)
Iris
Ciliary body
Posterior portion
Anterior portion
Vitreous humor (transparent)
Lens

FIGURE 34.10   Internal structures of the preserved beef eye after dissection. (J and J Photography)

Name _____

Date _____

Section _____

The ⒜ corresponds to the indicated Learning Outcome(s) found at the beginning of the laboratory exercise.

Eye Structure

♻ PART A ASSESSMENTS

Match the terms in column A with the descriptions in column B. Place the letter of your choice in the space provided. ⒜ ⒉

Column A	Column B
a. Aqueous humor	_____ **1.** Posterior five-sixths of middle (vascular) tunic
b. Choroid coat	_____ **2.** White part of outer (fibrous) tunic
c. Ciliary muscle	_____ **3.** Transparent anterior portion of outer tunic
d. Conjunctiva	_____ **4.** Inner lining of eyelid
e. Cornea	_____ **5.** Secretes tears
f. Iris	_____ **6.** Fills posterior cavity of eye
g. Lacrimal gland	_____ **7.** Area where optic nerve exits the eye
h. Optic disc	_____ **8.** Smooth muscle that controls the pupil size and light entering the eye
i. Retina	_____ **9.** Fills anterior and posterior chambers of the anterior cavity of the eye
j. Sclera	_____ **10.** Contains photoreceptor cells called rods and cones
k. Suspensory ligament	_____ **11.** Connects lens to ciliary body
l. Vitreous humor	_____ **12.** Cause lens to change shape

Complete the following:

13. List the structures and fluids through which light passes as it travels from the cornea to the retina. ⒊ _____

_____

_____

_____

14. List three ways in which rods and cones differ in structure or function. ⒉ _____

_____

_____

_____

PART B ASSESSMENTS

Complete the following:

1. Which tunic/layer of the eye was the most difficult to cut? **4** _____

2. What kind of tissue do you think is responsible for this quality of toughness? **4** _____

3. How do you compare the shape of the pupil in the dissected eye with your pupil? **4** _____
_____

4. Where was the aqueous humor in the dissected eye? **4** _____
_____

5. What is the function of the dark pigment in the choroid coat? **2** _____
_____

6. Describe the lens of the dissected eye. **4** _____
_____

7. Describe the vitreous humor of the dissected eye. **4** _____
_____

 PART C ASSESSMENTS

Locate the 23 anatomical terms pertaining to the "Eye Structure." **5**

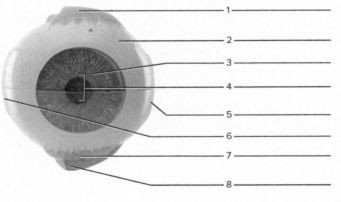

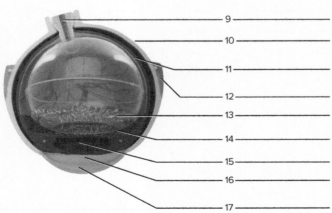

Terms:
Anterior chamber
Cornea
Inferior oblique muscle
Inferior rectus muscle
Iris
Lateral rectus muscle
Lens
Medial rectus muscle
Optic nerve
Pupil
Retina
Sclera
Superior rectus muscle
Suspensory ligaments

FIGURE 34.11   Using the terms provided label the structures indicated on the eye models pictured. Some terms can be used morethan once. (©3B Scientific GmbH, Germany, 2015 www.3bscientific.com)

35

Visual Tests and Demonstrations

MATERIALS NEEDED

Snellen eye chart
3" × 5" card (plain)
3" × 5" card with word typed in center
Astigmatism chart
Meterstick
Metric ruler
Pen flashlight
Ichikawa's or Ishihara's color plates for colorblindness test

PURPOSE OF THE EXERCISE

To conduct and interpret the tests for visual acuity, astigmatism, accommodation, color vision, the blind spot, and reflexes of the eye.

LEARNING OUTCOMES

After completing this exercise, you should be able to

1. Conduct and interpret the tests used to evaluate visual acuity, astigmatism, the ability to accommodate for close vision, and color vision.

2. Describe four conditions that can lead to defective vision.

3. Demonstrate and describe the blind spot, photopupillary reflex, accommodation pupillary reflex, and convergence reflex.

Normal vision (emmetropia) results when light rays from objects in the external environment are refracted by the cornea and lens of the eye and focused onto the photoreceptors of the retina. If a person has *nearsightedness (myopia)*, the image focuses in front of the retina because the eyeball is too long; nearby objects are clear, but distant objects are blurred. If a person has *farsightedness (hyperopia)*, the image focuses behind the retina because the eyeball is too short; distant objects are clear, but nearby objects are blurred. A concave lens is required to correct vision for nearsightedness; a convex lens is required to correct vision for farsightedness.

If a defect occurs from unequal curvatures of the cornea or lens, a condition called *astigmatism* results. Focusing in different planes cannot occur simultaneously. Corrective cylindrical lenses allow proper refraction of light onto the retina to compensate for the unequal curvatures.

As part of a natural aging process, the lens's elasticity decreases, which degrades our ability to focus on nearby objects. The near point of accommodation test is used to determine the ability to accommodate. Often a person will use "reading glasses" to compensate for this condition. Hereditary defects in color vision result from a lack of certain cones needed to absorb certain wavelengths of light. Special color test plates are used to diagnose colorblindness.

As you conduct this laboratory exercise, it does not matter in what order the visual tests or visual demonstrations are performed. The tests and demonstrations do not depend upon the results of any of the others. If you wear glasses, perform the tests with and without the corrective lenses. If you wear contact lenses, it is not necessary to run the tests under both conditions; however, indicate in the laboratory report that all tests were performed with contact lenses in place.

PRACTICE

PROCEDURE A—Visual Tests

Perform the following visual tests using your laboratory partner as a test subject. If your partner usually wears glasses, test each eye with and without the glasses.

1. *Visual acuity test.* Visual acuity (sharpness of vision) can be measured by using a Snellen eye chart (fig. 35.1). This chart consists of several sets of letters in different sizes printed on a white card. The letters near the top of the chart are relatively large, and those in each lower set become smaller. At one end of each set of letters is an acuity value in the form of a fraction. One of the sets near the bottom of the chart, for example, is marked 20/20. The normal eye can clearly see these letters from the standard distance of 20 feet and thus is said to have 20/20 vision. The letter at the top of the chart is marked 20/200. The normal eye can read letters of this size from a distance of 200 feet. Thus, an eye able to read only the top letter of the chart from a distance

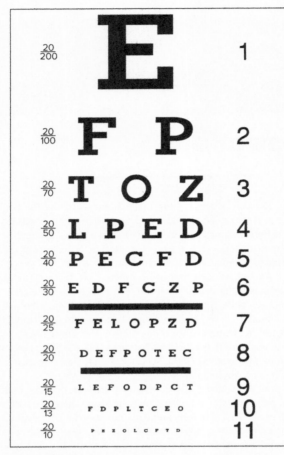

$\frac{20}{200}$	E	1
$\frac{20}{100}$	F P	2
$\frac{20}{70}$	T O Z	3
$\frac{20}{50}$	L P E D	4
$\frac{20}{40}$	P E C F D	5
$\frac{20}{30}$	E D F C Z P	6
$\frac{20}{25}$	F E L O P Z D	7
$\frac{20}{20}$	D E F P O T E C	8
$\frac{20}{15}$	L E F O D P C T	9
$\frac{20}{13}$	F D P L T C E O	10
$\frac{20}{10}$	P E Z O L C F T D	11

FIGURE 35.1   The Snellen eye chart looks similar to this but is somewhat larger.

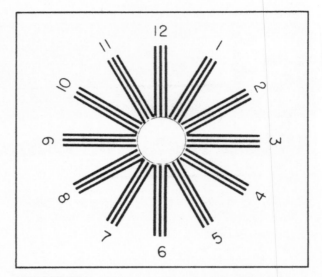

FIGURE 35.2   Astigmatism is evaluated using a chart such as this one.

of 20 feet is said to have 20/200 vision. This person has less than normal vision. A line of letters near the bottom of the chart is marked 20/15. The normal eye can read letters of this size from a distance of 15 feet, but a person might be able to read it from 20 feet. This person has better than normal vision.

To conduct the visual acuity test, follow these steps:

a. Hang the Snellen eye chart on a well-illuminated wall at eye level.

b. Have your partner stand 20 feet in front of the chart, gently cover your partner's left eye with a 3" × 5" card, and ask your partner to read the smallest set of letters possible using the right eye.

c. Record the visual acuity value for that set of letters in Part A of Laboratory Report 35.

d. Repeat the procedure, using the left eye for the test.

2. *Astigmatism test.* Astigmatism is a condition that results from a defect in the curvature of the cornea or lens. As a consequence, some portions of the image projected on the retina are sharply focused, but other portions are blurred. Astigmatism can be evaluated by using an

astigmatism chart (fig. 35.2). This chart consists of sets of black lines radiating from a central spot like the spokes of a wheel. To a normal eye, these lines appear sharply focused and equally dark; however, if the eye has an astigmatism, some sets of lines appear sharply focused and dark, whereas others are blurred and less dark.

To conduct the astigmatism test, follow these steps:

a. Hang the astigmatism chart on a well-illuminated wall at eye level.

b. Have your partner stand 20 feet in front of the chart, and gently cover your partner's left eye with a 3" × 5" card. Ask your partner to focus on the spot in the center of the radiating lines using the right eye and report which lines, if any, appear more sharply focused and darker.

c. Repeat the procedure, using the left eye for the test.

d. Record the results in Part A of the laboratory report.

3. *Accommodation test.* Accommodation is the changing of the shape of the lens that occurs when the normal eye is focused for near (close) vision. It involves a reflex in which muscles of the ciliary body are stimulated to contract, releasing tension on the suspensory ligaments that are fastened to the lens capsule. This allows the capsule to rebound elastically, causing the surface of the lens to become more spherical (convex). The ability to accommodate is likely to decrease with age because the tissues involved tend to lose their elasticity. This common aging condition, called *presbyopia,* can be corrected with reading glasses or bifocal lenses.

To evaluate the ability to accommodate, follow these steps:

a. Hold the end of a meterstick against your partner's chin so that the stick extends outward at a right angle to the plane of the face (fig. 35.3).

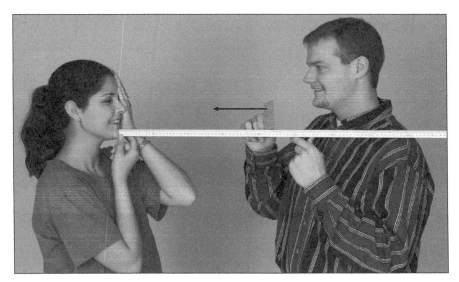

FIGURE 35.3   To determine the near point of accommodation, slide the 3" × 5" card along the meterstick toward your partner's open eye until it reaches the closest location where your partner can still see the word sharply focused. (J and J Photography)

b. Have your partner close the left eye.

c. Hold a 3" × 5" card with a word typed in the center at the distal end of the meterstick.

d. Slide the card along the stick toward your partner's open right eye, and locate the *point closest to the eye* where your partner can still see the letters of the word sharply focused. This distance is called the *near point of accommodation,* and it tends to increase with age (table 35.1).

e. Repeat the procedure with the right eye closed.

f. Record the results in Part A of the laboratory report.

4. *Color vision test.* Some people exhibit defective color vision because they lack certain cones, usually those sensitive to the reds or greens. This trait is an X-linked (sex-linked) inheritance, so the condition is more prevalent in males (7%) than in females (0.4%). People who lack or possess decreased sensitivity to the red-sensitive cones possess protanopia colorblindness; those who lack or are less sensitive to green-sensitive cones possess deuteranopia colorblindness. The colorblindness condition is often more of a deficiency or a weakness than a condition of blindness. Laboratory Exercise 53 describes the genetics for colorblindness.

To conduct the color vision test, follow these steps:

a. Examine the color test plates in Ichikawa's or Ishihara's book to test for any red-green color vision deficiency. Also examine figure 35.4.

b. Hold the test plates approximately 30 inches from the subject in bright light. All responses should occur within 3 seconds.

c. Compare your responses with the correct answers in Ichikawa's or Ishihara's book. Determine the percentage of males and females in your class who exhibit any color-deficient vision. If an individual exhibits

TABLE 35.1 Near Point of Accommodation

Age (years)	Average Near Point (cm)
10	7
20	10
30	13
40	20
50	45
60	90

color-deficient vision, determine if the condition is protanopia or deuteranopia.

d. Record the class results in Part A of the laboratory report.

5. Complete Part A of the laboratory report.

 **PRACTICE**

PROCEDURE B—Visual Demonstrations

Perform the following demonstrations with the help of your laboratory partner.

1. *Blind-spot demonstration.* There are no photoreceptors in the *optic disc,* located where the nerve fibers of the retina leave the eye and enter the optic nerve. Consequently, this region of the retina is commonly called the *blind spot.*

To demonstrate the blind spot, follow these steps:

a. Close your left eye, hold figure 35.5 about 35 cm away from your face, and stare at the "+" sign in the figure with your right eye.

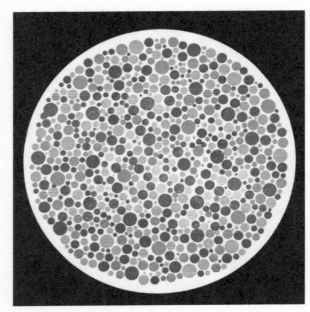

FIGURE 35.4 Sample color test plate to diagnose possible colorblindness. A person with normal color vision perceives the embedded pattern of the number 74; some colorblind individuals only see dots. Tests for colorblindness cannot be conducted with this sample test plate reproduction. For accurate testing, the original plates should be used. (Steve Allen/Getty Images)

b. Move the figure closer to your face as you continue to stare at the "+" until the dot on the figure suddenly disappears. This happens when the image of the dot is focused on the optic disc. Measure the right eye distance using a metric ruler or a meterstick.

c. Repeat the procedures with your right eye closed. This time, stare at the dot using your left eye, and the "+" will disappear when the image falls on the optic disc. Measure the distance.

d. Record the results in Part B of the laboratory report.

 ASSESS

CRITICAL THINKING

Explain why small objects are not lost from our vision under normal visual circumstances.

_____

_____

_____

_____

FIGURE 35.5 Blind-spot demonstration.

2. *Photopupillary reflex.* The smooth muscles of the iris control the size of the pupil. For example, when the intensity of light entering the eye increases, a photopupillary reflex is triggered, and the circular muscles of the iris are stimulated to contract. As a result, the size of the pupil decreases, and less light enters the eye. This reflex is an autonomic response and protects the photoreceptor cells of the retina from damage.

To demonstrate this photopupillary reflex, follow these steps:

a. Ask your partner to sit with his/her hands thoroughly covering the eyes for 2 minutes.

b. Position a pen flashlight close to one eye with the light shining on the hand that covers the eye.

c. Ask your partner to remove the hand quickly.

d. Observe the pupil and note any change in its size.

e. Have your partner remove the other hand, but keep that uncovered eye shielded from extra light.

f. Observe both pupils and note any difference in their sizes.

3. *Accommodation pupillary reflex.* The pupil constricts as a normal accommodation reflex to focusing on close objects. This reduces peripheral light rays and allows the refraction of light to occur closer to the center portion of the lens, resulting in a clearer image.

To demonstrate the accommodation pupillary reflex, follow these steps:

a. Have your partner stare for several seconds at a dimly illuminated object in the room that is more than 20 feet away.

b. Observe the size of the pupil of one eye. Then hold a pencil about 25 cm in front of your partner's face, and have your partner stare at it.

c. Note any change in the size of the pupil.

4. *Convergence reflex.* The eyes converge as a normal convergence reflex to focusing on close objects. The convergence reflex orients the eyeballs toward the object so the image falls on the fovea centralis of both eyes.

To demonstrate the convergence reflex, follow these steps:

a. Repeat the procedure outlined for the accommodation pupillary reflex.

b. Note any change in the position of the eyeballs as your partner changes focus from the distant object to the pencil.

5. Complete Part B of the laboratory report.

Name _____

Date _____

Section _____

The 🄰 corresponds to the indicated Learning Outcome(s) found at the beginning of the laboratory exercise.

Visual Tests and Demonstrations

🔄 PART A ASSESSMENTS

1. Visual acuity test results: 🄰1

Eye Tested	Acuity Values
Right eye	
Right eye with glasses (if applicable)	
Left eye	
Left eye with glasses (if applicable)	

2. Astigmatism test results: 🄰1

Eye Tested	Darker Lines
Right eye	
Right eye with glasses (if applicable)	
Left eye	
Left eye with glasses (if applicable)	

3. Accommodation test results: 🄰1

Eye Tested	Near Point (cm)
Right eye	
Right eye with glasses (if applicable)	
Left eye	
Left eye with glasses (if applicable)	

4. Color vision test results: $\boxed{1}$

Condition	Males			Females		
	Class Number	Class Percentage	Expected Percentage	Class Number	Class Percentage	Expected Percentage
Normal color vision			93			99.6
Deficient red-green color vision			7			0.4
Protanopia (lack red-sensitive cones)			Less frequent type			Less frequent type
Deuteranopia (lack green-sensitive cones)			More frequent type			More frequent type

5. Complete the following: $\boxed{2}$

   a. What is meant by 20/70 vision? _____

   _____

   b. What is meant by 20/10 vision? _____

   _____

   c. What visual problem is created by astigmatism? _____

   _____

   d. Why does the near point of accommodation often increase with age? ____

   _____

   e. Describe the eye defect that causes color-deficient vision. _____

   _____

♻ PART B ASSESSMENTS

1. Blind-spot results: $\boxed{3}$

   a. Right eye distance _____

   b. Left eye distance _____

2. Complete the following:

   a. Explain why an eye has a blind spot. $\boxed{3}$ _____

   _____

   b. Describe the photopupillary reflex. $\boxed{3}$ _____

   _____

   c. What difference did you note in the size of the pupils when one eye was exposed to bright light and the other eye was shielded from the light? $\boxed{3}$ _____

   _____

   d. Describe the accommodation pupillary reflex. $\boxed{3}$ _____

   _____

   e. Describe the convergence reflex. $\boxed{3}$ _____

   _____

Endocrine System

36

Endocrine Histology and Diabetic Physiology

MATERIALS NEEDED

For Procedure A—Endocrine Histology:
Textbook
Human torso model
Compound light microscope
Prepared microscope slides of the following:
    Pituitary gland
    Thyroid gland
    Parathyroid gland
    Adrenal gland
    Pancreas

For Procedure B—Insulin Shock:
500 mL beaker
Live small fish (1" to 1.5" goldfish, guppy, rosy red
    feeder, or other)
Small fish net
Insulin (regular U-100) (HumulinR in 10 mL vials has
    100 units/mL; store in refrigerator—do not freeze)
Syringes for U-100 insulin
10% glucose solution
Clock with second hand or timer
Compound light microscope
Prepared microscope slides of the following:
    Normal human pancreas (stained for alpha and beta
      cells)
    Human pancreas of diabetic

⚠ SAFETY

- Review all Laboratory Safety Guidelines in
  Appendix 1 of your laboratory manual.
- Wear disposable gloves when handling the fish
  and syringes.
- Dispose of the used syringe and needle in the
  puncture-resistant container.
- Wash your hands before leaving the laboratory.

PURPOSE OF THE EXERCISE

To examine microscopically the tissues of endocrine glands and to observe behavioral changes that occur during insulin shock.

LEARNING OUTCOMES APR

After completing this exercise, you should be able to

1. Name and locate the major endocrine glands.

2. Associate the principal hormones and functions of the major endocrine glands.

3. Sketch and label tissue sections from the pituitary gland, thyroid gland, parathyroid glands, adrenal glands, and pancreas.

4. Compare the causes, symptoms, and treatments for type 1 and type 2 diabetes mellitus.

5. Observe and record the behavioral changes caused by insulin shock.

6. Observe and record the recovery from insulin shock when sugar is provided to cells.

The endocrine system consists of ductless glands that act together with parts of the nervous system to help control body activities. The endocrine glands secrete hormones that are transported in body fluids and affect cells that have appropriate receptor molecules. In this way, hormones influence the rate of metabolic reactions, the transport of substances through cell membranes, the regulation of water and electrolyte balances, and many other functions. By controlling cellular activities, endocrine glands play important roles in the maintenance of homeostasis.

The primary "fuel" for cells is sugar (glucose). Insulin produced by the beta cells of the pancreatic islets has a primary regulation role in the transport of blood sugar across the cell membranes of most cells of the body. Some individuals do not produce enough insulin;

others do not have adequate transport of glucose into the body cells when these target cells lose insulin receptors. This rather common disease is called *diabetes mellitus*. Diabetes mellitus is a functional disease that can have its onset either during childhood or later in life.

The protein hormone insulin has several functions: it stimulates the liver and skeletal muscles to synthesize glycogen from glucose; it inhibits the conversion of noncarbohydrates into glucose; it promotes the facilitated diffusion of glucose across cell membranes of cells possessing insulin receptors (adipose tissue, liver, and skeletal muscle); it decreases blood sugar (glucose); it increases protein synthesis; and it promotes fat storage in adipose cells. In a normal person, as blood sugar increases during nutrient absorption after a meal, the rising glucose levels directly stimulate beta cells of the pancreas to secrete insulin. This increase in insulin prevents blood sugar from sudden surges (hyperglycemia) by promoting glycogen production in the liver and an increase in the entry of glucose into muscle and adipose cells.

Between meals and during sleep, blood glucose levels decrease and insulin decreases. When insulin concentration decreases, more glucose is available to enter cells that lack insulin receptors. Neurons, liver cells, kidney cells, and red blood cells either lack insulin receptors or express limited insulin receptors, and they absorb and utilize glucose without the need for insulin. These cells depend upon blood glucose concentrations to enable cellular respiration and ATP production. When insulin is given as a medication for diabetes mellitus, blood glucose levels may drop drastically (hypoglycemia) if the individual does not eat. This can have significant negative effects on the nervous system, resulting in insulin shock.

Type 1 diabetes mellitus usually has a rapid onset relatively early in life, but it can occur in adults. It occurs when insulin is no longer produced; thus, sometimes it is referred to as *insulin-dependent diabetes mellitus (IDDM)*. This autoimmune disorder occurs when the immune system destroys beta cells, resulting in a decrease in insulin production. This disorder affects about 10% to 15% of diabetics. The treatment for type 1 diabetes mellitus is to administer insulin, usually by injection.

Type 2 diabetes mellitus has a gradual onset, usually diagnosed in people over age forty. It is sometimes called *non-insulin-dependent diabetes mellitus (NIDDM)*. This disease results as body cells become less sensitive to insulin or lose insulin receptors and thus cannot respond to insulin even if insulin levels remain normal. This situation is called *insulin resistance*. Risk factors for the onset of type 2 diabetes mellitus, which afflicts about 85% to 90% of diabetics, include heredity, obesity, and lack of exercise. Obesity is a major indicator for predicting type 2 diabetes; a weight loss of a mere 10 pounds will increase the body's sensitivity to insulin as a means of prevention. The treatments for type 2 diabetes mellitus include a diet that avoids foods that stimulate insulin production, weight control, exercise, and medications.

(The causes and treatments for type 2 diabetes closely resemble those for coronary artery disease.)

PRACTICE

PROCEDURE A—Endocrine Histology

1. Review sections 13.4 through 13.9 titled "Pituitary Gland," "Thyroid Gland," "Parathyroid Glands," "Adrenal Glands," "Pancreas," and "Pineal, Thymus, and Other Endocrine Glands" in chapter 13 of the textbook.
2. As a review activity, label figure 36.1.
3. Complete Part A of Laboratory Report 36.
4. Examine the human torso model and locate the following:

   hypothalamus
   pituitary stalk (infundibulum)
   pituitary gland (hypophysis)
       anterior lobe (adenohypophysis)
       posterior lobe (neurohypophysis)
   thyroid gland
   parathyroid glands
   adrenal glands (suprarenal glands)
       adrenal medulla
       adrenal cortex
   pancreas
   pineal gland
   thymus
   ovaries
   testes

5. Examine the microscopic tissue sections of the following glands, and identify the features described:

   Pituitary gland. The pituitary gland is located in the sella turcica of the sphenoid bone. To examine the pituitary tissue, follow these steps:

   a. Observe the tissues using low-power magnification (fig. 36.2).
   b. Locate the *pituitary stalk (infundibulum)*, the *anterior lobe* (the largest part of the gland), and the *posterior lobe*. The anterior lobe makes hormones and appears glandular; the posterior lobe stores and releases hormones made by the hypothalamus and is nervous in appearance.
   c. Observe an area of the anterior lobe with high-power magnification. Locate a cluster of relatively large cells and identify some *acidophil cells,* which contain pink-stained granules, and some *basophil cells,* which contain blue-stained granules. These acidophil and basophil cells are the hormone-secreting cells. The hormones secreted include TSH, ACTH, prolactin, growth hormone, FSH, and LH.

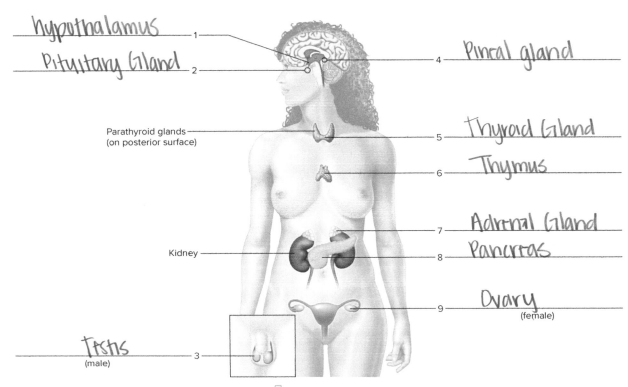

hypothalamus — 1

Pituitary Gland — 2

4 — Pineal gland

Parathyroid glands
(on posterior surface) — 5 — Thyroid Gland

6 — Thymus

7 — Adrenal Gland

Kidney — 8 — Pancreas

9 — Ovary
(female)

Testis
(male) — 3

FIGURE 36.1 Label the major endocrine glands.

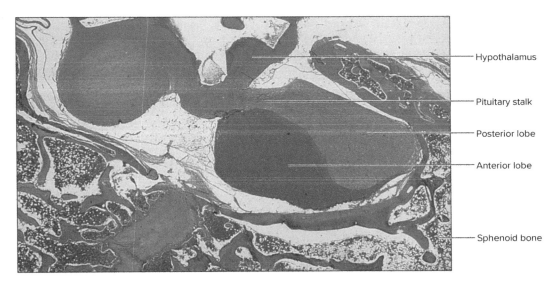

Hypothalamus

Pituitary stalk

Posterior lobe

Anterior lobe

Sphenoid bone

FIGURE 36.2 Micrograph of the pituitary gland (6×). **APR** (Biophoto Associates/Science Source)

d. Observe an area of the posterior lobe with high-power magnification. Note the numerous unmyelinated nerve fibers present in this lobe. Also locate some *pituicytes,* a type of neuroglia, scattered among the nerve fibers. The posterior lobe stores and releases ADH and oxytocin, which are synthesized in the hypothalamus.

e. Prepare labeled sketches of representative portions of the anterior and posterior lobes of the pituitary gland in Part B of the laboratory report.

Thyroid gland. To examine the thyroid tissue, follow these steps:

a. Use low-power magnification to observe the tissue (fig. 36.3). Note the numerous *follicles,* each of which consists of a layer of cells surrounding a colloid-filled cavity.

b. Observe the tissue using high-power magnification. The cells forming the wall of a follicle are simple cuboidal epithelial cells. These cells secrete thyroid hormone (T_3 and T_4). The extrafollicular (parafollicular) cells secrete calcitonin.

c. Prepare a labeled sketch of a representative portion of the thyroid gland in Part B of the laboratory report.

Parathyroid gland. Typically, four parathyroid glands are attached to the posterior surface of the thyroid gland. To examine the parathyroid tissue, follow these steps:

a. Use low-power magnification to observe the tissue (fig. 36.4). The gland consists of numerous tightly packed secretory cells.

b. Switch to high-power magnification and locate two types of cells—a smaller form (chief cells) arranged in cordlike patterns and a larger form (oxyphil cells) that have distinct cell boundaries and are present in clusters. *Chief cells* secrete parathyroid hormone, whereas the function of *oxyphil cells* is not clearly understood.

c. Prepare a labeled sketch of a representative portion of the parathyroid gland in Part B of the laboratory report.

Adrenal gland. To examine the adrenal tissue, follow these steps:

a. Use low-power magnification to observe the tissue (fig. 36.5). Note the thin capsule of connective tissue that covers the gland. Just beneath the capsule, there is a relatively thick *adrenal cortex.* The central portion of the gland is the *adrenal medulla.* The cells of the cortex are in three poorly defined layers. Those of the outer layer (zona glomerulosa) are arranged irregularly; those of the middle layer (zona fasciculata) are in long cords; and those of the inner layer (zona reticularis) are arranged in an interconnected network of cords. The most noted hormones secreted from the cortex include aldosterone from the zona glomerulosa, cortisol from the zona fasciculata, and estrogens and androgens from the zona reticularis. The cells of the medulla are relatively large and irregularly shaped, and they often occur in clusters. The medulla cells secrete epinephrine (adrenaline) and norepinephrine (noradrenaline).

b. Using high-power magnification, observe each of the layers of the cortex and the cells of the medulla.

c. Prepare labeled sketches of representative portions of the adrenal cortex and medulla in Part B of the laboratory report.

Pancreas. To examine the pancreas tissue, follow these steps:

a. Use low-power magnification to observe the tissue (fig. 36.6). The gland largely consists of deeply stained exocrine cells arranged in clusters around secretory ducts. These exocrine cells (acinar cells) secrete pancreatic juice rich in digestive enzymes.

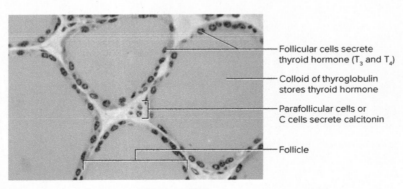

Follicular cells secrete thyroid hormone (T_3 and T_4)

Colloid of thyroglobulin stores thyroid hormone

Parafollicular cells or C cells secrete calcitonin

Follicle

FIGURE 36.3   Micrograph of the thyroid gland (300×).   **APR** (Biophoto Associates/Science Source)

There are circular masses of differently stained cells scattered throughout the gland. These clumps of cells constitute the *pancreatic islets (islets of Langerhans)*, and they represent the endocrine portion of the pancreas. The beta cells secrete insulin and the alpha cells secrete glucagon.

b. Examine an islet, using high-power magnification. Unless special stains are used, it is not possible to distinguish alpha and beta cells.

c. Prepare a labeled sketch of a representative portion of the pancreas in Part B of the laboratory report.

6. Complete Part C of the laboratory report.

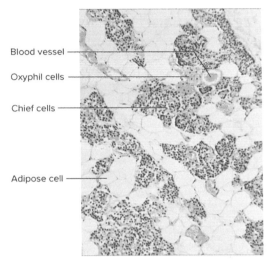

FIGURE 36.4 Micrograph of the parathyroid gland (65×). **APR** (Biophoto Associates/Science Source)

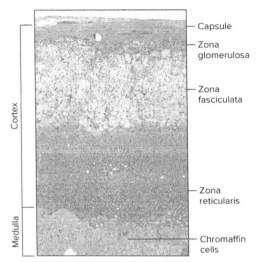

FIGURE 36.5 Micrograph of the adrenal cortex and the adrenal medulla (75×). **APR** (Ed Reschke)

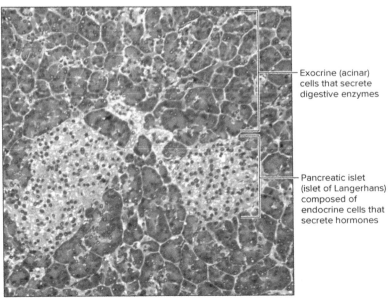

FIGURE 36.6 Micrograph of the pancreas (200×). **APR** (From Kent M. Van De Graaff and Stuart Ira Fox, Concepts of Human Anatomy and Physiology, 2/e. 1989 Wm. C. Brown Publisher, Dubuque, Iowa.)

PRACTICE

PROCEDURE B—Insulin Shock

If too much insulin is present relative to the amount of glucose available to the cells, insulin shock can occur. In this procedure, insulin shock will be created in a fish by introducing insulin into the water with a fish. Insulin will be absorbed into the blood of the gills of the fish. When insulin in the blood reaches levels above normal, rapid hypoglycemia occurs. As hypoglycemia happens from the increased insulin absorption, the brain cells obtain less of the much-needed glucose, and epinephrine is secreted. Consequently, a rapid heart rate, anxiety, sweating, mental disorientation, impaired vision, dizziness, convulsions, and possible unconsciousness are complications during insulin shock. Although these complications take place in humans, some of them are visually noticeable in a fish that is in insulin shock. The most observable components of insulin shock in a fish are related to behavioral changes, including rapid and irregular movements of the entire fish and increased gill cover and mouth movements.

1. Review section 13.8 titled "Pancreas" and the "Diabetes Mellitus—Clinical Application" in chapter 13 of the textbook, as well as the introduction to this laboratory exercise.
2. Complete the table in Part D of the laboratory report.
3. A fish can be used to observe normal behavior, induced insulin shock, and recovery from insulin shock. The pancreatic islets are located in the pyloric ceca or other scattered regions in fish. The behavioral changes of fish when they experience insulin shock and recovery mimic those of humans. Observe a fish in 200 mL of aquarium water in a 500 mL beaker. Make your observations of swimming, gill cover (operculum), and mouth movements for 5 minutes. Record your observations in Part E of the laboratory report.
4. Add 200 units of room-temperature insulin slowly, using the syringe, into the water with the fish to induce insulin shock. Record the time or start the timer when the insulin is added. Watch for changes of the swimming, gill cover, and mouth movements as insulin diffuses into the blood at the gills. Signs of insulin shock might include rapid and irregular movements. Record your observations of any behavioral changes and the time of the onset in Part E of the laboratory report.
5. After definite behavioral changes are observed, use the small fish net to move the fish into another 500 mL beaker containing 200 mL of a 10% room-temperature glucose solution. Record the time when the fish is transferred into this container. Make observations of any recovery of normal behaviors, including the amount of time involved. Record your observations in Part E of the laboratory report.
6. When the fish appears to be fully recovered, return it to its normal container. Dispose of the syringe according to directions from your laboratory instructor.
7. Complete Part E of the laboratory report.

LEARN: ACTIVITY

Examine a normal stained pancreas under high-power magnification (fig. 36.6). Locate the clumps of cells that constitute the *pancreatic islets (islets of Langerhans)* that represent the endocrine portions of the organ. A pancreatic islet contains four distinct cells that secrete different hormones: *alpha cells,* which secrete glucagon; *beta cells,* which secrete insulin; *delta cells,* which secrete somatostatin; and *F cells,* which secrete pancreatic polypeptide. Some of the specially stained pancreatic islets allow for distinguishing the different types of cells (figs. 36.7 and 36.8). The normal pancreatic islet has about 80% beta cells concentrated in the central region of the islet, about 15% alpha cells near the periphery of the islet, about 5% delta cells, and a small number of F cells. There are complex interactions among the four hormones affecting blood sugar, but in general, glucagon will increase blood sugar, whereas insulin will lower blood sugar; thus, these two hormones are considered antagonistic in their function. Somatostatin acts as a paracrine secretion to inhibit alpha and beta cell secretions, and pancreatic polypeptide inhibits secretions from delta cells and digestive enzyme secretions from the pancreatic acini.

Examine the pancreas of a diabetic under high-power magnification (fig. 36.9). Locate a pancreatic islet and note the number of beta cells as compared to a normal number of beta cells in the normal pancreatic islet. Prepare a labeled sketch of a pancreatic islet showing indications of diabetes mellitus in Part B of the laboratory report.

Describe the differences that you observed between a normal pancreatic islet and a pancreatic islet of a person with diabetes mellitus.

_____

_____

_____

_____

_____

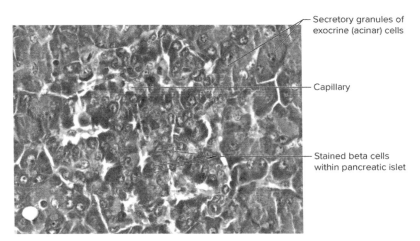

FIGURE 36.7   Micrograph of a specially stained pancreatic islet surrounded by exocrine (acinar) cells (500×). The insulin-containing beta cells are stained dark purple. **APR** (©J & J Photography)

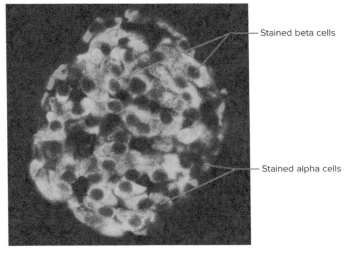

FIGURE 36.8   Micrograph of a pancreatic islet using immunofluorescence stains to visualize alpha and beta cells. Only the areas of interest are visible in the fluorescence microscope. **APR** (Al Telser/McGraw-Hill Education)

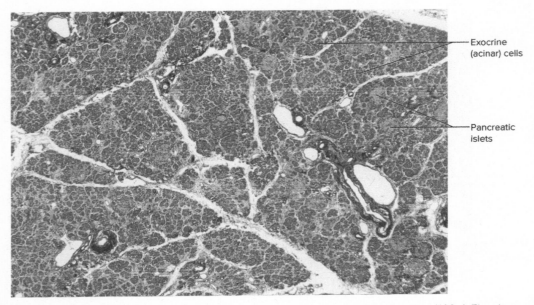

FIGURE 36.9 Micrograph of pancreas showing indications of changes from diabetes mellitus (100×). The changes from diabetes mellitus include areas of fibrosis and amyloid replacement of islet cells. (Michael Abbey/Science Source)

Cardiovascular System

LABORATORY EXERCISE

37

Blood Cells and Blood Typing

MATERIALS NEEDED

Textbook
Compound light microscope
Prepared microscope slides of human blood (Wright's stain)
Colored pencils
ABO blood-typing kit
Simulated blood-typing kits are suggested as a substitute for collected blood
Microscope slide
Alcohol swabs
Sterile blood lancet
Toothpicks
Anti-D serum
Slide warming table (Rh blood-typing box)
Disposable gloves
Prepared slides of pathological blood, such as eosinophilia, leukocytosis, leukopenia, and lymphocytosis

⚠ SAFETY

- It is important that students learn and practice correct procedures for handling body fluids. Consider using simulated blood-typing kits, mammal blood other than human, or contaminant-free blood that has been tested and is available from various laboratory supply houses. Some of the procedures might be accomplished as demonstrations only. If student blood is used, it is important that students handle only their own blood.
- Use an appropriate disinfectant to wash the laboratory tables before and after the procedures.
- Wear disposable gloves and safety goggles when handling blood samples.
- Clean the end of a finger with 70% alcohol before the puncture is performed.
- Use the sterile blood lancet only once.
- Dispose of used lancets and blood-contaminated items in an appropriate container (never use the wastebasket).
- Wash your hands before leaving the laboratory.

PURPOSE OF THE EXERCISE

To review the characteristics of blood cells, to examine them microscopically, to perform a differential white blood cell count, to determine the ABO blood type of a blood sample, and to observe an Rh blood-typing test.

LEARNING OUTCOMES APR

After completing this exercise, you should be able to

1. Describe the structure and function of red blood cells, white blood cells, and platelets.
2. Identify and sketch red blood cells, five types of white blood cells, and platelets.
3. Perform and interpret the results of a differential white blood cell count.
4. Analyze the basis of ABO blood typing.
5. Match the ABO type of a blood sample.
6. Explain the basis of Rh blood typing.
7. Interpret how the Rh type of a blood sample is determined.

Blood is a type of connective tissue whose cells are suspended in a liquid extracellular matrix. These cells are formed mainly in red bone marrow, and they include *red blood cells, white blood cells,* and some cellular fragments called *platelets* (fig. 37.1).

Red blood cells contain hemoglobin and transport gases between the body cells and the lungs, white blood cells defend the body against infections, and platelets play an important role in stoppage of bleeding (hemostasis).

Blood typing involves identifying protein substances called *antigens* present on the outer surface of red blood cell membranes. Although many different antigens are associated with human red blood cells, only a few of them are of clinical importance. These include the antigens of the ABO group and those of the Rh group.

To determine which antigens are present, a blood sample is mixed with blood-typing sera that contain known types of antibodies. If a particular antibody contacts a corresponding antigen, a reaction occurs, and the red blood cells clump together (agglutination). Thus, if blood cells are mixed with serum-containing antibodies that react with antigen A and the cells clump together, antigen A must be present in those cells.

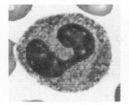
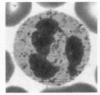
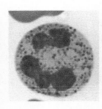
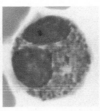
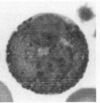

Neutrophils (three of many variations)
•Fine light-purple granules
•Nucleus single to five lobes (highly variable)
•Immature neutrophils, called bands, have a
 single C-shaped nucleus
•Mature neutrophils, called segs, have a lobed nucleus
•Often called polymorphonuclear leukocytes when older

Eosinophils (three of many variations)
•Coarse reddish granules
•Nucleus usually bilobed

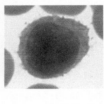
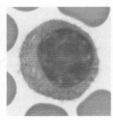

Basophils (three of many variations)
•Coarse deep blue to almost black granules
•Nucleus often almost hidden by granules

Lymphocytes (three of many variations)
•Slightly larger than RBCs
•Thin rim of nearly clear cytoplasm
•Nearly round nucleus appears to fill most of cell
 in smaller lymphocytes
•Larger lymphocytes hard to distinguish from monocytes

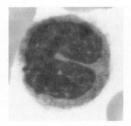
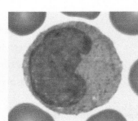
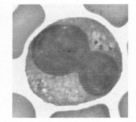

Monocytes (three of many variations)
•Largest WBC; 2–3× larger than RBCs
•Cytoplasm nearly clear
•Nucleus round, kidney-shaped, oval, or lobed

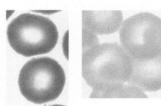

Platelets (several variations)
•Cell fragments
•Single to small clusters

Erythrocytes (several variations)
•Lack nucleus (mature cell)
•Biconcave discs
•Thin centers appear almost hollow

FIGURE 37.1   Micrographs of blood cells illustrating some of the numerous variations of each type. Appearance characteristics
for each cell pertain to a thin blood film using Wright's stain (1,000×).   APR (Al Telser/McGraw-Hill Education)

PRACTICE

PROCEDURE A—Types of Blood Cells

1. Review section 14.2 titled "Formed Elements" in chapter 14 of the textbook.
2. Complete Part A of Laboratory Report 37.
3. Refer to figure 37.1 as an aid in identifying the various types of blood cells. Study the functions of the blood cells listed in table 37.1. Use the prepared slide of blood and locate each of the following:

   red blood cell (erythrocyte)

   white blood cell (leukocyte)

   > granulocytes

   >> neutrophil

   >> eosinophil

   >> basophil

   > agranulocytes

   >> lymphocyte

   >> monocyte

   platelet (thrombocyte)

TABLE 37.1 Cellular Components of Blood

Component	Function
Red blood cell (erythrocyte)	Transports oxygen and carbon dioxide
White blood cell (leukocyte)	Destroys pathogenic microorganisms and parasites, removes worn cells, provides immunity
Granulocytes—Have Granular Cytoplasm	
Neutrophil	Phagocytizes small particles
Eosinophil	Destroys parasites, helps control inflammation and allergic reactions
Basophil	Releases heparin and histamine
Agranulocytes—Lack Granular Cytoplasm	
Lymphocyte	Provides immunity
Monocyte	Phagocytizes large particles
Platelet (thrombocyte)	Helps control blood loss from injured blood vessels

4. In Part B of the laboratory report, prepare sketches of single blood cells to illustrate each type. Pay particular attention to the relative size, nuclear shape, and color of granules in the cytoplasm (if present). The sketches should be accomplished using either the high-power objective or the oil immersion objective of the compound light microscope.

 PRACTICE

PROCEDURE B—Differential White Blood Cell Count

A differential white blood cell count is performed to determine the percentage of each of the various types of white blood cells present in a blood sample. The test is useful because the relative proportions of white blood cells may change in particular diseases, as indicated in table 37.2. Neutrophils, for example, usually increase during bacterial infections, whereas eosinophils may increase during certain parasitic infections and allergic reactions.

1. To make a differential white blood cell count, follow these steps:
   a. Using high-power magnification or an oil immersion objective, focus on the cells at one end of a prepared blood slide where the cells are well distributed.
   b. Slowly move the blood slide back and forth, following a path that avoids passing over the same cells twice (fig. 37.2).
   c. Each time you encounter a white blood cell, identify its type and record it in Part C of the laboratory report.
   d. Continue searching for and identifying white blood cells until you have recorded 100 cells in the data table. *Percent* means "parts of 100" for each type of white blood cell, so the total number observed is equal to its percentage in the blood sample.
2. Complete Part C of the laboratory report.

TABLE 37.2 Differential White Blood Cell Count

Cell Type	Normal Value (percent)	Elevated Levels May Indicate
Neutrophil	50–70	Bacterial infections, stress
Lymphocyte	25–33	Mononucleosis, whooping cough, viral infections
Monocyte	3–9	Malaria, tuberculosis, fungal infections
Eosinophil	1–3	Allergic reactions, autoimmune diseases, parasitic worms
Basophil	<1	Cancers, chicken pox, hypothyroidism

FIGURE 37.2 Move the blood slide back and forth to avoid passing the same cells twice.

 **LEARN: ACTIVITY**

Obtain a prepared slide of pathological blood that has been stained with Wright's stain. Perform a differential white blood cell count using this slide, and compare the results with the values for normal blood listed in table 37.2. What differences do you note?

_____

_____

_____

_____

_____

_____

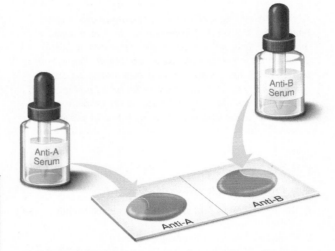

FIGURE 37.3   Slide prepared for ABO blood typing.

b. Place a small drop of blood on each half of the microscope slide. Work quickly so that the blood will not have time to clot.

c. Add a drop of anti-A serum to the blood on the left half and a drop of anti-B serum to the blood on the right half. Note the color coding of the anti-A and anti-B typing sera. To avoid contaminating the serum, avoid touching the blood with the serum while it is in the dropper; instead, allow the serum to fall from the dropper onto the blood.

d. Use separate toothpicks to stir each sample of serum and blood together, and spread each over an area about as large as a quarter. Dispose of toothpicks in an appropriate container.

e. Examine the samples for clumping of red blood cells (agglutination) after 2 minutes.

f. See table 37.3 and figure 37.4 for aid in interpreting the test results.

g. Discard contaminated materials as directed by the laboratory instructor.

 **PRACTICE**

PROCEDURE C—ABO Blood Typing

1. Review section 14.5 titled "Blood Groups and Transfusions" in chapter 14 of the textbook.
2. Complete Part D of the laboratory report.
3. Perform the ABO blood type test using the blood-typing kit. To do this, follow these steps:
   a. Obtain a clean microscope slide and mark across its center with a wax pencil to divide it into right and left halves. Also write "Anti-A" near the edge of the left half and "Anti-B" near the edge of the right half (fig. 37.3).

4. Complete Part E of the laboratory report.

TABLE 37.3   Possible Reactions of ABO Blood-Typing Sera and Donor Compatibilities

| REACTIONS | | | | Permissible | |
Anti-A Serum	Anti-B Serum	Blood Type	Preferred Donor Type	Donor in Limited Amounts	Incompatible Donor
Clumping	No clumping	Type A	A	O	B, AB
No clumping	Clumping	Type B	B	O	A, AB
Clumping	Clumping	Type AB (universal recipient)	AB	A, B, O	None
No clumping	No clumping	Type O (universal donor)	O	No alternative types	A, B, AB

Note: The inheritance of the ABO blood groups is described in Laboratory Exercise 53.

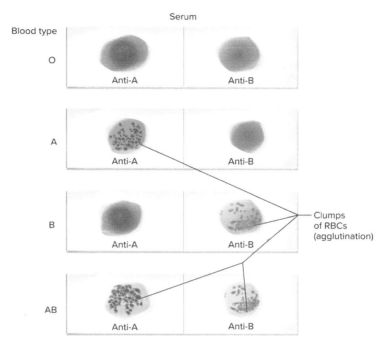

FIGURE 37.4   Four possible results of the ABO test.

ASSESS

CRITICAL THINKING

Judging from the observations of the blood-typing results, suggest which components in the anti-A and anti-B sera caused clumping (agglutination).

_____

_____

_____

_____

PRACTICE

DEMONSTRATION—Rh Blood Typing

1. Review section 14.5 titled "Blood Groups and Transfusions" in chapter 14 of the textbook.
2. Complete Part F of the laboratory report.
3. To determine the Rh blood type of a blood sample, follow these steps:
   a. Thoroughly wash hands with soap and water, and dry them with paper towels.
   b. Cleanse the end of the middle finger with an alcohol swab, and let the finger dry in the air.
   c. Lance the tip of a finger using a sterile disposable blood lancet. Place a small drop of blood in the center of a clean microscope slide. Cover the lanced finger location with a bandage.
   d. Add a drop of anti-D serum to the blood and mix them together with a clean toothpick.
   e. Place the slide on the plate of a slide warming table (Rh blood-typing box) prewarmed to 45°C (113°F) (fig. 37.5).
   f. Slowly rock the table (box) back and forth to keep the mixture moving, and watch for clumping (agglutination) of the blood cells. When clumping occurs in anti-D serum, the clumps usually are smaller than those that appear in anti-A or anti-B sera, so they may be less obvious. However, if clumping occurs, the blood is called Rh positive; if no clumping occurs _within 2 minutes,_ the blood is called Rh negative.
   g. Discard all contaminated materials in appropriate containers.
4. Complete Parts G and H of the laboratory report.

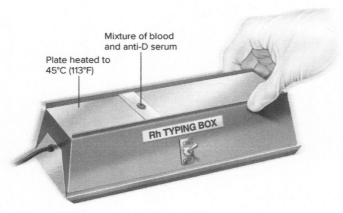

Mixture of blood
and anti-D serum

Plate heated to
45°C (113°F)

Rh TYPING BOX

FIGURE 37.5 Slide warming box (table) used for Rh blood typing.

LABORATORY EXERCISE

38

Heart Structure

MATERIALS NEEDED

Textbook
Dissectible human heart model
Preserved sheep or other mammalian heart
Dissecting tray
Dissecting instruments
Disposable gloves
Colored pencils (red and blue)

⚠ SAFETY

- Wear disposable gloves when working on the heart dissection.
- Save or dispose of the dissected heart as instructed.
- Wash the dissecting tray and instruments as instructed.
- Wash your laboratory table.
- Wash your hands before leaving the laboratory.

PURPOSE OF THE EXERCISE

To review the structural characteristics of the human heart and to examine the major features of a mammalian heart.

LEARNING OUTCOMES APR

After completing this exercise, you should be able to

1. Identify and label the major structural features of the human heart and closely associated blood vessels.

2. Match heart structures with appropriate locations and functions.

3. Compare the features of the human heart with those of another mammal.

The heart is a muscular pump located within the mediastinum and resting upon the diaphragm. It is enclosed by the lungs, thoracic vertebrae, and sternum, and attached at its superior end (the base) are several large blood vessels. Its

inferior end extends downward to the left and terminates as a bluntly pointed apex.

The heart and the proximal ends of the attached blood vessels are enclosed by a double-layered pericardium. The innermost layer of this membrane (visceral pericardium) consists of a thin covering closely applied to the surface of the heart, whereas the outer layer (parietal pericardium with fibrous pericardium) forms a tough, protective sac surrounding the heart. Between the parietal and visceral layers of the pericardium is a space, the pericardial cavity, that contains a small volume of serous (pericardial) fluid.

PRACTICE

PROCEDURE A—The Human Heart

1. Review section 15.2 titled "Structure of the Heart" in chapter 15 of the textbook.
2. As a review activity, label figures 38.1–38.3.
3. Complete Part A of Laboratory Report 38.
4. Examine the human heart model, and locate the following features:

heart
  base (superior region where blood vessels emerge)
  apex (inferior, rounded end)
pericardium (pericardial sac)
  fibrous pericardium (outer layer)
  parietal pericardium (inner lining of fibrous pericardium)
  visceral pericardium (epicardium)
pericardial cavity (between parietal and visceral pericardial membranes)
myocardium (cardiac muscle)
endocardium (lines heart chambers)
atria
  right atrium
  left atrium
  auricles
ventricles
  right ventricle
  left ventricle

atrioventricular orifices

atrioventricular valves (AV valves)

   tricuspid valve (right atrioventricular valve)

   mitral valve (bicuspid valve; left atrioventricular valve)

semilunar valves

   pulmonary valve

   aortic valve

chordae tendineae

papillary muscles

atrioventricular sulcus

interventricular sulci

   anterior sulcus

   posterior sulcus

superior vena cava

inferior vena cava

pulmonary trunk

pulmonary arteries

pulmonary veins

aorta

left coronary artery

   circumflex artery

   anterior interventricular (descending) artery

right coronary artery

   posterior interventricular artery

   marginal artery

cardiac veins

   great cardiac vein

   middle cardiac vein

   small cardiac vein

coronary sinus (for return of blood from cardiac veins into right atrium)

5. Label the human heart model in figure 38.4.

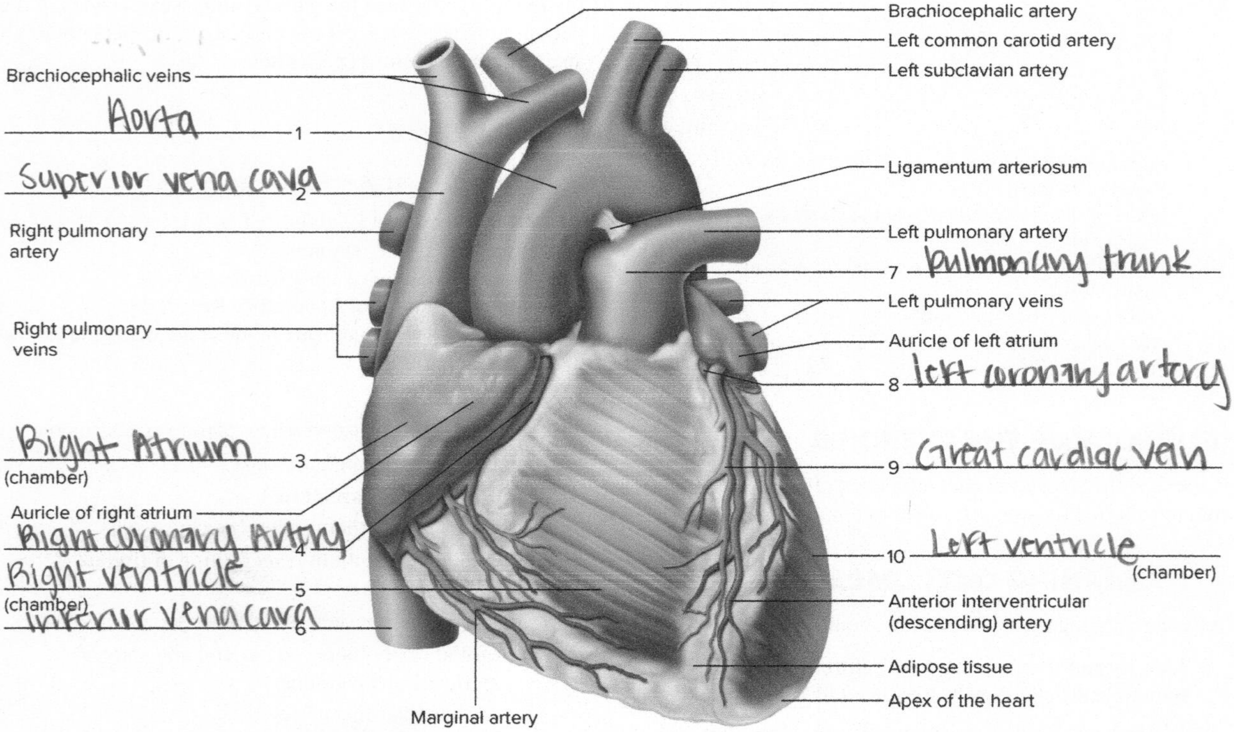

Brachiocephalic veins

Aorta — 1

Superior vena cava — 2

Right pulmonary artery

Right pulmonary veins

Right Atrium — 3 (chamber)

Auricle of right atrium

Right coronary Artry

Right ventricle — 5 (chamber)

Inferior vena cava — 6

Marginal artery

Brachiocephalic artery

Left common carotid artery

Left subclavian artery

Ligamentum arteriosum

Left pulmonary artery

7 — *pulmonary trunk*

Left pulmonary veins

Auricle of left atrium

8 — *left coronary artery*

9 — *Great cardiac vein*

10 — *Left ventricle* (chamber)

Anterior interventricular (descending) artery

Adipose tissue

Apex of the heart

FIGURE 38.1   Label this anterior view of the human heart.   [A] APR

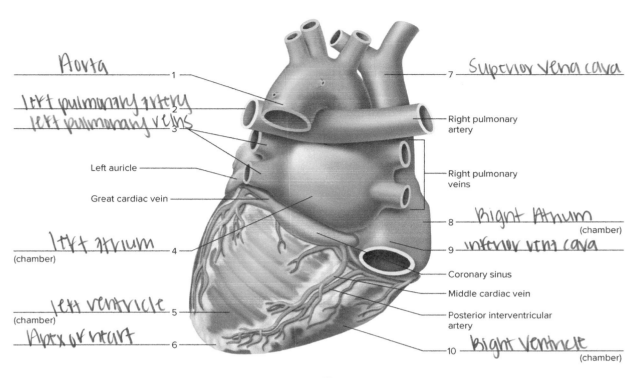

Aorta — 1

left pulmonary artery — 2
left pulmonary veins — 3

Left auricle —

Great cardiac vein —

left atrium — 4
(chamber)

left ventricle — 5
(chamber)

Apex of heart — 6

7 — Superior Vena cava

Right pulmonary artery

Right pulmonary veins

8 — Right Atrium
(chamber)

9 — Inferior vena cava

Coronary sinus

Middle cardiac vein

Posterior interventricular artery

10 — Right Ventricle
(chamber)

FIGURE 38.2   Label this posterior view of the human heart. 🅰

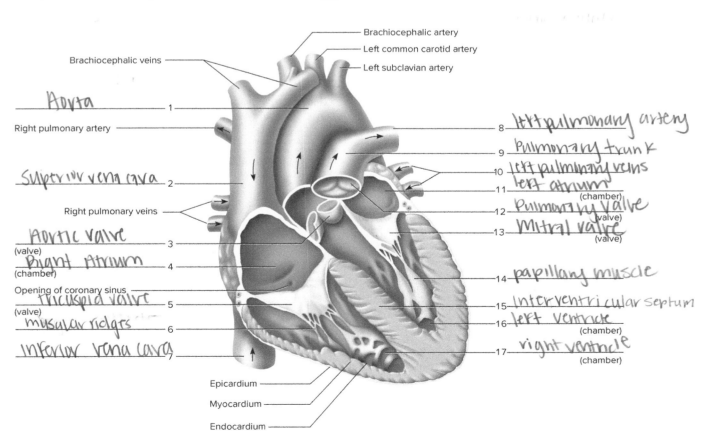

Brachiocephalic artery

Left common carotid artery

Left subclavian artery

Brachiocephalic veins —

Aorta — 1

Right pulmonary artery —

Superior vena cava — 2

Right pulmonary veins —

Aortic valve — 3
(valve)

Right Atrium — 4
(chamber)

Opening of coronary sinus —

tricuspid valve — 5
(valve)

muscular ridges — 6

Inferior vena cava — 7

Epicardium —

Myocardium —

Endocardium —

8 — left pulmonary artery

9 — Pulmonary trunk

10 — left pulmonary veins

11 — left atrium
(chamber)

12 — Pulmonary valve
(valve)

13 — Mitral valve
(valve)

14 — papillary muscle

15 — Interventricular septum

16 — left ventricle
(chamber)

17 — right ventricle
(chamber)

FIGURE 38.3   Label this frontal section of the human heart. The arrows indicate the direction of blood flow. 🅰 **APR**

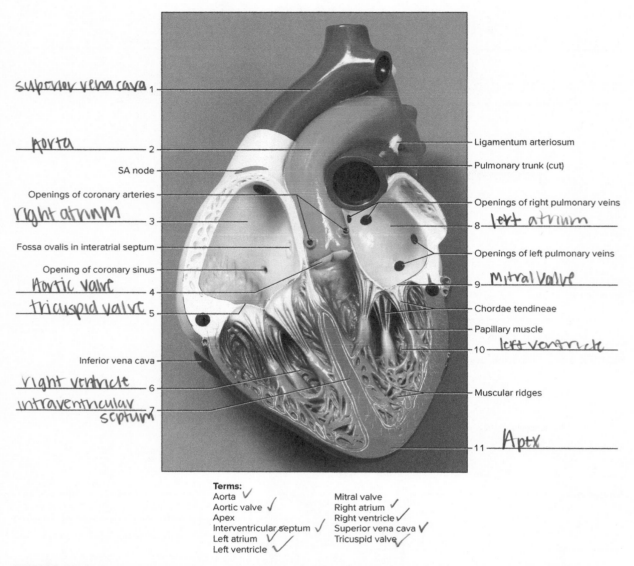

Labels (handwritten and printed):

- 1 — superior vena cava
- 2 — Aorta
- SA node
- Openings of coronary arteries
- 3 — right atrium
- Fossa ovalis in interatrial septum
- Opening of coronary sinus
- 4 — Aortic valve
- 5 — tricuspid valve
- Inferior vena cava
- 6 — right ventricle
- 7 — intraventricular septum
- Ligamentum arteriosum
- Pulmonary trunk (cut)
- Openings of right pulmonary veins
- 8 — left atrium
- Openings of left pulmonary veins
- 9 — Mitral valve
- Chordae tendineae
- Papillary muscle
- 10 — left ventricle
- Muscular ridges
- 11 — Apex

Terms:
Aorta ✓
Aortic valve ✓
Apex
Interventricular septum ✓
Left atrium ✓
Left ventricle ✓
Mitral valve ✓
Right atrium ✓
Right ventricle ✓
Superior vena cava ✓
Tricuspid valve ✓

FIGURE 38.4   Identify the features indicated on this anterior view of a frontal section of a human heart model, using the terms provided. (*Note*: The pulmonary valve is not shown on the portion of the model photographed.) **A** **APR** (Al Telser/McGraw-Hill Education)

 LEARN: ACTIVITY

Use red and blue colored pencils to color the blood vessels in figure 38.3. Use red to illustrate a blood vessel with oxygen-rich (oxygenated) blood, and use blue to illustrate a blood vessel with oxygen-poor (deoxygenated) blood. You can check your work by referring to the corresponding figures in the textbook, presented in full color.

 PRACTICE

PROCEDURE B—Dissection of a Sheep Heart

1. Obtain a preserved sheep heart. Rinse it in water thoroughly to remove as much of the preservative as possible. Also run water into the large blood vessels to force any blood clots out of the heart chambers. The structures of a sheep heart are similar to those of a human heart and will therefore be used for further exploration. Although sheep are quadrupeds, human terminology will be used for the sheep heart.

2. Place the heart in a dissecting tray with its anterior surface up (fig. 38.5), and proceed as follows:

   a. Although the relatively thick *pericardial sac* probably is missing, look for traces of this membrane around the origins of the large blood vessels.

   b. Locate the *visceral pericardium,* which appears as a thin, transparent layer on the surface of the heart. Use a scalpel to remove a portion of this layer and expose the *myocardium* beneath. Also note the abundance of fat along the paths of various blood vessels. This adipose tissue occurs in the loose connective tissue that underlies the visceral pericardium.

   c. Identify the following:

   right atrium

   right ventricle

   left atrium

   left ventricle

   atrioventricular (coronary) sulcus

   anterior interventricular sulcus

   d. Carefully remove the fat from the anterior interventricular sulcus, and expose the blood vessels that pass along this groove. They include a branch of the *left coronary artery (anterior interventricular artery)* and a *cardiac vein.*

3. Examine the posterior surface of the heart (fig. 38.6), and proceed as follows:

   a. Identify the *atrioventricular (coronary) sulcus* and the *posterior interventricular sulcus.*

   b. Locate the stumps of two relatively thin-walled veins that enter the right atrium. Demonstrate this connection by passing a slender probe through them. The upper vessel is the *superior vena cava,* and the lower one is the *inferior vena cava.*

4. Open the right atrium. To do this, follow these steps:

   a. Insert a blade of the scissors into the superior vena cava and cut downward through the atrial wall (fig. 38.6).

   b. Open the chamber, locate the *right atrioventricular valve (tricuspid valve),* and examine its cusps.

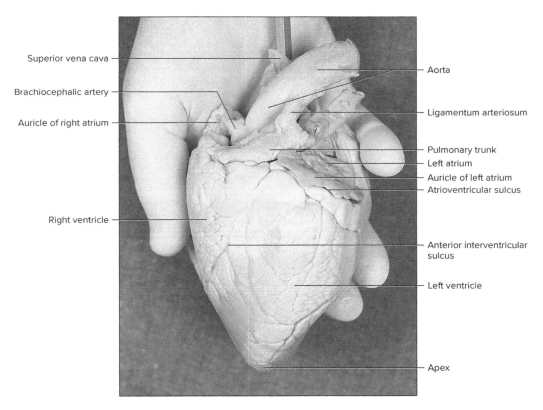

Superior vena cava

Brachiocephalic artery

Auricle of right atrium

Right ventricle

Aorta

Ligamentum arteriosum

Pulmonary trunk

Left atrium

Auricle of left atrium

Atrioventricular sulcus

Anterior interventricular sulcus

Left ventricle

Apex

FIGURE 38.5 Anterior surface of sheep heart. **APR** (Al Telser/McGraw-Hill Education)

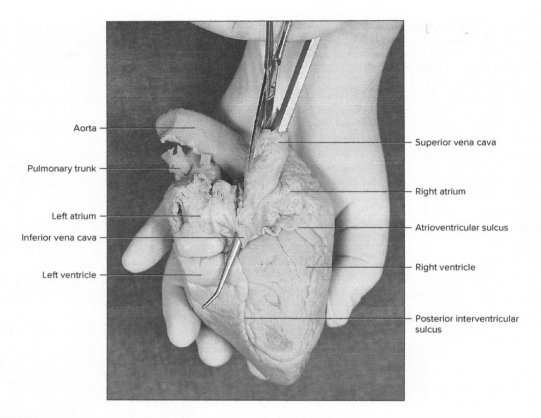

Aorta

Pulmonary trunk

Left atrium

Inferior vena cava

Left ventricle

Superior vena cava

Right atrium

Atrioventricular sulcus

Right ventricle

Posterior interventricular sulcus

FIGURE 38.6 Anterior surface of sheep heart. **APR** (Al Telser/McGraw-Hill Education)

c. Also locate the opening to the *coronary sinus* between the valve and the inferior vena cava.

d. Run some water through the right atrioventricular valve to fill the chamber of the right ventricle.

e. Gently squeeze the ventricles, and watch the cusps of the valve as the water moves up against them.

5. Open the right ventricle as follows:

a. Continue cutting downward through the right atrioventricular valve and the right ventricular wall until you reach the apex of the heart.

b. Locate the *chordae tendineae* and the *papillary muscles.*

c. Find the opening to the *pulmonary trunk,* and use the scissors to cut upward through the wall of the right ventricle. Follow the pulmonary trunk until you have exposed the *pulmonary valve.*

d. Examine the valve and its cusps.

6. Open the left side of the heart. To do this, follow these steps:

a. Insert the blade of the scissors through the wall of the left atrium and cut downward on the anterior portion to the apex of the heart.

b. Open the left atrium, and locate the four openings of the *pulmonary veins.* Pass a slender probe through each opening, and locate the stump of its vessel.

c. Examine the *left atrioventricular valve (bicuspid valve)* and its cusps.

d. Also examine the left ventricle, and compare the thickness of its wall with that of the right ventricle.

7. Locate the aorta, which leads away from the left ventricle, and proceed as follows:

a. Compare the thickness of the aortic wall with that of a pulmonary artery.

b. Use scissors to cut along the length of the aorta to expose the *aortic valve* at its base.

c. Examine the cusps of the valve, and locate the openings of the *coronary arteries* just distal to them.

8. As a review, locate and identify the stumps of each of the major blood vessels associated with the heart.

9. Discard or save the specimen as directed by the laboratory instructor.

10. Complete Part B of the laboratory report.

39

Cardiac Cycle

MATERIALS NEEDED

Textbook

For Procedure A—Heart Sounds:
Stethoscope
Alcohol swabs

For Procedure B—Electrocardiogram:
Electrocardiograph (or other instrument for recording an ECG)
Cot or table
Alcohol swabs
Electrode paste (gel)
Electrodes and cables
Lead selector switch

PURPOSE OF THE EXERCISE

To review the events of a cardiac cycle, to become acquainted with normal heart sounds, and to record an electrocardiogram.

 ## LEARNING OUTCOMES APR

After completing this exercise, you should be able to

1. Interpret the major events of a cardiac cycle.

2. Associate the sounds produced during a cardiac cycle with the valves closing.

3. Correlate the components of a normal ECG pattern with the phases of a cardiac cycle.

4. Record and interpret an electrocardiogram.

A set of atrial contractions while the ventricular walls relax, followed by ventricular contractions while the atrial walls relax, constitutes a *cardiac cycle.* Such a cycle is accompanied by blood pressure changes within the heart chambers, movement of blood into and out of the chambers, and opening and closing of heart valves. These closing valves produce vibrations in the tissues and thus create the sounds associated with the heartbeat. If backflow of blood occurs when a heart valve is closed, creating a turbulence noise, the condition is known as a murmur.

The regulation and coordination of the cardiac cycle involves the *cardiac conduction system.* The pathway of electrical signals originates from the *SA (sinoatrial) node* located in the right atrium near the entrance of the superior vena cava. The stimulation of the heartbeat and the heart rate originate from the SA node, so it is often called the *pacemaker.* As the signals pass through the atrial walls toward the *AV (atrioventricular) node,* contractions of the atria take place. When the AV node has been signaled, the rapid continuation of the electrical signals occurs through the *AV bundle,* passes through the *right and left bundle branches* within the interventricular septum, and terminates via the *Purkinje fibers* throughout the ventricular walls. Once the myocardium has been stimulated, ventricular contractions happen, completing one cardiac cycle. The sympathetic and parasympathetic subdivisions of the autonomic nervous system influence the activity of the pacemaker under various conditions. An increased rate results from sympathetic responses; a decreased rate results from parasympathetic responses.

A number of electrical changes also occur in the myocardium as it contracts and relaxes. These changes can be detected by using metal electrodes and an instrument called an *electrocardiograph.* The recording produced by the instrument is an *electrocardiogram,* or *ECG (EKG).*

 ## PRACTICE

PROCEDURE A—Heart Sounds

1. Review section 15.3 titled "Heart Actions" in chapter 15 of the textbook.
2. Complete Part A of Laboratory Report 39.
3. Listen to your heart sounds. To do this, follow these steps:
   a. Obtain a stethoscope, and clean its earpieces and the diaphragm by using alcohol swabs.
   b. Fit the earpieces into your ear canals so that the angles are positioned in the forward direction.
   c. Firmly place the diaphragm (bell) of the stethoscope on the chest over the fifth intercostal space near the apex of the heart (fig. 39.1) and listen to the sounds. This is a good location to hear the first sound (*lubb*) of a cardiac cycle when the AV valves are closing, which occurs during ventricular *systole* (contraction).

d. Move the diaphragm to the second intercostal space, just to the left of the sternum, and listen to the sounds from this region. You should be able to hear the second sound (*dupp*) of the cardiac cycle clearly when the semilunar valves are closing, which occurs during ventricular *diastole* (relaxation).

e. It is possible to hear sounds associated with the aortic and pulmonary valves by listening from the second intercostal space on either side of the sternum. The aortic valve sound comes from the right and the pulmonary valve sound from the left. The sound associated with the mitral valve can be heard from the fifth intercostal space at the nipple line on the left. The sound of the tricuspid valve can be heard at the fourth intercostal space just to the left of the sternum (fig. 39.1). The four points for listening (auscultating) the four heart valve sounds when closing are not directly superficial to the valve locations. This is because the sounds associated with the closing valves travel in a slanting direction to the surface of the chest wall where the stethoscope placements occur.

4. Inhale slowly and deeply, and exhale slowly while you listen to the heart sounds from each of the locations as before. Note any changes that have occurred in the sounds.

5. Exercise moderately outside the laboratory for a few minutes so that other students listening to heart sounds will not be disturbed. After the exercise period, listen to the heart sounds and note any changes that have occurred in them. *This exercise should be avoided by anyone with health risks.*

6. Complete Part B of the laboratory report.

 PRACTICE

PROCEDURE B—Electrocardiogram

1. Review the concept headings titled "Cardiac Conduction System" and "Electrocardiogram" of section 15.3 and the Clinical Application "Arrhythmias" in chapter 15 of the textbook.

2. Complete Part C of the laboratory report.

3. The laboratory instructor will demonstrate the proper adjustment and use of the instrument available to record an electrocardiogram.

4. Record your laboratory partner's ECG. To do this, follow these steps:

a. Have your partner lie on a cot or table close to the electrocardiograph, remaining as relaxed and still as possible.

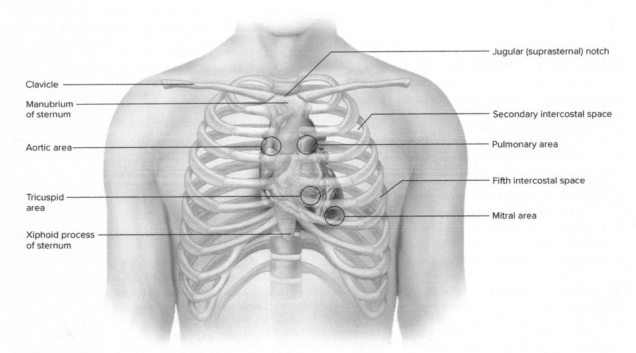

Clavicle

Manubrium of sternum

Aortic area

Tricuspid area

Xiphoid process of sternum

Jugular (suprasternal) notch

Secondary intercostal space

Pulmonary area

Fifth intercostal space

Mitral area

FIGURE 39.1   The first sound (lubb) of a cardiac cycle can be heard by placing the diaphragm of a stethoscope over the fifth intercostal space near the apex of the heart. The second sound (dupp) can be heard over the second intercostal space, just left of the sternum. The thoracic regions circled indicate where the sounds of each heart valve are most easily heard. Note that where the sound is heard may differ slightly from the anatomical location of the structure. **APR**

b. Scrub the electrode placement locations with alcohol swabs (fig. 39.2). Apply a small quantity of electrode paste to the skin on the insides of the wrists and ankles. (Any jewelry on the wrists or ankles should be removed.)

c. Spread some electrode paste over the inner surfaces of four electrodes and attach one to each of the prepared skin areas (fig. 39.2). Make sure there is good contact between the skin and the metal of the electrodes. The electrode on the right ankle is the grounding system.

d. Attach the electrodes to the corresponding cables of a lead selector switch. When an ECG recording is made, only two electrodes are used at a time, and the selector switch allows various combinations of electrodes (leads) to be activated. Three standard limb leads placed on the two wrists and the left ankle are used for an ECG. This arrangement has become known as *Einthoven's triangle,*[1] which permits the recording of the potential difference between any two of the electrodes.

The standard leads I, II, and III are called bipolar leads because they are the potential difference between two electrodes (a positive and a negative). Lead I measures the potential difference between the right wrist (negative) and the left wrist (positive). Lead II measures the potential difference between the right wrist and the left ankle, and lead III measures the potential difference between the left wrist and the left ankle. The right ankle is always the ground.

e. Turn on the recording instrument and adjust it as previously demonstrated by the laboratory instructor. The paper speed should be set at 2.5 cm/second. This is the standard speed for ECG recordings.

f. Set the lead selector switch to lead I (right wrist, left wrist electrodes), and record the ECG for 1 minute.

g. Set the lead selector switch to lead II (right wrist, left ankle electrodes), and record the ECG for 1 minute.

h. Set the lead selector switch to lead III (left wrist, left ankle electrodes), and record the ECG for 1 minute.

i. Remove the electrodes and clean the paste from the metal and skin.

j. Use figure 39.3 to label the ECG components of the results from leads I, II, and III. The PQ interval is often called the PR interval because the Q wave is often small or absent. The normal PQ interval is 0.12–0.20 seconds. The normal QRS complex duration is less than 0.10 second.

5. Complete Part D of the laboratory report.

---

[1]Willem Einthoven (1860–1927), a Dutch physiologist, received the Nobel Prize for Physiology or Medicine for his work with electrocardiograms.

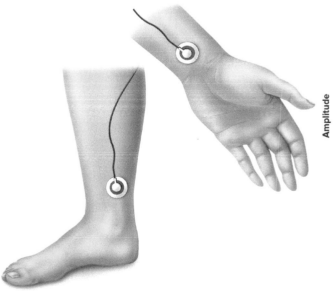

FIGURE 39.2 To record an ECG, attach electrodes to the wrists and ankles.

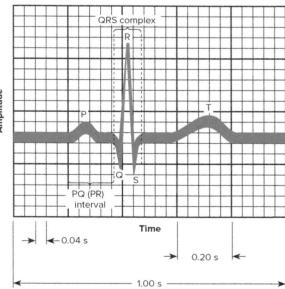

FIGURE 39.3 Components of a normal ECG pattern with a time scale. **APR**

Notes

Lead II

Lead III

2. What differences do you find in the ECG patterns of these leads? 🄰 _____

_____

_____

_____

_____

_____

3. How much time passed from the beginning of the P wave to the beginning of the QRS complex (PQ interval, or PR interval) in the ECG from lead I? 🔺 _____

4. What is the significance of this PQ (PR) interval? 🔺 _____

_____

_____

_____

_____

5. How can you determine heart rate from an electrocardiogram? 🔺 _____

_____

_____

6. What was your heart rate as determined from the ECG? 🔺 _____

ASSESS

CRITICAL THINKING

If a person's heart rate were 72 beats per minute, determine the number of QRS complexes that would have appeared on an ECG during the first 30 seconds. 🔺

_____

_____

_____

40

Blood Vessel Structure, Arteries, and Veins

MATERIALS NEEDED

Textbook
Dissectible human heart model
Human torso model
Anatomical charts of the cardiovascular system
Compound light microscope
Prepared microscope slides:
    Artery cross section
    Vein cross section
Live frog or goldfish
Frog Ringer's solution
Paper towel
Rubber bands
Frog board or heavy cardboard (with a 1-inch hole cut
    in one corner)
Dissecting pins
Thread
Masking tape
Ice
Hot plate
Thermometer

⚠ SAFETY

- Wear disposable gloves when handling the live frogs.
- Return the frogs to the location indicated after the experiment.
- Wash your hands before leaving the laboratory.

PURPOSE OF THE EXERCISE

To review the structure and functions of blood vessels, to trace major circulatory pathways, and to locate the major arteries and veins.

 ## LEARNING OUTCOMES APR

After completing this exercise, you should be able to

1. Distinguish the structures and functions of arteries, veins, and capillaries.
2. Identify the types of blood vessels in the web of a frog's foot.
3. Distinguish and trace the pulmonary and systemic circuits.
4. Locate the major arteries in these circuits on a diagram, chart, or model.
5. Locate the major veins in these circuits on a diagram, chart, or model.
6. Recognize the anatomical terms pertaining to the blood vessel structure, arteries, and veins.

The blood vessels form a closed system of tubes that carry blood to and from the heart, lungs, and body cells. These tubes include arteries and arterioles that transport blood away from the heart; capillaries in which exchanges of substances occur between the blood and surrounding tissues; and venules and veins that return blood to the heart.

The blood vessels of the cardiovascular system can be divided into two major pathways—the pulmonary circuit and the systemic circuit. Within each circuit, arteries transport blood away from the heart. After exchanges of gases, nutrients, and wastes have occurred between the blood and the surrounding tissues, veins return the blood to the heart.

Variations exist in anatomical structures among humans, especially in arteries and veins. The illustrations in this laboratory manual represent normal (meaning the most common variation) anatomy.

Artery

Vein

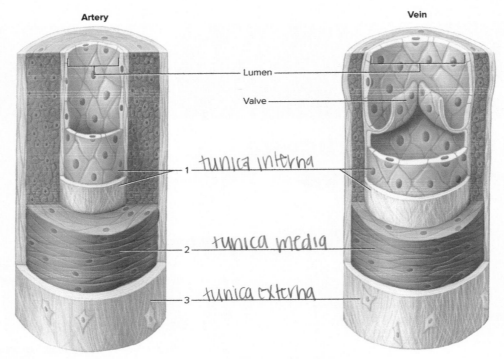

Lumen

Valve

1 — tunica interna

2 — tunica media

3 — tunica externa

FIGURE 40.1   Label the tunics of the walls of this artery and vein. **APR**

 PRACTICE

PROCEDURE A—Blood Vessel Structure

1. Review section 15.4 titled "Blood Vessels" in chapter 15 of the textbook.
2. As a review activity, label figure 40.1.
3. Complete Part A of Laboratory Report 40.
4. Obtain a microscope slide of an artery cross section, and examine it using low-power and high-power magnification. Identify the three distinct layers (tunics) of the arterial wall. The inner layer **(tunica interna)** is composed of an endothelium (simple squamous epithelium) and appears as a wavy line due to an abundance of elastic fibers that have recoiled just beneath it. The middle layer **(tunica media)** consists of numerous concentrically arranged smooth muscle cells with elastic fibers scattered among them. The outer layer **(tunica externa)** contains connective tissue rich in collagen fibers (fig. 40.2).
5. Prepare a labeled sketch of the arterial wall in Part B of the laboratory report.
6. Obtain a slide of a vein cross section and examine it as you did the artery cross section. Note the thinner wall and larger lumen relative to an artery of comparable locations. Identify the three layers of the wall, and prepare a labeled sketch in Part B of the laboratory report.
7. Complete Part B of the laboratory report.
8. Observe the blood vessels in the webbing of a frog's foot. (As a substitute for a frog, a live goldfish could be used to observe circulation in the tail.) To do this, follow these steps:
   a. Obtain a live frog. Wrap its body in a moist paper towel, leaving one foot extending outward. Secure the towel with rubber bands, but be careful not to wrap the animal so tightly that it could be injured. Try to keep the nostrils exposed.
   b. Place the frog on a frog board or on a piece of heavy cardboard with the foot near the hole in one corner.
   c. Fasten the wrapped body to the board with masking tape.
   d. Carefully spread the web of the foot over the hole and secure it to the board with dissecting pins and thread (fig. 40.3). Keep the web moist with frog Ringer's solution.
   e. Secure the board on the stage of a microscope with heavy rubber bands, and position it so that the web is beneath the objective lens.

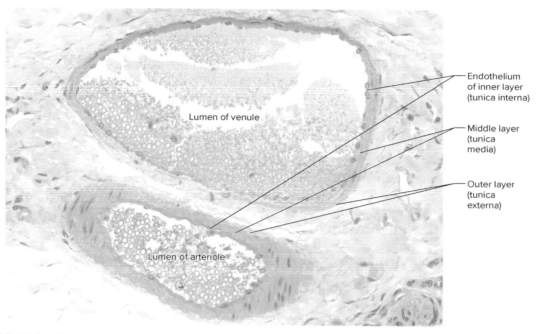

FIGURE 40.2   Cross section of a small artery (arteriole) and a small vein (venule) (200×). **APR** (Biophoto Associates/Science Source)

 LEARN: ACTIVITY

Investigate the effect of temperature change on the blood vessels of the frog's foot by flooding the web with a small quantity of ice water. Observe the blood vessels with low-power magnification and note any changes in their diameters or the rate of blood flow. Remove the ice water and replace it with water heated to about 35°C (95°F). Repeat your observations. What do you conclude from this experiment?

_____

_____

_____

    f. Focus on the web, using low-power magnification, and locate some blood vessels. Note the movement of the blood cells and the direction of the blood flow. You might notice that red blood cells of frogs are nucleated. Identify an arteriole, a capillary, and a venule.

    g. Examine each of these vessels with high-power magnification.

    h. When finished, return the frog to the location indicated by your instructor. The microscope lenses and stage will likely need cleaning after the experiment.

9. Complete Part C of the laboratory report.

 PRACTICE

PROCEDURE B—Paths of Circulation

1. Review section 15.6 titled "Paths of Circulation" in chapter 15 of the textbook.
2. As a review activity, label figure 40.4.
3. Locate the following blood vessels on the available anatomical charts, the dissectible human heart model, and the human torso model:

pulmonary circuit
    pulmonary trunk 1
    pulmonary arteries 2
    pulmonary veins 4

systemic circuit
    aorta 1
    superior vena cava 1
    inferior vena cava 1

 ASSESS

CRITICAL THINKING

Why is the left ventricle wall thicker than the right ventricle wall?

_____

_____

_____

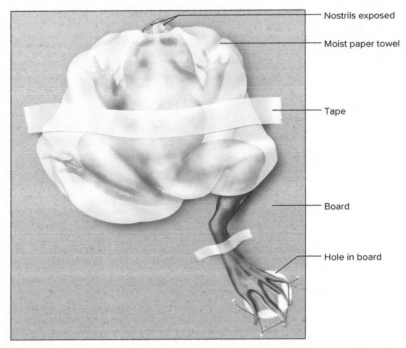

Nostrils exposed

Moist paper towel

Tape

Board

Hole in board

FIGURE 40.3 Spread the web of the foot over the hole and secure it to the board with pins and thread.

 PRACTICE

PROCEDURE C—Arterial System

1. Review section 15.7 titled "Arterial System" in chapter 15 of the textbook.
2. As a review activity, label and study figures 40.5, 40.6, 40.7, 40.8, and 40.9.
3. Locate the following arteries of the systemic circuit on the charts and human torso model:

aorta
   ascending aorta
   aortic arch (arch of aorta)
   thoracic aorta
   abdominal aorta
branches of the aorta
   coronary artery
   brachiocephalic trunk (artery)
   left common carotid artery
   left subclavian artery
   celiac trunk (artery)
   superior mesenteric artery
   renal artery
   inferior mesenteric artery
   common iliac artery
arteries to neck, head, and brain
   vertebral artery
   common carotid artery

   external carotid artery
   internal carotid artery
   facial artery
arteries of base of brain (fig. 40.7)
   vertebral artery
   basilar artery
   internal carotid artery
   cerebral arterial circle (circle of Willis)
arteries to shoulder and upper limb
   subclavian artery
   axillary artery
   brachial artery
   deep brachial artery
   ulnar artery
   radial artery
arteries to pelvis and lower limb
   common iliac artery
   internal iliac artery
   external iliac artery
   femoral artery
   popliteal artery
   anterior tibial artery
   dorsalis pedis artery (dorsal artery of foot)
   posterior tibial artery

4. Complete Part D of the laboratory report.

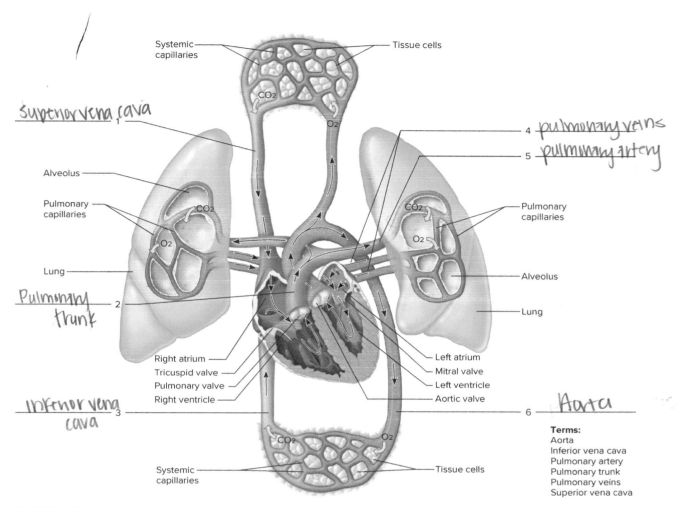

Systemic capillaries

Tissue cells

CO_2

O_2

Superior vena cava 1

Alveolus

Pulmonary capillaries

CO_2

O_2

Lung

Pulmonary trunk 2

4 pulmonary veins

5 pulmonary artery

CO_2

O_2

Pulmonary capillaries

Alveolus

Lung

Right atrium
Tricuspid valve
Pulmonary valve
Right ventricle

Left atrium
Mitral valve
Left ventricle
Aortic valve

Inferior vena cava 3

6 Aorta

CO_2

O_2

Systemic capillaries

Tissue cells

Terms:
Aorta
Inferior vena cava
Pulmonary artery
Pulmonary trunk
Pulmonary veins
Superior vena cava

FIGURE 40.4 Label the major blood vessels associated with the pulmonary and systemic circuits, using the terms provided. The red colors indicate locations of oxygen-rich (oxygenated) blood; the blue colors indicate locations of oxygen-poor (deoxygenated) blood. (Structures are not drawn to scale.) **3** **APR**

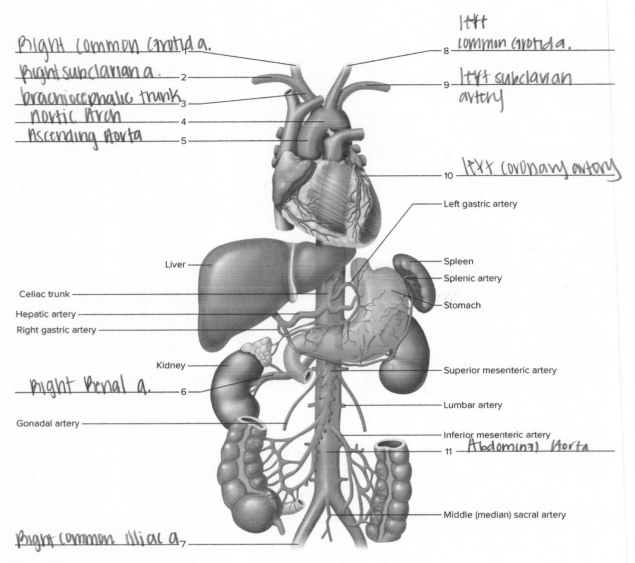

Right common Carotid a.
Right subclavian a.   2
brachiocephalic trunk   3
Aortic Arch   4
Ascending Aorta   5

left
common Carotid.   8
left subclavian   9
artery

left coronary artery   10

Left gastric artery

Liver

Celiac trunk

Hepatic artery

Right gastric artery

Kidney

Right Renal a.   6

Gonadal artery

Spleen
Splenic artery

Stomach

Superior mesenteric artery

Lumbar artery

Inferior mesenteric artery   Abdominal Aorta   11

Middle (median) sacral artery

Right common Illiac a.   7

FIGURE 40.5   Label the portions of the aorta and its principal branches. Ⓐ APR

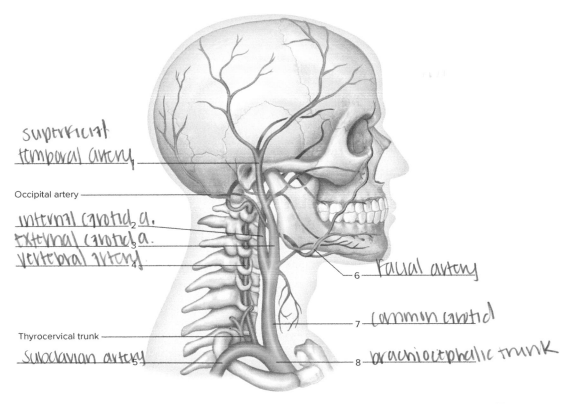

superficial
temporal artery
Occipital artery
internal carotid a.
external carotid a.
vertebral artery
facial artery
Thyrocervical trunk
common carotid
subclavian artery
brachiocephalic trunk

FIGURE 40.6   Label the arteries supplying the right side of the neck and head. (The clavicle has been removed.) **4 APR**

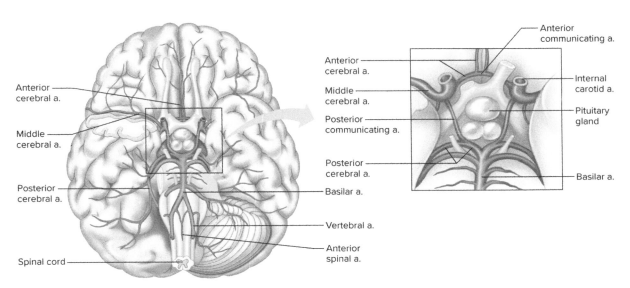

Anterior
cerebral a.

Middle
cerebral a.

Posterior
cerebral a.

Spinal cord

Anterior
cerebral a.

Middle
cerebral a.

Posterior
communicating a.

Posterior
cerebral a.

Basilar a.

Vertebral a.

Anterior
spinal a.

Anterior
communicating a.

Internal
carotid a.

Pituitary
gland

Basilar a.

FIGURE 40.7   View of inferior surface of the brain. The cerebral arterial circle (circle of Willis) is formed by the anterior and posterior cerebral arteries, which join the internal carotid arteries (a. stands for artery). **APR**

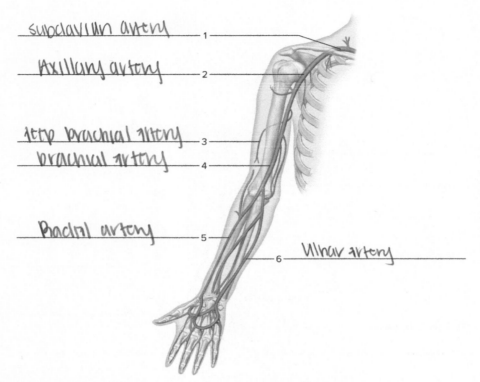

subclavian artery _____ 1 _____

Axillary artery _____ 2 _____

1ttp brachial artery __ 3 _____
brachial artery _____ 4 _____

Brachial artery _____ 5 _____

_____ 6 ____ Ulnar artery _____

FIGURE 40.8   Label the major arteries of the shoulder and upper limb.   APR

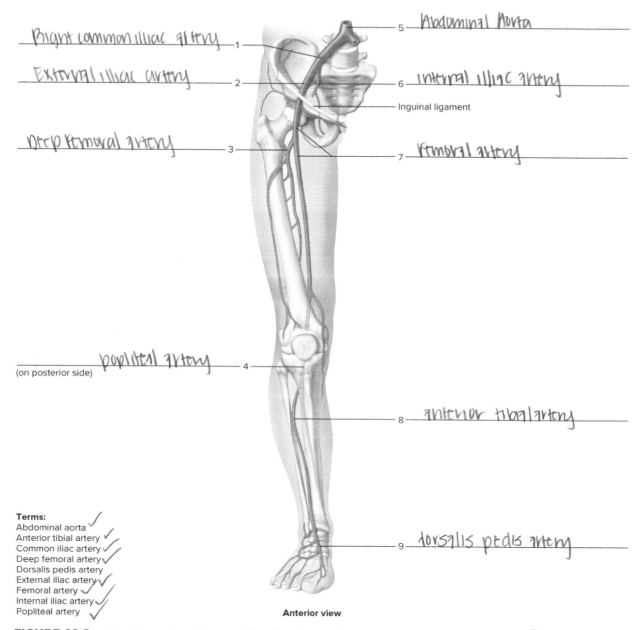

Right common illiac artery — 1

External illiac artery — 2

Deep femoral artery — 3

_____ popliteal artery — 4
(on posterior side)

5 — Abdominal Aorta

6 — internal illiac artery

— Inguinal ligament

7 — Femoral artery

8 — anterior tibial artery

9 — dorsalis pedis artery

Terms:
Abdominal aorta ✓
Anterior tibial artery ✓
Common iliac artery ✓
Deep femoral artery ✓
Dorsalis pedis artery
External iliac artery ✓
Femoral artery ✓
Internal iliac artery ✓
Popliteal artery ✓

Anterior view

FIGURE 40.9   Label the major arteries supplying the pelvis and lower limb, using the terms provided. **A** **APR**

PRACTICE

PROCEDURE D—Venous System

1. Review section 15.8 titled "Venous System" in chapter 15 of the textbook.
2. As a review activity, label figures 40.10, 40.11, 40.12, and 40.13.
3. Locate the following veins of the systemic circuit on the charts and human torso model:

veins from brain, head, and neck
    external jugular vein
    internal jugular vein
    vertebral vein
    subclavian vein
    brachiocephalic vein
    superior vena cava

veins from upper limb and shoulder
    radial vein
    ulnar vein
    brachial vein
    basilic vein
    cephalic vein
    median cubital vein (antecubital vein)

    axillary vein
    subclavian vein

veins of abdominal viscera
    hepatic portal vein
    gastric vein
    superior mesenteric vein
    splenic vein
    inferior mesenteric vein
    hepatic vein
    renal vein

veins from lower limb and pelvis
    anterior tibial vein
    posterior tibial vein
    popliteal vein
    femoral vein
    great (long) saphenous vein
    small (short) saphenous vein
    external iliac vein
    internal iliac vein
    common iliac vein
    inferior vena cava

4. Complete Parts E and F of the laboratory report.

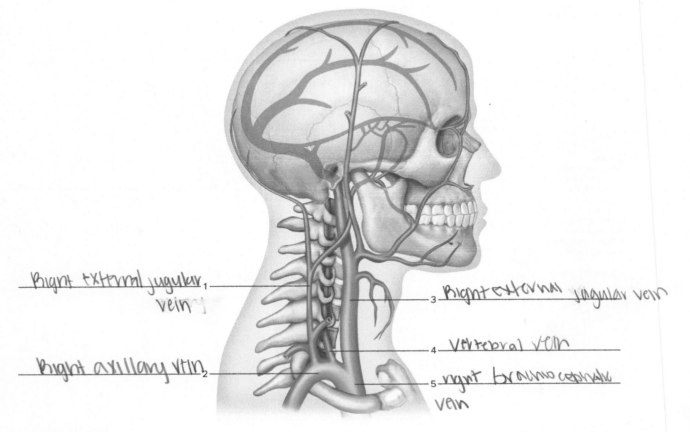

1. Right external jugular vein
2. Right axillary vein
3. Right external jugular vein
4. Vertebral vein
5. right brachiocephalic vein

FIGURE 40.10   Label the major veins associated with the head and neck. (The clavicle has been removed.) **APR**

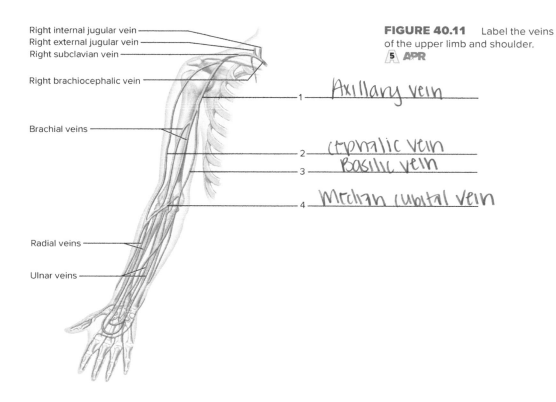

Right internal jugular vein
Right external jugular vein
Right subclavian vein
Right brachiocephalic vein
Brachial veins
Radial veins
Ulnar veins

FIGURE 40.11   Label the veins of the upper limb and shoulder. Ⓐ APR

1 _Axillary vein_
2 _cephalic vein_
3 _Basilic vein_
4 _Median cubital vein_

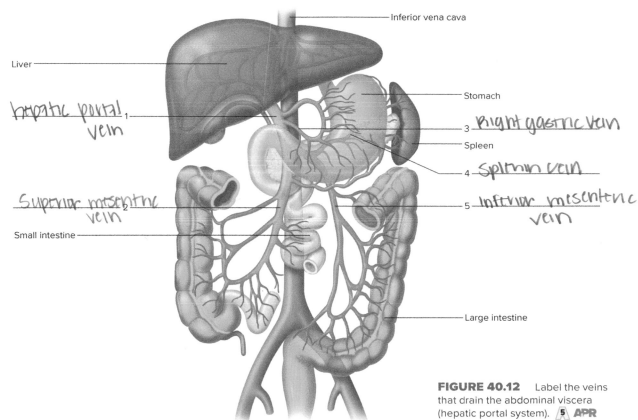

Inferior vena cava
Liver
Stomach
Right gastric vein — 3
Spleen
Splenic vein — 4
Inferior mesenteric vein — 5
hepatic portal vein — 1
Superior mesenteric vein — 2
Small intestine
Large intestine

FIGURE 40.12   Label the veins that drain the abdominal viscera (hepatic portal system). Ⓐ APR

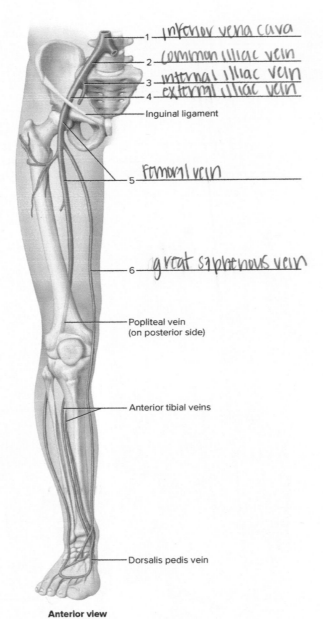

1 _Inferior vena cava_____

2 _common illiac vein_____

3 _internal illiac vein_____

4 _external illiac vein_____

— Inguinal ligament

5 _Femoral vein_____

6 _great saphenous vein_____

— Popliteal vein
(on posterior side)

— Anterior tibial veins

— Dorsalis pedis vein

Anterior view

FIGURE 40.13   Label the veins of the pelvis and lower limb. 5 APR

LABORATORY EXERCISE

41

Pulse Rate and Blood Pressure

MATERIALS NEEDED

Textbook
Clock with second hand
Sphygmomanometer
Stethoscope
Alcohol swabs

PURPOSE OF THE EXERCISE

To examine the pulse, to determine the pulse rate, to measure blood pressure, and to investigate the effects of body position and exercise on pulse rate and blood pressure.

 LEARNING OUTCOMES

After completing this exercise, you should be able to

(1) Determine pulse rate and correlate pulse characteristics.

(2) Test the effects of various factors on pulse rate.

(3) Describe and measure blood pressure using a sphygmomanometer.

(4) Test the effects of various factors on blood pressure.

The surge of blood that enters the arteries each time the ventricles of the heart contract causes the elastic walls of these vessels to swell. Then, as the ventricles relax, the walls recoil. This alternate expanding and recoiling of an arterial wall can be felt as a pulse in vessels that run close to the surface of the body.

The force exerted by the blood pressing against the inner walls of arteries also creates blood pressure. This pressure reaches a maximum during ventricular contraction and then drops to its lowest level while the ventricles are relaxed.

 PRACTICE

PROCEDURE A—Pulse Rate

1. Review section 15.5 titled "Blood Pres-sure" in chapter 15 of the textbook.

2. Complete Part A of Laboratory Report 41.
3. Examine your laboratory partner's radial pulse. To do this, follow these steps:
   a. Have your partner sit quietly, remaining as relaxed as possible.
   b. Locate the pulse by placing your index and middle fingers over the radial artery on the anterior surface near the lateral side of the wrist. Do not use your thumb for sensing the pulse because you may feel a pulse coming from an artery in the thumb. The radial artery is most often used because it is easy and convenient to locate.
   c. To determine the pulse rate, count the number of pulses that occur in 1 minute. This can be accomplished by counting pulses in 30 seconds and multiplying that number by 2. (A pulse count for a full minute, although taking a little longer, would give a better opportunity to detect pulse characteristics and irregularities.)
   d. Note the characteristics of the pulse. That is, can it be described as regular or irregular, strong or weak, hard or soft? The pulse should be regular and the amplitude (magnitude) will decrease as the distance from the left ventricle increases. The strength of the pulse reveals some indications of blood pressure. Under high blood pressure, the pulse feels very hard and strong; under low blood pressure, the pulse feels weak and can be easily compressed.
   e. Record the pulse rate and pulse characteristics while sitting in Part B of the laboratory report.
4. Repeat the procedure and determine the pulse rate and pulse characteristics in each of the following conditions:
   a. Immediately after lying down
   b. 3–5 minutes after lying down
   c. Immediately after standing
   d. 3–5 minutes after standing quietly
   e. Immediately after 3 minutes of moderate exercise *(omit if the person has health problems)*
   f. 3–5 minutes after exercise has ended
   Record the pulse rate and pulse characteristics in Part B of the laboratory report.
5. Complete Part B of the laboratory report.

⟳ LEARN: LAB IN MOTION

Determine the pulse rate and pulse characteristics in two additional locations on your laboratory partner. Locate and record the pulse rate from the common carotid artery and the dorsalis pedis artery. ⚠1

Common carotid artery pulse rate _____

Dorsalis pedis artery pulse rate _____

Compare the amplitude of the pulse characteristics between these two pulse locations and interpret any of the variations noted. _____

_____

_____

_____

_____

⟳ PRACTICE

PROCEDURE B—Blood Pressure

1. Although blood pressure is present in all blood vessels, the standard location to record blood pressure is the brachial artery.

2. Measure your laboratory partner's arterial blood pressure. To do this, follow these steps:

   a. Obtain a sphygmomanometer and a stethoscope.

   b. Clean the earpieces and the diaphragm of the stethoscope with alcohol swabs.

   c. Have your partner sit quietly with upper limb resting on a table at heart level. Have the person remain as relaxed as possible.

   d. Locate the brachial artery at the antecubital space. Wrap the cuff of the sphygmomanometer around the arm so that its lower border is about 2.5 cm above the bend of the elbow. Center the bladder of the cuff in line with the *brachial pulse* (fig. 41.1).

   e. Palpate the *radial pulse.* Close the valve on the neck of the rubber bulb connected to the cuff, and pump air from the bulb into the cuff. Inflate the cuff while watching the sphygmomanometer, and note the pressure when the pulse disappears. (This is a rough estimate of the systolic pressure.) Immediately deflate the cuff. *Do not leave the cuff inflated for more than 1 minute.*

   f. Position the stethoscope over the brachial artery. Reinflate the cuff to a level 30 mm Hg higher than the point where the pulse disappeared during palpation.

   g. Slowly open the valve of the bulb until the pressure in the cuff drops at a rate of about 2 or 3 mm Hg per second.

   h. Listen for sounds (Korotkoff sounds) from the brachial artery. When the first loud tapping sound is heard, record the reading as the systolic pressure.

This indicates the pressure exerted against the arterial wall during systole.

   i. Continue to listen to the sounds as the pressure drops, and note the level when the last sound is heard. Record this reading as the diastolic pressure, which measures the constant arterial resistance.

   j. Release all of the pressure from the cuff.

   k. Repeat the procedure until you have two blood pressure measurements from each arm, allowing 2–3 minutes of rest between readings.

   l. Average your readings and enter them in the table in Part C of the laboratory report.

3. Measure and record your partner's blood pressure in each of the following conditions:

   a. 3–5 minutes after lying down

   b. 3–5 minutes after standing quietly

   c. Immediately after 3 minutes of moderate exercise *(omit if the person has health problems)*

   d. 3–5 minutes after exercise has ended

   Record the blood pressures in Part C of the laboratory report.

4. Complete Part C of the laboratory report.

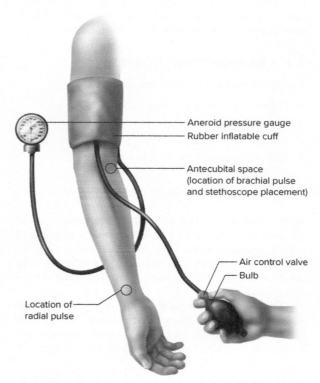

FIGURE 41.1 Blood pressure is commonly measured by using a sphygmomanometer (blood pressure cuff). The use of a column of mercury is the most accurate measurement, but due to environmental concerns, it has been replaced by alternative gauges and digital readouts.

Labels: Aneroid pressure gauge; Rubber inflatable cuff; Antecubital space (location of brachial pulse and stethoscope placement); Air control valve; Bulb; Location of radial pulse

Lymphatic System

LABORATORY EXERCISE

42

Lymphatic System

MATERIALS NEEDED

Textbook
Human torso model
Anatomical chart of the lymphatic system
Compound light microscope
Prepared microscope slides:
    Lymph node section
    Human thymus section
    Human spleen section

PURPOSE OF THE EXERCISE

To review the structure of the lymphatic system and to observe the microscopic structure of a lymph node, the thymus, and the spleen.

 LEARNING OUTCOMES APR

After completing this exercise, you should be able to

1. Trace the major lymphatic pathways in an anatomical chart or model.

2. Locate and identify the major clusters of lymph nodes (lymph glands) in an anatomical chart or a model.

3. Describe the structure and function of a lymphatic vessel, a lymph node, the thymus, and the spleen.

4. Locate and sketch the major microscopic structures of a lymph node, the thymus, and the spleen.

The lymphatic system is a vast collection of cells and biochemicals that travel in lymphatic capillaries and vessels and the organs and glands that produce them. The system is closely associated with the cardiovascular system and includes a network of vessels that assist in the circulation of body fluids. These vessels provide pathways through which excess fluid can be transported away from interstitial spaces within most tissues and returned to the bloodstream. Without the lymphatic system, this fluid would accumulate in tissue spaces, producing edema.

The organs of the lymphatic system also help defend the tissues against infections by filtering particles from lymph and by supporting the activities of lymphocytes that furnish immunity against specific disease-causing agents or pathogens.

 PRACTICE

PROCEDURE A—Lymphatic Pathways APR

1. Review section 16.2 titled "Lymphatic Pathways" in chapter 16 of the textbook.
2. As a review activity, label figure 42.1.
3. Complete Part A of Laboratory Report 42.
4. Observe the human torso model and the anatomical chart of the lymphatic system, and locate the following features:

lymphatic vessels
lymph nodes
lymphatic trunks
    lumbar trunk
    intestinal trunk
    intercostal trunk
    bronchomediastinal trunk
    subclavian trunk
    jugular trunk
cisterna chyli
collecting ducts
    thoracic (left lymphatic) duct
    right lymphatic duct
internal jugular veins
subclavian veins

 PRACTICE

PROCEDURE B—Lymph Nodes

1. Review the concept heading titled "Lymph Nodes" in section 16.4 in chapter 16 of the textbook.
2. As a review activity, label figure 42.2.
3. Complete Part B of the laboratory report.

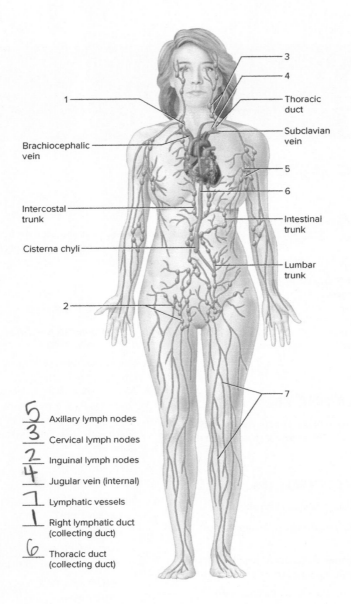

FIGURE 42.1   Label the diagram by placing the correct numbers in the spaces provided.

5   Axillary lymph nodes
3   Cervical lymph nodes
2   Inguinal lymph nodes
4   Jugular vein (internal)
7   Lymphatic vessels
1   Right lymphatic duct (collecting duct)
6   Thoracic duct (collecting duct)

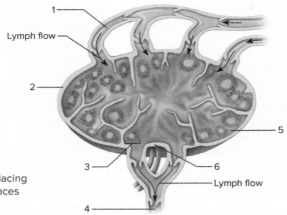

FIGURE 42.2   Label this diagram of a lymph node by placing the correct numbers in the spaces provided.

1   Afferent lymphatic vessel
2   Capsule
4   Efferent lymphatic vessel
6   Hilum
3   Lymphatic nodule
5   Lymphatic sinus

4. Observe the anatomical chart of the lymphatic system and the human torso model, and locate the clusters of lymph nodes in the following regions:

cervical region
axillary region
inguinal region
pelvic cavity
abdominal cavity
thoracic cavity

5. Palpate the lymph nodes in your cervical region. They are located along the lower border of the mandible and between the ramus of the mandible and the sternocleidomastoid muscle. They feel like small, firm lumps.

6. Observe a human lymph node in figure 42.3*a*.

7. Obtain a prepared microscope slide of a lymph node and observe it, using low-power magnification (fig. 42.3*b*). Identify the *capsule* that surrounds the node and is mainly composed of collagen fibers, the *lymphatic nodules* that appear as dense masses near the surface of the node, and the *lymphatic sinuses* that appear as narrow spaces where lymph circulates between the nodules and the capsule.

8. Using high-power magnification, examine a nodule within the lymph node. The nodule contains densely packed *lymphocytes.*

9. Prepare a labeled sketch of a representative section of a lymph node in Part D of the laboratory report.

 PRACTICE

PROCEDURE C—Thymus and Spleen

1. Review the concept headings titled "Thymus" and "Spleen" in section 16.4 in chapter 16 of the textbook.

2. Locate the thymus and spleen in the anatomical chart of the lymphatic system and on the human torso model.

3. Complete Part C of the laboratory report.

4. Obtain a prepared microscope slide of human thymus and observe it, using low-power magnification (fig. 42.4). Note how the thymus is subdivided into *lobules* by *septa* of connective tissue that contain blood vessels. Identify the *capsule* of loose connective tissue that surrounds the thymus, the outer *cortex* of a lobule composed of densely packed cells and deeply stained, and the inner *medulla* of a lobule composed of loosely packed *lymphocytes* and epithelial cells and lightly stained.

5. Examine the cortex tissue of a lobule using high-power magnification. The cells of the cortex are composed of densely packed lymphocytes among some epithelial cells and *macrophages.* Some of these cortical cells may be undergoing mitosis, so their chromosomes may be visible.

6. Prepare a labeled sketch of a representative section of the thymus in Part D of the laboratory report.

7. Obtain a prepared slide of the human spleen and observe it, using low-power magnification (fig. 42.5).

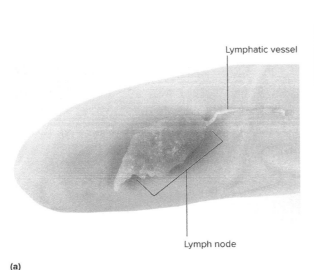

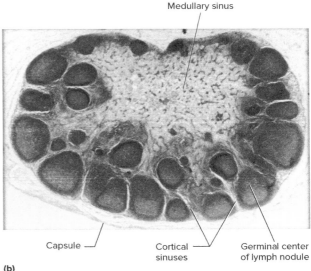

Lymphatic vessel

Lymph node

(a)

Medullary sinus

Capsule

Cortical sinuses

Germinal center of lymph nodule

(b)

FIGURE 42.3 Lymph nodes: (*a*) photograph of a human lymph node on tip of finger; (*b*) micrograph of a lymph node (20×).
APR ((*a*) J and J Photography; (*b*) McGraw-Hill Education/Greg Reeder)

Identify the *capsule* of dense connective tissue that surrounds the spleen. The tissues of the spleen include circular *nodules* of *white pulp* enclosed in a matrix of *red pulp.*

8. Using high-power magnification, examine a nodule of white pulp and red pulp. The cells of the white pulp are mainly *lymphocytes.* Also, there may be an arteriole centrally located in the nodule. The cells of the red pulp are mostly red blood cells with many lymphocytes and macrophages. The macrophages engulf and destroy cellular debris and bacteria.

9. Prepare a labeled sketch of a representative section of the spleen in Part D of the laboratory report.

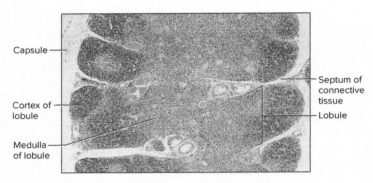

FIGURE 42.4   Micrograph of a section of the thymus (15×).   **APR** (Biophoto Associates/Science Source)

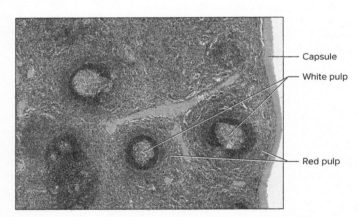

FIGURE 42.5   Micrograph of a section of the spleen (40×).   **APR** (Biophoto Associates/Science Source)

Digestive System

LABORATORY EXERCISE

43

Digestive Organs

MATERIALS NEEDED

Textbook
Human torso model
Head model, sagittal section
Skull with teeth
Teeth, sectioned
Tooth model, sectioned
Paper cup
Compound light microscope
Prepared microscope slides of the following:
    Salivary gland
    Esophagus
    Stomach (fundus)
    Pancreas (exocrine portion)
    Liver
    Small intestine (jejunum)
    Large intestine

PURPOSE OF THE EXERCISE APR

To review the structure and function of the digestive organs
and to examine the tissues of these organs.

 LEARNING OUTCOMES APR

After completing this exercise, you should be able to

1. Locate and label the major digestive organs and their
major structures.

2. Describe the functions of these organs.

3. Examine and sketch the structures of a section of the
small intestine.

The digestive system includes the organs associated with
the alimentary canal and several accessory structures.
The alimentary canal, a muscular tube, passes through the
body from the opening of the mouth to the anus. It includes
the mouth, pharynx, esophagus, stomach, small intestine, and
large intestine. The function of the canal is to move substances
throughout its length. It is specialized in various regions to
store, digest, and absorb food materials and to eliminate the
residues. The accessory organs, which include the salivary

glands, liver, gallbladder, and pancreas, secrete products into
the alimentary canal that aid digestive functions.

 PRACTICE

PROCEDURE A—Mouth and Salivary Glands

1. Review sections 17.2 and 17.3 titled "Mouth" and "Salivary Glands" in chapter 17 of the textbook.
2. As a review activity, label figures 43.1–43.3.
3. Examine the head model (sagittal section) and a skull.
Locate the following structures:
mouth
oral cavity
vestibule
tongue
    lingual frenulum
    papillae
palate
    hard palate
    soft palate
    uvula
palatine tonsils
gums (gingivae)
teeth
    incisors
    canines (cuspids)
    premolars (bicuspids)
    molars

4. Examine a sectioned tooth and a tooth model. Locate
the following features:
crown
    enamel
    dentin
neck
root
    pulp cavity
    cementum
    root canal

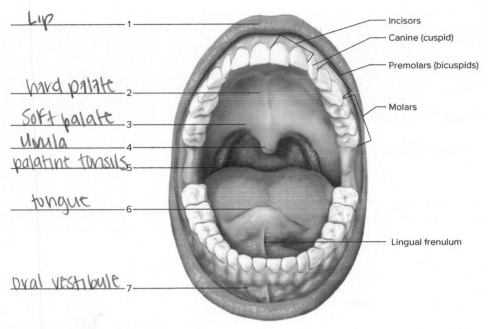

Lip — 1

Incisors

Canine (cuspid)

Premolars (bicuspids)

hard palate — 2

Molars

soft palate — 3

Uvula — 4

palatine tonsils — 5

tongue — 6

Lingual frenulum

oral vestibule — 7

FIGURE 43.1   Label the features of the mouth.

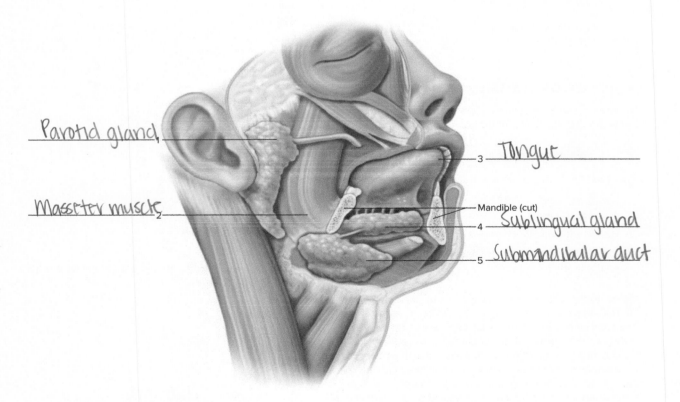

Parotid gland — 1

Tongue — 3

Masseter muscle — 2

Mandible (cut)

Sublingual gland — 4

Submandibular duct — 5

FIGURE 43.2   Label the features associated with the major salivary glands.   APR

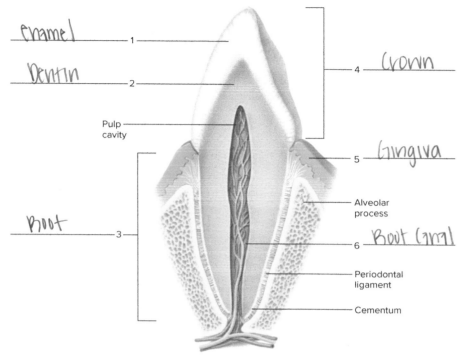

Labels (handwritten):
- enamel — 1
- Dentin — 2
- 4 — Crown
- Pulp cavity
- 5 — Gingiva
- Alveolar process
- Root — 3
- 6 — Root (gm)
- Periodontal ligament
- Cementum

FIGURE 43.3 Label the features of this canine (cuspid) tooth. Ⓐ

---

5. Observe the head of the human torso model, and locate the following:

   parotid salivary gland
   parotid duct (Stensen's duct)
   submandibular salivary gland
   submandibular duct (Wharton's duct)
   sublingual salivary gland
   sublingual ducts (10–12)

6. Examine a microscopic section of a salivary gland, using low- and high-power magnification. Note that the mucous cells that produce mucus and the serous cells that produce enzymes are clustered around small ducts. Also note a larger secretory duct surrounded by lightly stained cuboidal epithelial cells (fig. 43.4).

7. Complete Part A of Laboratory Report 43.

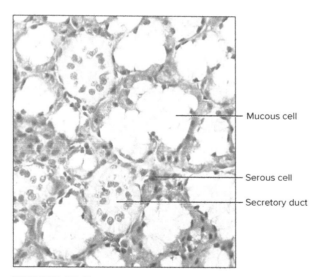

- Mucous cell
- Serous cell
- Secretory duct

FIGURE 43.4 Micrograph of the salivary gland (300×). (Biophoto Associates/Science Source)

Flower-like lots of little circles
Clustered & small

PRACTICE

PROCEDURE B—Pharynx and Esophagus

1. Review section 17.4 titled "Pharynx and Esophagus" in chapter 17 of the textbook.
2. As a review activity, label figure 43.5.
3. Observe the human torso model, and locate the following features:

pharynx
  nasopharynx
  oropharynx
  laryngopharynx
  constrictor muscles
epiglottis
esophagus
lower esophageal sphincter (cardiac sphincter)

4. Have your laboratory partner take a swallow from a cup of water. Carefully watch the movements in the anterior region of the neck. What steps in the swallowing process did you observe?

_____

_____

_____

5. Examine a microscopic section of esophagus wall, using low-power magnification (fig. 43.6). The inner lining is composed of stratified squamous epithelium, and there are layers of muscle tissue in the wall. Locate some mucous glands in the submucosa. They appear as clusters of lightly stained cells.
6. Complete Part B of the laboratory report.

tongue: Innes & groove w/ (steak) duts

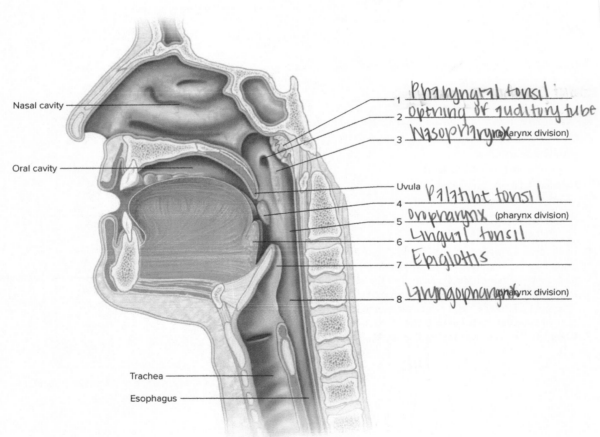

Nasal cavity

Oral cavity

Uvula

Trachea

Esophagus

1 *Pharyngeal tonsil*
2 *Opening of auditory tube*
3 *Nasopharynx* (pharynx division)

4 *Palatine tonsil*
5 *Oropharynx* (pharynx division)
6 *Lingual tonsil*
7 *Epiglottis*
8 *Laryngopharynx* (pharynx division)

FIGURE 43.5   Label the features associated with the pharynx.   **APR**

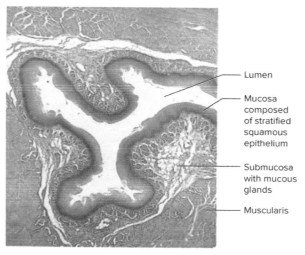

FIGURE 43.6   Micrograph of a cross section of the esophagus (10×). **APR** (Ed Reschke)

Lumen

Mucosa composed of stratified squamous epithelium

Submucosa with mucous glands

Muscularis

 **PRACTICE**

PROCEDURE C—Stomach

1. Review section 17.5 entitled "Stomach" in chapter 17 of the textbook.
2. As a review activity, label figure 43.7.
3. Observe the human torso model, and locate the following features of the stomach:

cardia
fundus
body
pylorus
   pyloric antrum
   pyloric canal
pyloric sphincter (valve)
lesser curvature
greater curvature
gastric folds (rugae)

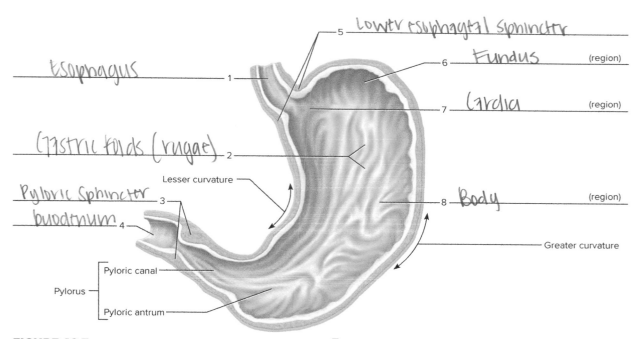

5 — *Lower esophageal sphincter*
6 — *Fundus* (region)
7 — *Cardia* (region)
8 — *Body* (region)

1 — *esophagus*
2 — *Gastric folds (rugae)*
3 — *Pyloric sphincter*
4 — *duodenum*

Lesser curvature
Greater curvature
Pyloric canal
Pylorus
Pyloric antrum

FIGURE 43.7   Label the stomach and associated structures. **A APR**

4. Examine a microscopic section of the stomach wall, using low-power magnification. Note how the inner lining of simple columnar epithelium dips inward to form gastric pits. The gastric glands are tubular structures that open into the gastric pits. Near the deep ends of these glands, you should be able to locate some intensely stained (bluish) chief cells and some lightly stained (pinkish) parietal cells (fig. 43.8). What are the functions of these cells?

_____

_____

_____

_____

5. Complete Part C of the laboratory report.

 PRACTICE _____

PROCEDURE D—Pancreas and Liver

1. Review sections 17.6, 17.7, and 17.8 titled "Pancreas," "Liver and Gallbladder," and "Small Intestine" in chapter 17 of the textbook.
2. As a review activity, label figure 43.9.
3. Observe the human torso model, and locate the following structures:

pancreas
pancreatic duct

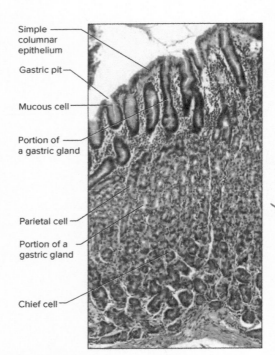

FIGURE 43.8   Micrograph of the mucosa of the stomach wall (60×). **APR** (Ed Reschke)

Simple columnar epithelium
Gastric pit
Mucous cell
Portion of a gastric gland
Parietal cell
Portion of a gastric gland
Chief cell

liver
    right lobe
    quadrate lobe
    caudate lobe
    left lobe
falciform ligament
gallbladder
hepatic ducts
common hepatic duct
cystic duct
bile duct (common bile duct)
hepatopancreatic sphincter

4. Examine the pancreas slide, using low-power magnification. Observe the exocrine (acinar) cells that secrete pancreatic juice (fig. 43.10).
5. Examine the liver slide, using low-power magnification. Observe the hepatocytes within the hepatic lobules. Each lobule has a central vein and numerous hepatic sinusoids. A hepatic triad would possess a small bile duct and branches of a hepatic artery and a hepatic portal vein (fig. 43.11).

 PRACTICE _____

PROCEDURE E—Small and Large Intestines

1. Review sections 17.8 and 17.9 titled "Small Intestine" and "Large Intestine" in chapter 17 of the textbook.
2. As a review activity, label figure 43.12 and examine figure 43.13.
3. Observe the human torso model, and locate each of the following features:

small intestine
    duodenum
    jejunum
    ileum
mesentery
ileocecal sphincter (valve)
large intestine
    large intestinal wall
        haustra
        taeniae coli
        epiploic (fatty) appendages
    cecum
    appendix (vermiform appendix)
    ascending colon
    right colic (hepatic) flexure
    transverse colon
    left colic (splenic) flexure
    descending colon
    sigmoid colon
    rectum

[handwritten margin note: 4 layers / grooves / MJC]

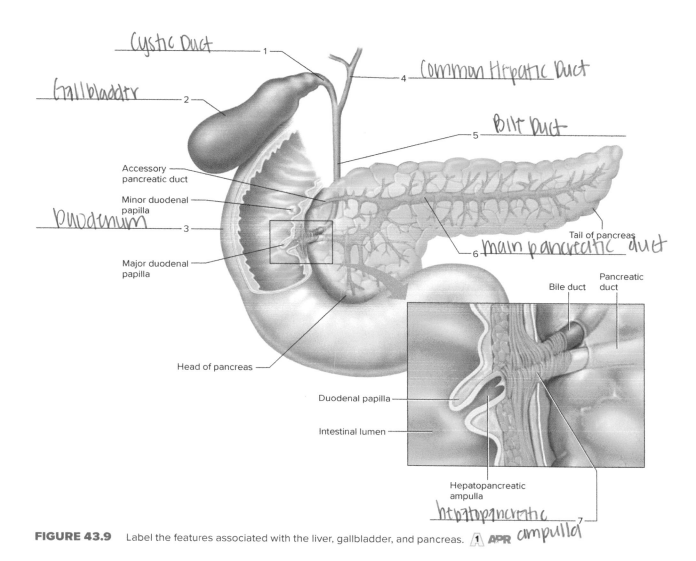

Cystic Duct — 1

Gallbladder — 2

Common Hepatic Duct — 4

Bile Duct — 5

Accessory pancreatic duct

Minor duodenal papilla

Duodenum — 3

Major duodenal papilla

Tail of pancreas

main pancreatic duct — 6

Head of pancreas

Bile duct

Pancreatic duct

Duodenal papilla

Intestinal lumen

Hepatopancreatic ampulla

hepatopancreatic ampulla — 7

FIGURE 43.9 Label the features associated with the liver, gallbladder, and pancreas. Ⓐ APR

anal canal

anal columns

anal sphincter muscles

external anal sphincter

internal anal sphincter

anus

4. Using low-power magnification, examine a microscopic section of the small intestine wall. Identify the mucosa, submucosa, muscularis, and serosa. Note the villi that extend into the lumen of the tube. Study a single villus, using high-power magnification. Note the core of connective tissue and the covering of simple columnar epithelium that contains some lightly stained goblet cells (fig. 43.14). What is the function of these villi?

_____

_____

_____

5. Prepare a labeled sketch of the wall of the small intestine in Part D of the laboratory report.

6. Examine a microscopic section of large intestine wall. Note the lack of villi. Also note the tubular mucous glands that open on the surface of the inner lining and the numerous lightly stained goblet cells. Locate the four layers of the wall (fig. 43.15). What is the function of the mucus secreted by these glands?

_____

_____

_____

7. Complete Part E of the laboratory report.

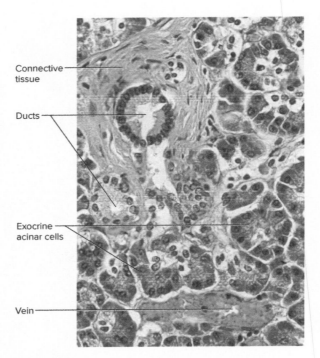

Connective tissue

Ducts

Exocrine acinar cells

Vein

FIGURE 43.10   Micrograph of exocrine portion of pancreas showing acinar cells and small pancreatic ducts (200×). The acinar cells secrete an alkaline pancreatic juice into small ducts that converge into the larger pancreatic duct. (Dennis Strete/McGraw-Hill Education)

 **ASSESS**

CRITICAL THINKING

How is the structure of the small intestine better adapted for absorption than that of the large intestine? ②

_____

_____

_____

_____

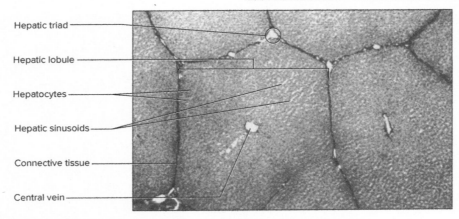

Hepatic triad

Hepatic lobule

Hepatocytes

Hepatic sinusoids

Connective tissue

Central vein

FIGURE 43.11   Micrograph of hepatic lobules of the liver (40×). (Victor P. Eroschenko)

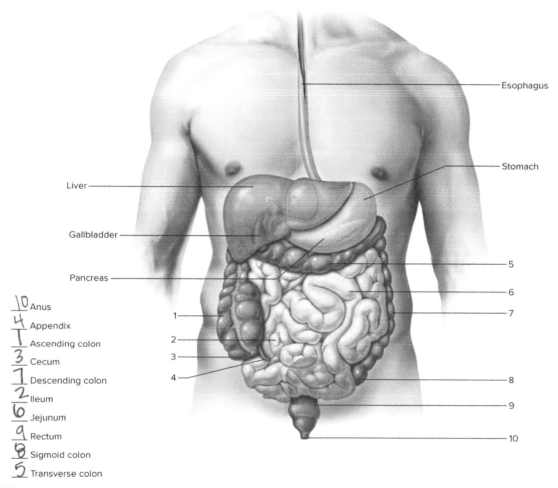

Esophagus

Stomach

Liver

Gallbladder

Pancreas

5

6

7

1

2

3

4

8

9

10

10 Anus
4 Appendix
1 Ascending colon
3 Cecum
7 Descending colon
2 Ileum
6 Jejunum
9 Rectum
8 Sigmoid colon
5 Transverse colon

FIGURE 43.12   Label this diagram of the small and large intestines by placing the correct numbers in the spaces provided.
🅐 APR

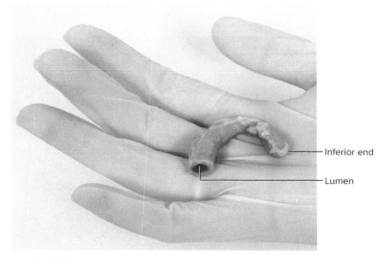

Inferior end

Lumen

FIGURE 43.13   Normal appendix. APR (© J & J Photography)

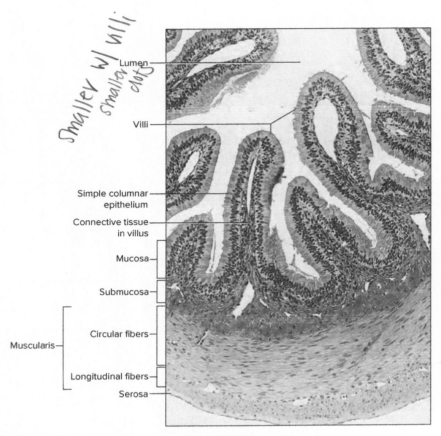

Smaller w/ villi
smaller dots

FIGURE 43.14 Micrograph of the small intestine wall (40×). **APR** (©Ed Reschke)

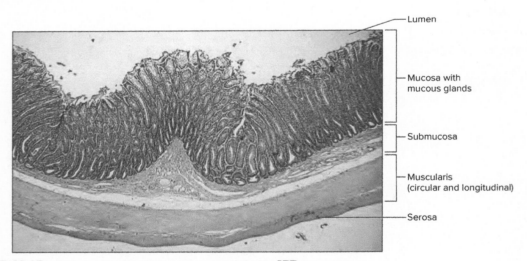

FIGURE 43.15 Micrograph of the large intestine wall (64×). **APR** (©Dennis Strete)

goblet cells
bigger cells

LABORATORY EXERCISE

44

Action of a Digestive Enzyme

MATERIALS NEEDED

0.5% amylase solution*
Beakers (50 and 500 mL)
Distilled water
Funnel
Pipets (1 and 10 mL)
Pipet rubber bulbs
0.5% starch solution
Graduated cylinder (10 mL)
Test tubes
Test-tube clamps

Wax marker
Iodine-potassium-iodide
    solution
Medicine dropper
Ice
Water bath, 37°C (98.6°F)
Porcelain test plate
Benedict's solution
Hot plates
Test-tube rack
Thermometer

For Alternative Procedure:
Small, disposable cups
Distilled water

*The amylase should be free of sugar for best results; a low-maltose solution of amylase yields good results.

⚠ SAFETY

- Review the Laboratory Safety Guidelines in Appendix 1.
- Use only a mechanical pipetting device (never your mouth). Use pipets with rubber bulbs or dropping pipets.
- Wear safety glasses when working with acids and when heating test tubes.
- Use test-tube clamps when handling hot test tubes.
- If an open flame is used for heating the test solutions, keep clothes and hair away from the flame.
- Dispose of chemicals according to appropriate directions.
- Use appropriate disinfectant to wash laboratory tables before and after the procedures.
- If student saliva is used as a source of amylase, it is important that students wear disposable gloves and handle only their own materials.
- Wash your hands before leaving the laboratory.

PURPOSE OF THE EXERCISE

To investigate the action of amylase and the effect of heat on its enzymatic activity.

 LEARNING OUTCOMES APR

After completing this exercise, you should be able to

1. Test a solution for the presence of starch or the presence of sugar.
2. Explain the action of amylase.
3. Test the effects of varying temperatures on the activity of amylase.

The digestive enzyme in salivary secretions is called *salivary amylase*. Pancreatic amylase is secreted among several other pancreatic enzymes. A bacterial extraction of amylase is available for laboratory experiments. This enzyme catalyzes the reaction of changing starch molecules into sugar (disaccharide) molecules, which is the first step in the digestion of complex carbohydrates.

As in the case of other enzymes, amylase is a protein catalyst. Its activity is affected by exposure to certain environmental factors, including various temperatures, pH, radiation, and electricity. As temperatures increase, faster chemical reactions occur as the collisions of molecules happen at a greater frequency. Eventually, temperatures increase to a point that the enzyme is denatured and the rate of the enzyme activity rapidly declines. As temperatures decrease, enzyme activity also decreases due to fewer collisions of the molecules; however, the colder temperatures do not denature the enzyme. Normal body temperature provides an environment for enzyme activity near the optimum for enzymatic reactions.

The pH of the environment where enzymes are secreted also has a major influence on the enzyme reactions. The optimum pH for amylase activity is between 6.8 and 7.0, typical of salivary secretions. When salivary amylase arrives in the stomach, hydrochloric acid deactivates the enzyme, diminishing any further chemical digestion of remaining starch in the stomach. Pancreatic amylase is secreted into the small intestine, where an optimum pH for amylase is once more provided. Other enzymes in the digestive system have different optimum pH ranges of activity compared to amylase. For example, pepsin from stomach secretions has an optimum activity around pH 2, whereas trypsin from pancreatic secretions operates best around pH 7–8.

PRACTICE

PROCEDURE A—Amylase Activity

1. Study the lock-and-key model of amylase action on starch digestion (fig. 44.1).
2. Examine the digestive locations where starch is converted to glucose (fig. 44.2).
3. Mark three clean test tubes as *tubes 1, 2,* and *3,* and prepare the tubes as follows (fig. 44.3):

   Tube 1: **Add 6 mL of amylase solution.**

   Tube 2: **Add 6 mL of starch solution.**

   Tube 3: **Add 5 mL of starch solution and 1 mL of amylase solution.**

ALTERNATIVE PROCEDURE

Human saliva can be used as a source of amylase solutions instead of bacterial amylase preparations. Collect about 5 mL of saliva in a small, disposable cup. Add an equal amount of distilled water and mix together for the amylase solutions during the laboratory procedures. Be sure to follow all of the safety guidelines.

4. Shake the tubes well to mix the contents, and place them in a warm water bath, 37°C (98.6°F), for 10 minutes.
5. At the end of the 10 minutes, test the contents of each tube for the presence of starch. To do this, follow these steps:
   a. Place 1 mL of the solution to be tested in a depression of a porcelain test plate.
   b. Next add 1 drop of iodine-potassium-iodide solution, and note the color of the mixture. If the solution becomes blue-black, starch is present.
   c. Record the results in Part A of Laboratory Report 44.
6. Test the contents of each tube for the presence of sugar (disaccharides, in this instance). To do this, follow these steps:
   a. Place 1 mL of the solution to be tested in a clean test tube.
   b. Add 1 mL of Benedict's solution.
   c. Place the test tube with a test-tube clamp in a beaker of boiling water for 2 minutes.
   d. Note the color of the liquid. If the solution becomes green, yellow, orange, or red, sugar is present. Blue indicates a negative test, whereas green indicates a

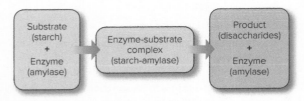

FIGURE 44.1   Lock-and-key model of enzyme action of amylase on starch digestion. **APR**

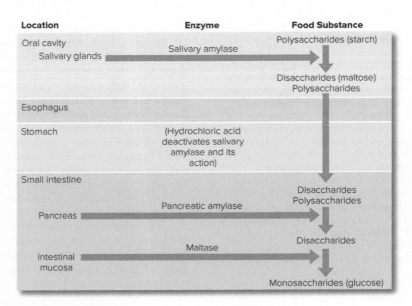

FIGURE 44.2   Flowchart of starch digestion.

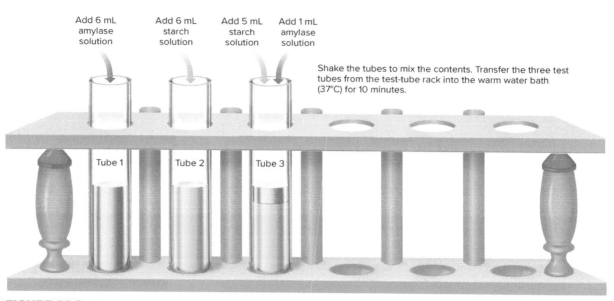

Add 6 mL amylase solution

Add 6 mL starch solution

Add 5 mL starch solution

Add 1 mL amylase solution

Shake the tubes to mix the contents. Transfer the three test tubes from the test-tube rack into the warm water bath (37°C) for 10 minutes.

Tube 1

Tube 2

Tube 3

FIGURE 44.3 Test tubes prepared for testing amylase activity.

positive test with the least amount of sugar, and red indicates the greatest amount of sugar present.

    e. Record the results in Part A of the laboratory report.

7. Complete Part A of the laboratory report.

 PRACTICE

PROCEDURE B—Effect of Heat

1. Mark three clean test tubes as *tubes 4, 5,* and *6.*
2. Add 1 mL of amylase solution to each of the tubes, and expose each solution to a different test temperature for 3 minutes as follows:

    Tube 4: **Place in beaker of ice water (about 0°C [32°F]).**

    Tube 5: **Place in warm water bath (about 37°C [98.6°F]).**

    Tube 6: **Place in beaker of boiling water (about 100°C [212°F]). Use a test-tube clamp.**

3. *It is important that the 5 mL of starch solution added to tube 4 be at ice-water temperature before it is added to the 1 mL of amylase solution.* Add 5 mL of starch solution to each tube, shake to mix the contents, and return the tubes to their respective test temperatures for 10 minutes.

4. At the end of the 10 minutes, test the contents of each tube for the presence of starch and the presence of sugar by following the same directions as in steps 5 and 6 of Procedure A.
5. Complete Part B of the laboratory report.

 LEARN: LAB IN MOTION

Devise an experiment to test the effect of some other environmental factor on amylase activity. For example, you might test the effect of a strong acid by adding a few drops of concentrated hydrochloric acid to a mixture of starch and amylase solutions. Be sure to include a control in your experimental plan. That is, include a tube containing everything except the factor you are testing. Then you will have something with which to compare your results. *Carry out your experiment only if it has been approved by the laboratory instructor.*

Notes

Name _____

Date _____

Section _____

The A corresponds to the indicated Learning Outcome(s) found at the beginning of the laboratory exercise.

Action of a Digestive Enzyme

PART A ASSESSMENTS

1. Test results: A 1

Tube	Starch	Sugar
1 Amylase solution		
2 Starch solution		
3 Starch-amylase solution		

2. Complete the following:

   a. Explain the reason for including tube 1 in this experiment. A 2 _____

_____

_____

_____

   b. What is the importance of tube 2? A 2 _____

_____

_____

_____

   c. What do you conclude from the results of this experiment? A 2 _____

_____

_____

_____

PART B ASSESSMENTS

1. Test results: △3

Tube	Starch	Sugar
4 0°C (32°F)		
5 37°C (98.6°F)		
6 100°C (212°F)		

2. Complete the following:

   a. What do you conclude from the results of this experiment? △3 _____

   _____

   _____

   _____

   _____

   _____

   b. If digestion failed to occur in one of the tubes in this experiment, how can you tell whether the amylase was destroyed by the factor being tested or the amylase activity was simply inhibited by the test treatment? △3 _____

   _____

   _____

   _____

   _____

   _____

ASSESS

CRITICAL THINKING

What test result would occur if the amylase you used contained sugar? _____
Would your results be as valid? Explain your answer. △2

_____

_____

_____

_____

Respiratory System

LABORATORY EXERCISE

45

Respiratory Organs

MATERIALS NEEDED

Textbook
Human skull (sagittal section)
Human torso model
Larynx model
Thoracic organs model
Compound light microscope
Prepared microscope slides of the following:
    Trachea (cross section)
    Lung, human (normal)
Animal lung with trachea (fresh or preserved)
Bicycle pump
Prepared microscope slides of the following:
    Lung tissue (smoker)
    Lung tissue (emphysema)

SAFETY

- Wear disposable gloves when working on the fresh or preserved animal lung demonstration.
- Wash your hands before leaving the laboratory.

PURPOSE OF THE EXERCISE

To review the structure and function of the respiratory organs and to examine the tissues of some of these organs.

 ## LEARNING OUTCOMES APR

After completing this exercise, you should be able to

① Locate the major organs and structural features of the respiratory system.

② Describe the functions of these organs.

③ Sketch and label features of tissue sections of the trachea and lung.

The organs of the respiratory system include the nose, nasal cavity, paranasal sinuses, pharynx, larynx, trachea, bronchial tree, and lungs. They mainly process incoming air and transport it to and from the atmosphere outside the body and the air sacs of the lungs.

In the alveoli (air sacs), gas exchanges take place between the air and the blood of nearby capillaries. The blood, in turn, transports gases to and from the alveoli and the body cells. This entire process of transporting and exchanging gases between the atmosphere and the body cells is called *respiration*.

 ## PRACTICE

PROCEDURE A—Respiratory Organs

1. Review section 19.2 titled "Organs of the Respiratory System" in chapter 19 of the textbook.
2. As a review activity, label figures 45.1–45.4.
3. Examine the sagittal section of the human skull, and locate the following features:

nasal cavity
    nostril (naris)
    nasal septum
    nasal conchae
    meatuses
paranasal sinuses
    maxillary sinus
    frontal sinus
    ethmoidal sinus (ethmoidal air cells)
    sphenoidal sinus

4. Observe the larynx model, the thoracic organs model, and the human torso model. Locate the features listed in step 3. Also locate the following features:

pharynx
    nasopharynx
    oropharynx
    laryngopharynx
larynx (palpate your larynx)
    vocal cords
        false vocal cords (vestibular folds)
        true vocal cords (vocal folds)
    thyroid cartilage ("Adam's apple")
    cricoid cartilage

cricothyroid ligament
epiglottis
epiglottic cartilage
arytenoid cartilages
corniculate cartilages
cuneiform cartilages
glottis
trachea (palpate your trachea)
bronchi
right and left main (primary) bronchi
lobar (secondary) bronchi
segmental (tertiary) bronchi
bronchioles
lung
hilum
lobes
superior lobe
middle lobe (right only)
inferior lobe
visceral pleura

parietal pleura
pleural cavity

5. Study figure 45.5 of the bronchial tree.
6. Complete Part A of Laboratory Report 45.

 LEARN: LAB IN MOTION

Observe the animal lung and the attached trachea. Identify the larynx, major laryngeal cartilages, trachea, and incomplete cartilaginous rings of the trachea. Open the larynx and locate the vocal folds. Examine the visceral pleura on the surface of a lung. A bicycle pump can be used to demonstrate lung inflation. Section the lung and locate some bronchioles and alveoli. Squeeze a portion of a lung between your fingers. How do you describe the texture of the lung?

_____

_____

_____

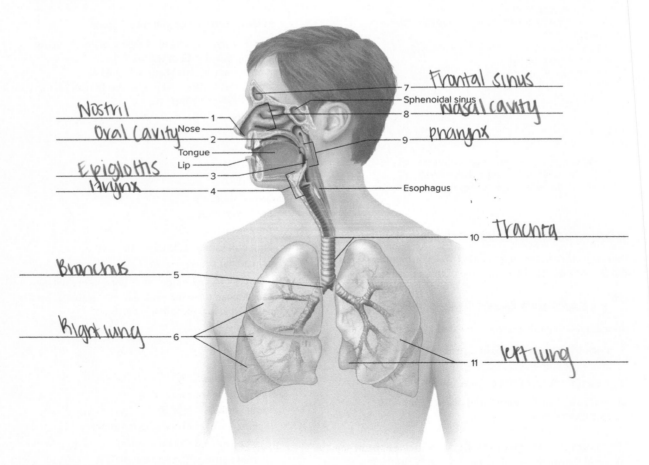

1 — Nostril
Oral Cavity — Nose
2 —
Tongue
Lip
Epiglottis — 3 —
larynx — 4 —
7 — Frontal sinus
Sphenoidal sinus
8 — Nasal cavity
9 — pharynx
Esophagus
10 — Trachea
Bronchus — 5 —
Right lung — 6 —
11 — left lung

FIGURE 45.1   Label the major features of the respiratory system. **APR**

FIGURE 45.2 Label the features of this sagittal section of the upper respiratory tract. A APR

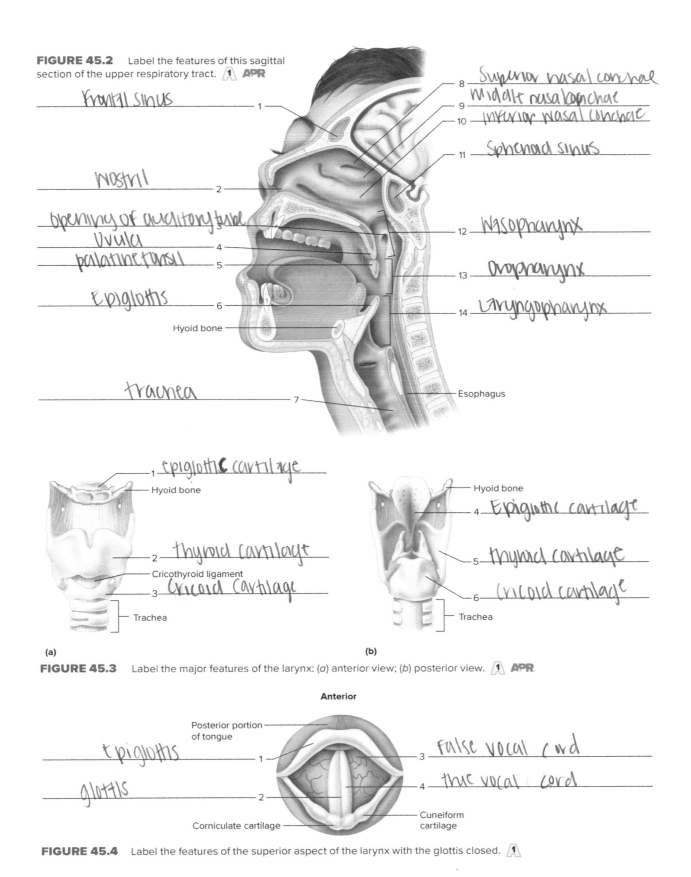

1 — frontal sinus

2 — nostril

3 — opening of auditory tube

4 — uvula

5 — palatine tonsil

6 — epiglottis

Hyoid bone

7 — trachea

8 — Superior nasal conchae

9 — middle nasal conchae

10 — inferior nasal conchae

11 — Sphenoid sinus

12 — Nasopharynx

13 — Oropharynx

14 — Laryngopharynx

Esophagus

FIGURE 45.3 Label the major features of the larynx: (a) anterior view; (b) posterior view. A APR

(a)

1 — epiglottic cartilage
Hyoid bone

2 — thyroid cartilage
Cricothyroid ligament

3 — Cricoid Cartilage
Trachea

(b)

Hyoid bone

4 — Epiglottic cartilage

5 — thyroid cartilage

6 — cricoid cartilage

Trachea

Anterior

Posterior portion of tongue

epiglottis — 1

glottis — 2

3 — False vocal cord

4 — true vocal cord

Corniculate cartilage

Cuneiform cartilage

FIGURE 45.4 Label the features of the superior aspect of the larynx with the glottis closed. A

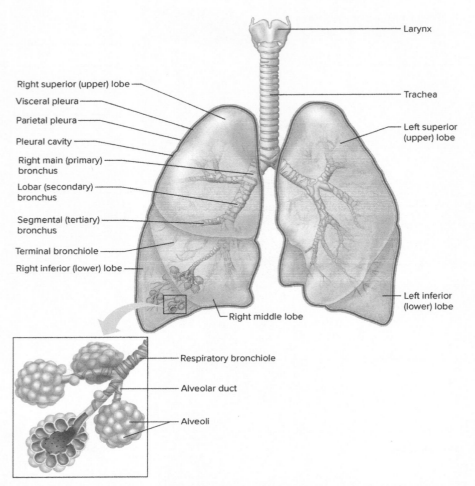

Larynx

Right superior (upper) lobe

Visceral pleura

Parietal pleura

Pleural cavity

Right main (primary) bronchus

Lobar (secondary) bronchus

Segmental (tertiary) bronchus

Terminal bronchiole

Right inferior (lower) lobe

Trachea

Left superior (upper) lobe

Left inferior (lower) lobe

Right middle lobe

Respiratory bronchiole

Alveolar duct

Alveoli

FIGURE 45.5   The bronchial tree consists of the passageways that connect the trachea and the alveoli. The alveolar ducts and alveoli are enlarged to show their locations. **APR**

 PRACTICE

PROCEDURE B—Respiratory Tissues

1. Obtain a prepared microscope slide of a trachea, and use low-power magnification to examine it. Notice the inner lining of ciliated pseudostratified columnar epithelium and the deep layer of hyaline cartilage, which represents a portion of an incomplete (C-shaped) tracheal ring (fig. 45.6).

2. Use high-power magnification to observe the cilia on the free surface of the epithelial lining. Locate the wine-glass-shaped goblet cells, which secrete a protective mucus, in the epithelium (fig. 45.7).

3. Prepare a labeled sketch of a representative portion of the tracheal wall in Part B of the laboratory report.

4. Study figure 45.8 of the respiratory tubes and the alveoli.

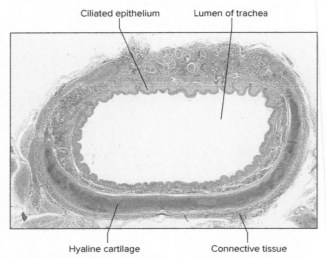

Ciliated epithelium   Lumen of trachea

Hyaline cartilage   Connective tissue

FIGURE 45.6   Micrograph of a section of the tracheal wall (100×). **APR** (©Ed Reschke)

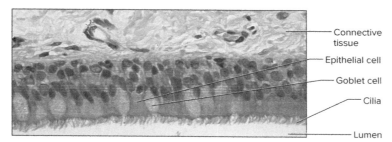

FIGURE 45.7   Micrograph of ciliated pseudostratified epithelium in the respiratory tract (275×).   **APR** (Biophoto Associates/ Science Source)

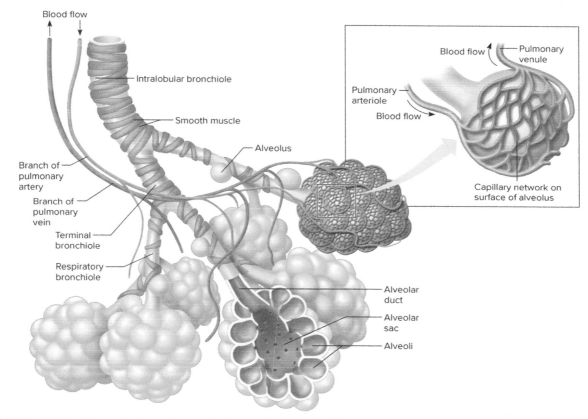

FIGURE 45.8   The respiratory tubes end in tiny alveoli, each of which is surrounded by a capillary network.   **APR**

5. Obtain a prepared microscope slide of human lung tissue. Examine it, using low-power magnification, and note the many open spaces of the air sacs (alveoli). Look for a bronchiole—a tube with a relatively thick wall and a wavy inner lining. Locate the smooth muscle tissue in the wall of this tube (fig. 45.9). You also may see a section of cartilage as part of the bronchiole wall.

6. Use high-power magnification to examine the alveoli (fig. 45.10). Their walls are composed of simple squamous epithelium. The thin walls of the alveoli and the associated capillaries allow efficient gas exchange of oxygen and carbon dioxide. You also may see sections of blood vessels containing blood cells.

7. Prepare a labeled sketch of a representative portion of the lung in Part B of the laboratory report.

8. Complete Part C of the laboratory report.

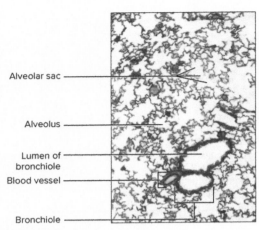

Alveolar sac

Alveolus

Lumen of
bronchiole

Blood vessel

Bronchiole

FIGURE 45.9   Micrograph of human lung tissue (70×).
APR (©Victor B. Eichler)

LEARN: ACTIVITY

Examine the prepared microscope slides of the lung tissue of a smoker and a person with emphysema, using low-power magnification. How does the smoker's lung tissue compare with that of the normal lung tissue that you examined previously?

_____

_____

_____

_____

_____

How does the emphysema patient's lung tissue compare with the normal lung tissue?

_____

_____

_____

_____

_____

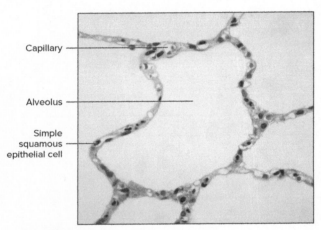

Capillary

Alveolus

Simple
squamous
epithelial cell

FIGURE 45.10   Micrograph of human lung tissue
(250×). **APR** (©Dwight Kuhn)

46

Breathing and Respiratory Volumes

MATERIALS NEEDED

Textbook
Spirometer, handheld (dry portable)
Alcohol swabs
Disposable mouthpieces
Nose clips
Meterstick
Lung function model

⚠ SAFETY

- Clean the spirometer with an alcohol swab before each use.
- Place a new disposable mouthpiece on the stem of the spirometer before each use.
- Dispose of the alcohol swabs and mouthpieces according to directions from your laboratory instructor.

PURPOSE OF THE EXERCISE

To review the mechanisms of breathing and to measure or calculate certain respiratory air volumes and respiratory capacities.

 ## LEARNING OUTCOMES APR

After completing this exercise, you should be able to

 Differentiate between the mechanisms responsible for inspiration and expiration.

② Match the respiratory air volumes and respiratory capacities with their definitions.

③ Measure respiratory air volumes using a spirometer.

④ Calculate respiratory capacities using the data obtained from respiratory air volumes.

Breathing involves the movement of air from outside the body through the bronchial tree and into the alveoli and the reversal of this air movement to allow gas (oxygen and carbon dioxide) exchange between air and blood. These movements are caused by changes in the size of the thoracic cavity that result from skeletal muscle contractions and from the elastic recoil of stretched tissues.

The volumes of air that move into and out of the lungs during various phases of breathing are called *respiratory air volumes* and *capacities*. A *volume* is a single measurement, whereas a *capacity* represents two or more combined volumes. Respiratory volumes can be measured by using an instrument called a *spirometer*. The values obtained vary with a person's age, sex, height, weight, stress, and physical fitness. Various respiratory capacities can be calculated by combining two or more of the respiratory volumes.

 ## PRACTICE

PROCEDURE A—Breathing Mechanisms APR

1. Review the concept headings titled "Inspiration" and "Expiration" in section 19.3 in chapter 19 of the textbook.
2. Complete Part A of Laboratory Report 46.

LEARN: LAB IN MOTION

Observe the mechanical lung function model (fig. 46.1). It consists of a heavy plastic bell jar with rubber sheeting clamped over its wide open end. Its narrow upper opening is plugged with a rubber stopper through which a Y tube is passed. Small rubber balloons are fastened to the arms of the Y. What happens to the balloons when the rubber sheeting is pulled downward? _____

_____

_____

What happens when the sheeting is pushed upward?

_____

_____

How do you explain these changes?

_____

_____

_____

What part of the respiratory system is represented by the

rubber sheeting? _____

The bell jar? _____

The Y tube? _____

The balloons? _____

Air moves into and out of the lungs in much the same way, as can be demonstrated using a syringe (fig. 46.2).

WARNING *If you begin to feel dizzy or light-headed while performing Procedure B, stop the exercise and breathe normally.*

 PRACTICE

PROCEDURE B—Respiratory Volumes and Capacities

1. Review the concept heading titled "Respiratory Volumes and Capacities" in section 19.3 in chapter 19 of the textbook.
2. Complete Part B of the laboratory report.
3. Get a handheld spirometer (fig. 46.3). Point the needle to zero by rotating the adjustable dial. Before using the instrument, clean it with an alcohol swab and place a new disposable mouthpiece over its stem. The instrument should be held with the dial upward, and the air should be blown only into the disposable mouthpiece (fig. 46.4). If air tends to exit the nostrils during exhalation, use a nose clip or pinch your nose as a prevention when exhaling into the spirometer. Movement of the needle indicates the air volume that leaves the lungs. The exhalations should be slowly and forcefully performed. Too rapid an exhalation can result in erroneous data or damage to the spirometer.
4. *Tidal volume (TV)* (about 500 mL) is the volume of air that enters (or leaves) the lungs during a *respiratory cycle* (one inspiration plus the following expiration). *Resting tidal volume* is the volume of air that enters (or leaves) the lungs during normal, quiet breathing (fig. 46.5). To measure this volume, follow these steps:
   a. Sit quietly for a few moments.
   b. Position the spirometer dial so that the needle points to zero.
   c. Place the mouthpiece between your lips and exhale three ordinary expirations into it after inhaling through the nose each time. *Do not force air out of your lungs; exhale normally.*

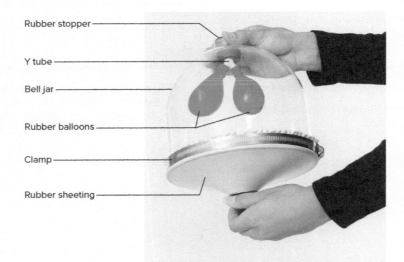

Rubber stopper

Y tube

Bell jar

Rubber balloons

Clamp

Rubber sheeting

FIGURE 46.1   A lung function model. (© J & J Photography)

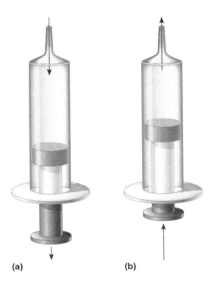

FIGURE 46.2 Moving the plunger of a syringe as in (*a*) results in air movements into the barrel of the syringe; moving the plunger of the syringe as in (*b*) results in air movements out of the barrel of the syringe. Air movements in and out of the lung function model and the lungs during breathing occur for similar reasons.

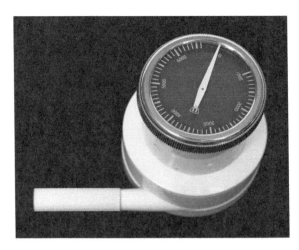

FIGURE 46.3 A handheld spirometer can be used to measure respiratory air volumes. (© J & J Photography)

    d. Divide the total value indicated by the needle by 3, and record this amount as your resting tidal volume on the table in Part C of the laboratory report.

5. *Expiratory reserve volume (ERV)* (about 1,100 mL) is the volume of air in addition to the tidal volume that leaves the lungs during forced expiration (fig. 46.5). To measure this volume, follow these steps:

    a. Breathe normally for a few moments. Set the needle to zero.

    b. At the end of an ordinary expiration, place the mouthpiece between your lips and exhale all of the air you can force from your lungs through the spirometer. Use a nose clip or pinch your nose to prevent air from exiting your nostrils.

    c. Record the results as your expiratory reserve volume in Part C of the laboratory report.

6. *Vital capacity (VC)* (about 4,600 mL) is the maximum volume of air that can be exhaled after taking the deepest breath possible (fig. 46.5). To measure this volume, follow these steps:

    a. Breathe normally for a few moments. Set the needle at zero.

    b. Breathe in and out deeply a couple of times; then take the deepest breath possible.

    c. Place the mouthpiece between your lips and exhale all the air out of your lungs, slowly and forcefully. Use a nose clip or pinch your nose to prevent air from exiting your nostrils.

FIGURE 46.4 Demonstration of a handheld spirometer. Air should be blown slowly and forcefully only into a disposable mouthpiece. *Use a nose clip or pinch your nose when measuring expiratory reserve volume and vital capacity volume if air exits the nostrils.* (© J & J Photography)

    d. Record the value as your vital capacity in Part C of the laboratory report. Compare your result with that expected for a person of your sex, age, and height listed in tables 46.1 and 46.2. Use the meterstick to determine your height in centimeters if necessary or multiply your height in inches by 2.54 to calculate your height in centimeters. Considerable individual variations from the expected will be noted due to parameters other than sex, age, and height, which can include fitness level, health, medications, and others.

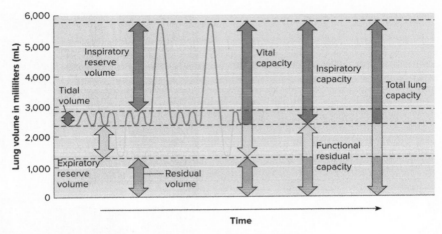

FIGURE 46.5 Graphic representation of respiratory volumes and capacities.

 ASSESS

CRITICAL THINKING

It can be noted from the data in tables 46.1 and 46.2 that vital capacities gradually decrease with age. Propose an explanation for this normal correlation.

_____

_____

_____

_____

_____

7. *Inspiratory reserve volume (IRV)* (about 3,000 mL) is the volume of air in addition to the tidal volume that enters the lungs during forced inspiration (fig. 46.5). Calculate your inspiratory reserve volume by subtracting your tidal volume (TV) and your expiratory reserve volume (ERV) from your vital capacity (VC):

$$IRV = VC - (TV + ERV)$$

8. *Inspiratory capacity (IC)* (about 3,500 mL) is the maximum volume of air a person can inhale following exhalation of the tidal volume (fig. 46.5). Calculate your inspiratory capacity by adding your tidal volume (TV) and your inspiratory reserve volume (IRV):

$$IC = TV + IRV$$

9. *Functional residual capacity (FRC)* (about 2,300 mL) is the volume of air that remains in the lungs following exhalation of the tidal volume (fig. 46.5). Calculate your functional residual capacity (FRC) by adding your expiratory reserve volume (ERV) and your residual volume (RV), which you can assume is 1,200 mL:

$$FRC = ERV + 1,200$$

10. *Residual volume (RV)* (about 1,200 mL) is the volume of air that always remains in the lungs after the most forceful expiration. Although it is part of *total lung capacity* (about 5,800 mL), it cannot be measured with a spirometer. The residual air allows gas exchange and the alveoli to remain open during the respiratory cycle (fig. 46.5).

11. Complete Part C of the laboratory report.

 LEARN: LAB IN MOTION

Determine your *minute respiratory volume.* To do this, follow these steps:

1. Sit quietly for a while, and then to establish your breathing rate, count the number of times you breathe in 1 minute. This might be inaccurate because conscious awareness of breathing rate can alter the results. You might ask a laboratory partner to record your breathing rate sometime when you are not expecting it to be recorded. A normal breathing rate is about 12–15 breaths per minute.

2. Calculate your minute respiratory volume by multiplying your breathing rate by your tidal volume:

_____ × _____ × _____
(breathing   (tidal   (minute respiratory
rate)   volume)   volume)

3. This value indicates the total volume of air that moves into your respiratory passages during each minute of ordinary breathing.

TABLE 46.1 Predicted Vital Capacities (in milliliters) for Females

Age	HEIGHT IN CENTIMETERS

Age	146	148	150	152	154	156	158	160	162	164	166	168	170	172	174	176	178	180	182	184	186	188	190	192	194
16	2950	2990	3030	3070	3110	3150	3190	3230	3270	3310	3350	3390	3430	3470	3510	3550	3590	3630	3670	3715	3755	3800	3840	3880	3920
18	2920	2960	3000	3040	3080	3120	3160	3200	3240	3280	3320	3360	3400	3440	3480	3520	3560	3600	3640	3680	3720	3760	3800	3840	3880
20	2890	2930	2970	3010	3050	3090	3130	3170	3210	3250	3290	3330	3370	3410	3450	3490	3525	3565	3605	3645	3695	3720	3760	3800	3840
22	2860	2900	2940	2980	3020	3060	3095	3135	3175	3215	3255	3290	3330	3370	3410	3450	3490	3530	3570	3610	3650	3685	3725	3765	3800
24	2830	2870	2910	2950	2985	3025	3065	3100	3140	3180	3220	3260	3300	3335	3375	3415	3455	3490	3530	3570	3610	3650	3685	3725	3765
26	2800	2840	2880	2920	2960	3000	3035	3070	3110	3150	3190	3230	3265	3300	3340	3380	3420	3455	3495	3530	3570	3610	3650	3685	3725
28	2775	2810	2850	2890	2930	2965	3000	3040	3070	3115	3155	3190	3230	3270	3305	3345	3380	3420	3460	3495	3535	3570	3610	3650	3685
30	2745	2780	2820	2860	2895	2935	2970	3010	3045	3085	3120	3160	3195	3235	3270	3310	3345	3385	3420	3460	3495	3535	3570	3610	3645
32	2715	2750	2790	2825	2865	2900	2940	2975	3015	3050	3090	3125	3160	3200	3235	3275	3310	3350	3385	3425	3460	3495	3535	3570	3610
34	2685	2725	2760	2795	2835	2870	2910	2945	2980	3020	3055	3090	3130	3165	3200	3240	3275	3310	3350	3385	3425	3460	3495	3535	3570
36	2655	2695	2730	2765	2805	2840	2875	2910	2950	2985	3020	3060	3095	3130	3165	3205	3240	3275	3310	3350	3385	3420	3460	3495	3530
38	2630	2665	2700	2735	2770	2810	2845	2880	2915	2950	2990	3025	3060	3095	3130	3170	3205	3240	3275	3310	3350	3385	3420	3455	3490
40	2600	2635	2670	2705	2740	2775	2810	2850	2885	2920	2955	2990	3025	3060	3095	3135	3170	3205	3240	3275	3310	3345	3380	3420	3455
42	2570	2605	2640	2675	2710	2745	2780	2815	2850	2885	2920	2955	2990	3025	3060	3100	3135	3170	3205	3240	3275	3310	3345	3380	3415
44	2540	2575	2610	2645	2680	2715	2750	2785	2820	2855	2890	2925	2960	2995	3030	3060	3095	3130	3165	3200	3235	3270	3305	3340	3375
46	2510	2545	2580	2615	2650	2685	2715	2750	2785	2820	2855	2890	2925	2960	2995	3030	3060	3095	3130	3165	3200	3235	3270	3305	3340
48	2480	2515	2550	2585	2620	2650	2685	2715	2750	2785	2820	2855	2890	2925	2960	2995	3030	3060	3095	3130	3160	3195	3230	3265	3300
50	2455	2485	2520	2555	2590	2625	2655	2690	2715	2750	2785	2820	2855	2890	2925	2955	2990	3025	3060	3090	3125	3155	3190	3225	3260
52	2425	2455	2490	2525	2555	2590	2625	2655	2690	2720	2755	2790	2820	2855	2890	2925	2955	2990	3020	3055	3090	3125	3155	3190	3220
54	2395	2425	2460	2495	2530	2560	2590	2625	2655	2690	2720	2755	2790	2820	2855	2885	2920	2950	2985	3020	3050	3085	3115	3150	3180
56	2365	2400	2430	2460	2495	2525	2560	2590	2625	2655	2690	2720	2755	2790	2820	2855	2885	2920	2950	2980	3015	3045	3080	3110	3145
58	2335	2370	2400	2430	2460	2495	2525	2560	2590	2625	2655	2690	2720	2750	2785	2815	2850	2880	2920	2945	2975	3010	3040	3075	3105
60	2305	2340	2370	2400	2430	2460	2495	2525	2560	2590	2625	2655	2685	2720	2750	2780	2810	2845	2875	2915	2940	2970	3000	3035	3065
62	2280	2310	2340	2370	2405	2435	2465	2495	2525	2560	2590	2620	2655	2685	2715	2745	2775	2810	2840	2870	2900	2935	2965	2995	3025
64	2250	2280	2310	2340	2370	2400	2430	2465	2495	2525	2555	2585	2620	2650	2680	2710	2740	2770	2805	2835	2865	2895	2920	2955	2990
66	2220	2250	2280	2310	2340	2370	2400	2430	2460	2495	2525	2555	2585	2615	2645	2675	2705	2735	2765	2800	2825	2860	2890	2920	2950
68	2190	2220	2250	2280	2310	2340	2370	2400	2430	2460	2490	2520	2550	2580	2610	2640	2670	2700	2730	2760	2795	2820	2850	2880	2910
70	2160	2190	2220	2250	2280	2310	2340	2370	2400	2425	2455	2485	2515	2545	2575	2605	2635	2665	2695	2725	2755	2780	2810	2840	2870
72	2130	2160	2190	2220	2250	2280	2310	2335	2365	2395	2425	2455	2480	2510	2540	2570	2600	2630	2660	2685	2715	2745	2775	2805	2830
74	2100	2130	2160	2190	2220	2245	2275	2305	2335	2360	2390	2420	2450	2475	2505	2535	2565	2590	2620	2650	2680	2710	2740	2765	2795

From E. DeF. Baldwin and E. W. Richards, Jr., "Pulmonary Insufficiency I. Physiologic Classification, Clinical Methods of Analysis, Standard Values in Normal Subjects" in Medicine 27:243.

TABLE 46.2　Predicted Vital Capacities (in milliliters) for Males

HEIGHT IN CENTIMETERS

Age	146	148	150	152	154	156	158	160	162	164	166	168	170	172	174	176	178	180	182	184	186	188	190	192	194
16	3765	3820	3870	3920	3975	4025	4075	4130	4180	4230	4285	4335	4385	4440	4490	4540	4590	4645	4695	4745	4800	4850	4900	4955	5005
18	3740	3790	3840	3890	3940	3995	4045	4095	4145	4200	4250	4300	4350	4405	4455	4505	4555	4610	4660	4710	4760	4815	4865	4915	4965
20	3710	3760	3810	3860	3910	3960	4015	4065	4115	4165	4215	4265	4320	4370	4420	4470	4520	4570	4625	4675	4725	4775	4825	4875	4930
22	3680	3730	3780	3830	3880	3930	3980	4030	4080	4135	4185	4235	4285	4335	4385	4435	4485	4535	4585	4635	4685	4735	4790	4840	4890
24	3635	3685	3735	3785	3835	3885	3935	3985	4035	4085	4135	4185	4235	4285	4330	4380	4430	4480	4530	4580	4630	4680	4730	4780	4830
26	3605	3655	3705	3755	3805	3855	3905	3955	4000	4050	4100	4150	4200	4250	4300	4350	4395	4445	4495	4545	4595	4645	4695	4740	4790
28	3575	3625	3675	3725	3775	3820	3870	3920	3970	4020	4070	4115	4165	4215	4265	4310	4360	4410	4460	4510	4555	4605	4655	4705	4755
30	3550	3595	3645	3695	3740	3790	3840	3890	3935	3985	4035	4080	4130	4180	4230	4275	4325	4375	4425	4470	4520	4570	4615	4665	4715
32	3520	3565	3615	3665	3710	3760	3810	3855	3905	3950	4000	4050	4095	4145	4195	4240	4290	4340	4385	4435	4485	4530	4580	4625	4675
34	3475	3525	3570	3620	3665	3715	3760	3810	3855	3905	3950	4000	4045	4095	4140	4190	4225	4285	4330	4380	4425	4475	4520	4570	4615
36	3445	3495	3540	3585	3635	3680	3730	3775	3825	3870	3920	3965	4010	4060	4105	4155	4200	4250	4295	4340	4390	4435	4485	4530	4580
38	3415	3465	3510	3555	3605	3650	3695	3745	3790	3840	3885	3930	3980	4025	4070	4120	4165	4210	4260	4305	4350	4400	4445	4495	4540
40	3385	3435	3480	3525	3575	3620	3665	3710	3760	3805	3850	3900	3945	3990	4035	4085	4130	4175	4220	4270	4315	4360	4410	4455	4500
42	3360	3405	3450	3495	3540	3590	3635	3680	3725	3770	3820	3865	3910	3955	4000	4050	4095	4140	4185	4230	4280	4325	4370	4415	4460
44	3315	3360	3405	3450	3495	3540	3585	3630	3675	3725	3770	3815	3860	3905	3950	3995	4040	4085	4130	4175	4220	4270	4315	4360	4405
46	3285	3330	3375	3420	3465	3510	3555	3600	3645	3690	3735	3780	3825	3870	3915	3960	4005	4050	4095	4140	4185	4230	4275	4320	4365
48	3255	3300	3345	3390	3435	3480	3525	3570	3615	3655	3700	3745	3790	3835	3880	3925	3970	4015	4060	4105	4150	4190	4235	4280	4325
50	3210	3255	3300	3345	3390	3430	3475	3520	3565	3610	3650	3695	3740	3785	3830	3870	3915	3960	4005	4050	4090	4135	4180	4225	4270
52	3185	3225	3270	3315	3355	3400	3445	3490	3530	3575	3620	3660	3705	3750	3795	3835	3880	3925	3970	4010	4055	4100	4140	4185	4230
54	3155	3195	3240	3285	3325	3370	3415	3455	3500	3540	3585	3630	3670	3715	3760	3800	3845	3890	3930	3975	4020	4060	4105	4145	4190
56	3125	3165	3210	3255	3295	3340	3380	3425	3465	3510	3550	3595	3640	3680	3725	3765	3800	3850	3895	3940	3980	4025	4065	4110	4150
58	3080	3125	3165	3210	3250	3295	3335	3380	3420	3460	3500	3545	3585	3630	3670	3715	3755	3800	3840	3880	3925	3965	4010	4050	4095
60	3050	3095	3135	3175	3220	3260	3300	3345	3385	3430	3470	3500	3555	3595	3635	3680	3720	3760	3805	3845	3885	3930	3970	4015	4055
62	3020	3060	3110	3150	3190	3230	3270	3310	3350	3390	3440	3480	3520	3560	3600	3640	3680	3730	3770	3810	3850	3890	3930	3970	4020
64	2990	3030	3080	3120	3160	3200	3240	3280	3320	3360	3400	3440	3490	3530	3570	3610	3650	3690	3730	3770	3810	3850	3900	3940	3980
66	2950	2990	3030	3070	3110	3150	3190	3230	3270	3310	3350	3400	3430	3470	3510	3550	3600	3640	3680	3720	3760	3800	3840	3880	3920
68	2920	2960	3000	3040	3080	3120	3160	3200	3240	3280	3320	3360	3400	3440	3480	3520	3560	3600	3640	3680	3720	3760	3800	3840	3880
70	2890	2930	2970	3010	3050	3090	3130	3170	3210	3250	3290	3330	3370	3410	3450	3480	3520	3560	3600	3640	3680	3720	3760	3800	3840
72	2860	2900	2940	2980	3020	3060	3100	3140	3180	3210	3250	3290	3330	3370	3410	3450	3490	3530	3570	3610	3650	3680	3720	3760	3800
74	2820	2860	2900	2930	2970	3010	3050	3090	3130	3170	3200	3240	3280	3320	3360	3400	3440	3470	3510	3550	3590	3630	3670	3710	3740

From E. DeF. Baldwin and E. W. Richards, Jr., "Pulmonary Insufficiency 1, Physiologic Classification, Clinical Methods of Analysis, Standard Values in Normal Subjects" in Medicine 27:243.

47

Control of Breathing

MATERIALS NEEDED

Textbook
Clock with second hand
Paper bags, small
Flasks
Glass tubing
Rubber stoppers, two-hole
Calcium hydroxide solution (limewater)
Pneumograph
Physiological recording apparatus

PURPOSE OF THE EXERCISE

To review the muscles that control breathing, and to describe
the mechanisms that regulate the rate and depth of breathing.

 ## LEARNING OUTCOMES APR

After completing this exercise, you should be able to

1. Locate the respiratory areas in the brainstem.

2. Describe the mechanisms that control and influence
breathing.

3. Select the respiratory muscles involved in inspiration
and forced expiration.

4. Test and record the effect of various factors on the rate
and depth of breathing.

Several skeletal muscles contract during breathing. The
principal muscles involved during inspiration are the diaphragm and external intercostals. The sternocleidomastoid,
scalenes, and pectoralis minor are synergistic during forceful inhalation. During quiet respiration, there is minimal
involvement of the expiratory muscles. During forced expiration, the principal muscles are the internal intercostals,
but the abdominal wall muscles can provide extra force.

Breathing is controlled from regions of the brainstem
called the *respiratory areas,* which control both inspiration
and expiration. These areas initiate and regulate impulses

that travel to various breathing muscles, causing rhythmic
breathing movements and adjustments to the rate and depth of
breathing to meet various cellular needs of oxygen supply and
carbon dioxide removal (fig. 47.1). The *medullary respiratory
center* is composed of two bilateral groups of neurons in the
medulla oblongata. They are called the *ventral respiratory group*
and the *dorsal respiratory group.* The ventral respiratory group
involves regulation of the basic rhythm of breathing. The
dorsal respiratory group primarily stimulates the diaphragm
contraction, resulting in inspiration, and helps process respiratory sensory information involving the cardiovascular system.
Neurons in another part of the brainstem, the pons, compose
the bilateral *pontine respiratory group* (formerly the pneumo-
taxic center). These neurons may contribute to the rhythm of
breathing by modifying the respirations during such situations
as exercise and sleep (fig. 47.2).

Various factors can influence the respiratory areas and
thus affect the rate and depth of breathing. These factors
include stretch of the lung tissues, emotional state, and presence in the blood of certain chemicals, such as carbon dioxide, hydrogen ions, and oxygen. The breathing rate increases
primarily as the blood concentration of carbon dioxide or
hydrogen ions increases, but much less often as the concentration of oxygen decreases.

 ## PRACTICE

PROCEDURE A—Control of Breathing

1. Review section 19.4 titled "Control of Breathing" in
chapter 19 of the textbook.

2. Respiratory areas in the brainstem control the cycle of
breathing. The dorsal respiratory group of neurons is
located in the medulla oblongata. Impulses from the
dorsal respiratory group stimulate the muscles of inspiration, especially the diaphragm (see fig. 47.1). This
results in a normal respiratory rate of 12–15 breaths per
minute. The ventral respiratory group of neurons will
fire during inspiration and expiration to control the
appropriate muscles of inspiration and expiration and
the basic rhythm of breathing (fig. 47.2).

3. Complete Part A of Laboratory Report 47.

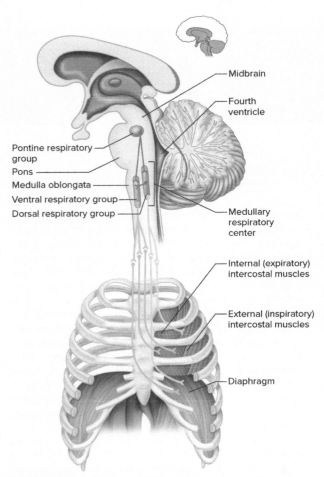

FIGURE 47.1   The respiratory areas are located in the pons and the medulla oblongata. **APR**

Midbrain

Fourth ventricle

Pontine respiratory group

Pons

Medulla oblongata

Ventral respiratory group

Dorsal respiratory group

Medullary respiratory center

Internal (expiratory) intercostal muscles

External (inspiratory) intercostal muscles

Diaphragm

LEARN: ACTIVITY

When a solution of calcium hydroxide is exposed to carbon dioxide, a chemical reaction occurs and a white precipitate of calcium carbonate is formed, as indicated by the following reaction:

$$Ca(OH)_2 + CO_2 \longrightarrow CaCO_3 + H_2O$$

Thus, a clear water solution of calcium hydroxide (lime-water) can be used to detect the presence of carbon dioxide because the solution becomes cloudy if this gas is bubbled through it.

The laboratory instructor will demonstrate this test for carbon dioxide by drawing some air through limewater in an apparatus such as that shown in figure 47.3. Then the instructor will blow an equal volume of expired air through a similar apparatus. (*Note*: A new sterile mouth-piece should be used each time the apparatus is demonstrated.) Watch for the appearance of a precipitate that causes the limewater to become cloudy. Was there any carbon dioxide in the atmospheric air drawn through the limewater?

_____

If so, how did the amount of carbon dioxide in the atmospheric air compare with the amount in the expired air?

_____

_____

_____

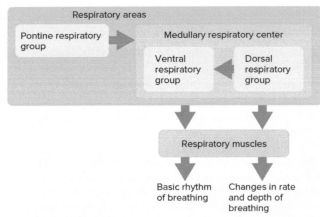

FIGURE 47.2   The medullary respiratory center and the pontine respiratory group control breathing.

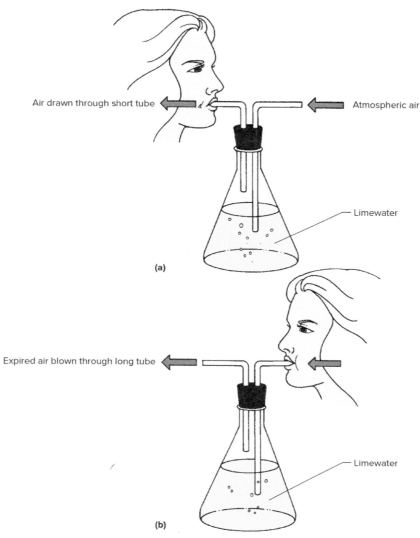

Air drawn through short tube

Atmospheric air

Limewater

(a)

Expired air blown through long tube

Limewater

(b)

FIGURE 47.3 Apparatus used to demonstrate the presence of carbon dioxide in air: (*a*) atmospheric air is drawn through limewater; (*b*) expired air is blown through limewater.

 PRACTICE

PROCEDURE B—Factors Affecting Breathing

Perform each of the following tests, using your laboratory partner as a test subject.

1. *Normal breathing.* To determine the subject's normal breathing rate and depth, follow these steps:
   a. Have the subject sit quietly for a few minutes.
   b. After the rest period, ask the subject to count backward mentally, beginning with 500.
   c. While the subject is distracted by counting, watch the subject's chest movements, and count the breaths taken in a minute. Use this value as the normal breathing rate (breaths per minute).

   d. Note the relative depth of the breathing movements.
   e. Record your observations in the table in Part B of the laboratory report.

2. *Effect of hyperventilation.* To test the effect of hyperventilation on breathing, follow these steps:
   a. Seat the subject and *guard to prevent the possibility of the subject falling over.*
   b. Have the subject breathe rapidly and deeply for a maximum of 20 seconds. *If the subject begins to feel dizzy, the hyperventilation should be halted immediately to prevent the subject from fainting from complications of alkalosis. The increased blood pH causes vasoconstriction of cerebral arterioles, which decreases circulation and oxygen to the brain.*

c. After the period of hyperventilation, determine the subject's breathing rate, and judge the breathing depth as before.

d. Record the results in Part B of the laboratory report.

3. *Effect of rebreathing air.* To test the effect of rebreathing air on breathing, follow these steps:

a. Have the subject sit quietly (approximately 5 minutes) until the breathing rate returns to normal.

b. Have the subject breathe deeply into a small paper bag that is held tightly over the nose and mouth. *If the subject begins to feel light-headed or like fainting, the rebreathing air should be halted immediately to prevent further acidosis and fainting.*

c. After 2 minutes of rebreathing air, determine the subject's breathing rate, and judge the depth of breathing.

d. Record the results in Part B of the laboratory report.

4. *Effect of breath holding.* To test the effect of breath holding on breathing, follow these steps:

a. Have the subject sit quietly (approximately 5 minutes) until the breathing rate returns to normal.

b. Have the subject hold his/her breath as long as possible. *If the subject begins to feel light-headed or like fainting, breath holding should be halted immediately to prevent further acidosis and fainting.*

c. As the subject begins to breathe again, determine the rate of breathing, and judge the depth of breathing.

d. Record the results in Part B of the laboratory report.

5. *Effect of exercise.* To test the effect of exercise on breathing, follow these steps:

a. Have the subject sit quietly (approximately 5 minutes) until breathing rate returns to normal.

b. Have the subject exercise by moderately running in place for 3–5 minutes. *This exercise should be avoided by anyone with health risks.*

c. After the exercise, determine the breathing rate, and judge the depth of breathing.

d. Record the results in Part B of the laboratory report.

6. Complete Part B of the laboratory report.

 LEARN: LAB IN MOTION

A *pneumograph* is a device that can be used together with some type of recording apparatus to record breathing movements. The laboratory instructor will demonstrate the use of this equipment to record various movements, such as those that accompany coughing, laughing, yawning, and speaking.

Devise an experiment to test the effect of some factor, such as hyperventilation, rebreathing of air, or exercise, on the length of time a person can hold his/her breath. *After the laboratory instructor has approved your plan,* carry out the experiment, using the pneumograph and recording equipment. What conclusion can you draw from the results of your experiment?

_____

_____

_____

_____

_____

Urinary System

48

Urinary Organs

MATERIALS NEEDED

Textbook
Human torso model
Kidney model
Preserved pig (or sheep) kidney
Dissecting tray
Dissecting instruments
Long knife
Compound light microscope
Prepared microscope slides of the following:
    Kidney section
    Ureter
    Urinary bladder

SAFETY

- Wear disposable gloves when working on the kidney dissection.
- Dispose of the kidney and gloves as directed by your laboratory instructor.
- Wash the dissecting tray and instruments as instructed.
- Wash your laboratory table.
- Wash your hands before leaving the laboratory.

PURPOSE OF THE EXERCISE

To review the structure of the urinary organs, to dissect a kidney, and to observe the major structures of a nephron.

LEARNING OUTCOMES APR

After completing this exercise, you should be able to

1. Locate and identify the major organs of the urinary system.
2. Locate and identify the major structures of a kidney.
3. Identify and sketch the structures of a nephron.
4. Trace the path of filtrate through a nephron.

5. Trace the path of blood and blood pressure changes through the renal blood vessels.
6. Identify and sketch the structures of a ureter and the urinary bladder.

The two kidneys are the primary organs of the urinary system. They are located in the abdominal cavity, against the posterior wall and behind the parietal peritoneum (retroperitoneal). Masses of adipose tissue associated with the kidneys hold them in place at a vertebral level between T12 and L3. The right kidney is slightly more inferior due to the large mass of the liver near its superior border. The movements of the muscular walls of the ureters force urine by means of peristaltic waves into the urinary bladder, which temporarily stores urine. The urethra conveys urine to the outside of the body.

Each kidney contains about 1 million nephrons, which serve as the basic structural and functional units of the kidney. A glomerular capsule, proximal tubule, nephron loop, and distal tubule compose the microscopic, multicellular structure of a relatively long nephron tubule. Several nephrons merge and drain into a common collecting duct. Approximately 80% of the nephrons are cortical nephrons with short nephron loops, whereas the remaining represent juxtamedullary nephrons, with long nephron loops extending deeper into the renal medulla. An elaborate network of blood vessels surrounds the entire nephron. Glomerular filtration, tubular reabsorption, and tubular secretion represent three processes resulting in urine as the final product.

A variety of functions occur in the kidneys. They remove metabolic wastes from the blood; help regulate blood volume, blood pressure, and pH of blood; control water and electrolyte concentrations; and secrete renin and erythropoietin.

PRACTICE

PROCEDURE A—Kidney

1. Review the concept heading titled "Kidney Structure" in section 20.2 in chapter 20 of the textbook.
2. As a review activity, label figures 48.1 and 48.2.

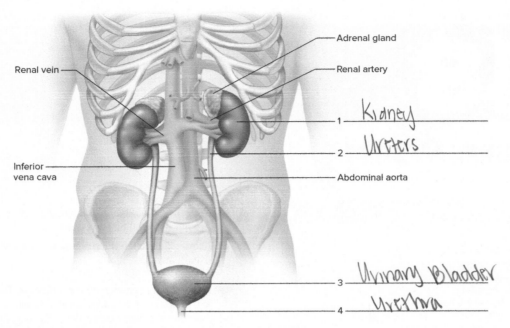

Adrenal gland

Renal vein

Renal artery

Inferior
vena cava

Abdominal aorta

1 _Kidney_

2 _Urrters_

3 _Urinary Bladder_

4 _Urethra_

FIGURE 48.1   Label the major organs of the urinary system. 🅰 APR

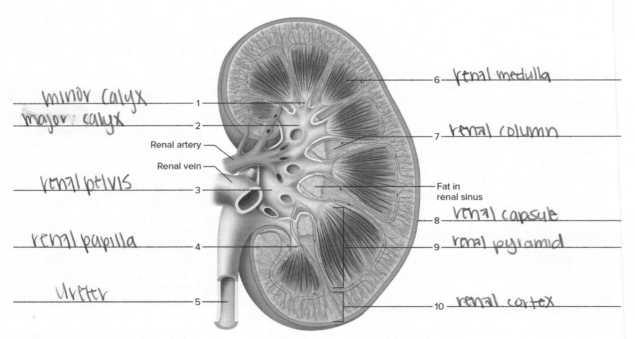

minor calyx   1

major calyx   2

Renal artery

Renal vein

renal pelvis   3

renal papilla   4

Ureter   5

6 _renal medulla_

7 _renal column_

Fat in
renal sinus

8 _renal capsule_

9 _renal pyramid_

10 _renal cortex_

FIGURE 48.2   Label the major structures in the longitudinal section of a kidney. 🄰 APR

3. Complete Part A of Laboratory Report 48.
4. Observe the human torso model and the kidney model. Locate the following:

   kidneys
   ureters
   urinary bladder
   urethra
   renal sinus
   renal pelvis
   > major calyces
   > minor calyces

   renal medulla
   > renal pyramids
   > renal papillae

   renal cortex
   renal columns
   nephrons
   > cortical nephrons (80% of nephrons)
   > juxtamedullary nephrons (20% of nephrons)

5. Obtain a pig or sheep kidney along with a dissecting tray, dissecting instruments, and disposable gloves, and follow these steps:
   a. Rinse the kidney with water to remove as much of the preserving fluid as possible and place it in a dissecting tray.
   b. Carefully remove any adipose tissue from the surface of the specimen.
   c. Locate the following external anatomical features:

   renal (fibrous) capsule
   hilum of kidney
   renal artery
   renal vein
   ureter

   d. Use a long knife to cut the kidney in half longitudinally along the frontal plane, beginning on the convex border.
   e. Rinse the interior of the kidney with water, and using figure 48.3 as a reference, locate the following:

   renal pelvis
   > major calyces
   > minor calyces

   renal cortex
   renal columns (extensions of renal cortical tissue between renal pyramids)
   renal medulla
   > renal pyramids
   > renal papillae

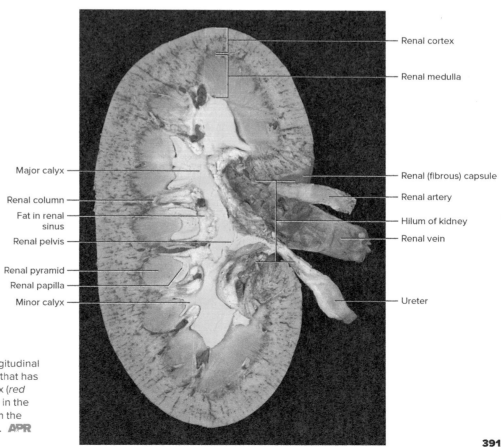

Major calyx
Renal column
Fat in renal sinus
Renal pelvis
Renal pyramid
Renal papilla
Minor calyx

Renal cortex
Renal medulla
Renal (fibrous) capsule
Renal artery
Hilum of kidney
Renal vein
Ureter

FIGURE 48.3 Longitudinal section of a pig kidney that has a triple injection of latex (*red* in the renal artery, *blue* in the renal vein, and *yellow* in the ureter and renal pelvis). APR
(J & J Photography)

PRACTICE

PROCEDURE B—Renal Blood Vessels and Nephrons

1. Review the concept headings titled "Renal Blood Supply" and "Nephrons" in section 20.2 in chapter 20 of the textbook.
2. As a review activity, label figure 48.4.
3. Complete Part B of the laboratory report.
4. Obtain a microscope slide of a kidney section, and examine it using low-power magnification. Locate the *renal capsule,* the *renal cortex* (which appears somewhat granular and may be more darkly stained than the other renal tissues), and the *renal medulla* (fig. 48.5).
5. Examine the renal cortex, using high-power magnification. Locate a *renal corpuscle.* These structures appear as isolated circular areas. Identify the *glomerulus,*

the capillary cluster inside the corpuscle, and the *glomerular capsule,* which appears as a clear area surrounding the glomerulus. A glomerulus and a glomerular capsule compose a *renal corpuscle.* Also note the numerous sections of renal tubules that occupy the spaces between renal corpuscles (fig. 48.5a). The renal cortex contains the proximal and distal convoluted tubules.

6. Prepare a labeled sketch of a representative section of the renal cortex in Part C of the laboratory report.
7. Examine the renal medulla, using high-power magnification. Identify longitudinal sections of various collecting ducts. These ducts are lined with simple epithelial cells, which vary in shape from squamous to cuboidal (fig. 48.5b).
8. Prepare a labeled sketch of a representative section of the renal medulla in Part C of the laboratory report.

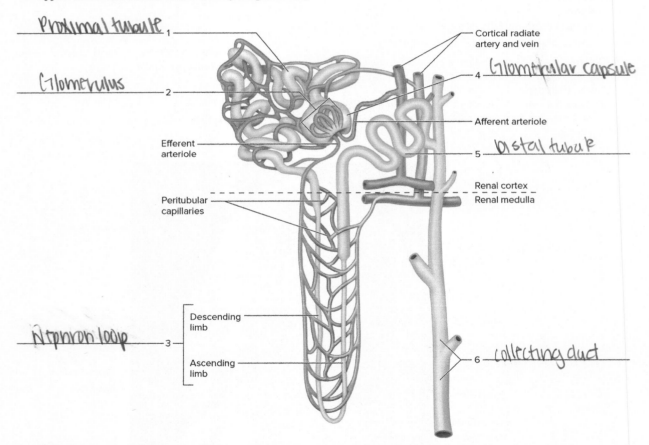

Proximal tubule — 1

Glomerulus — 2

Efferent arteriole

Peritubular capillaries

Nephron loop — 3

Descending limb

Ascending limb

Cortical radiate artery and vein

4 — **Glomerular capsule**

Afferent arteriole

5 — **Distal tubule**

Renal cortex
Renal medulla

6 — **collecting duct**

FIGURE 48.4   Label the major structures of the nephron and associated structures. [3] APR

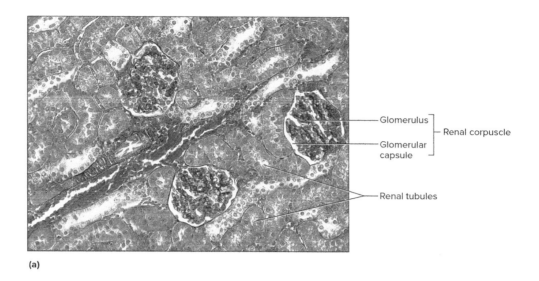

Glomerulus ⎫
⎬ Renal corpuscle
Glomerular ⎭
capsule

Renal tubules

(a)

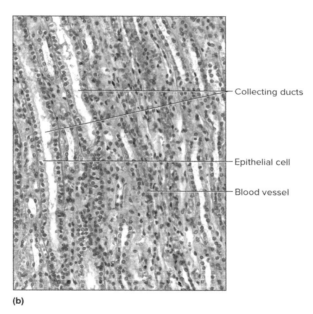

Collecting ducts

Epithelial cell

Blood vessel

(b)

FIGURE 48.5 (*a*) Micrograph of a section of the renal cortex (220×). (*b*) Micrograph of a section of the renal medulla (80×). (*Note*: With about a million nephrons in a kidney, there are many different orientations visible in micrograph sections of renal cortex and renal medulla.) **APR** ((*a*) Biophoto Associates/Science Source; (*b*) Al Telser/McGraw-Hill Education)

PRACTICE

PROCEDURE C—Ureter and Urinary Bladder

1. Review the concept heading titled "Ureters" in section 20.4 in chapter 20 of the textbook.
2. Obtain a microscope slide of a cross section of a ureter, and examine it using low-power magnification. Locate the *mucous coat* layer next to the lumen. Examine the middle *muscular coat* composed of longitudinal and circular smooth muscle cells, responsible for the peristaltic waves that propel urine from the kidneys to the urinary bladder. The outer *fibrous coat,* composed of connective tissue, secures the ureter in the retroperitoneal position (fig. 48.6).
3. Examine the mucous coat using high-power magnification. The specialized tissue is transitional epithelium, which allows changes in its thickness when unstretched and stretched.
4. Prepare a labeled sketch of a ureter in Part D of the laboratory report.

5. Review the concept heading titled "Urinary Bladder" in section 20.4 in chapter 20 of the textbook.
6. Obtain a microscope slide of a segment of the wall of a urinary bladder, and examine it using low-power magnification. Examine the *mucous coat* next to the lumen and the *submucous coat* composed of connective tissue just beneath the mucous coat. Examine the *muscular coat* composed of bundles of smooth muscle cells interlaced in many directions. This thick, muscular layer is called the *detrusor muscle* and functions in the elimination of urine. An outer *serous coat* is composed of connective tissue (fig. 48.7). The serous coat on the upper surface of the urinary bladder consists of the parietal peritoneum.
7. Examine the mucous coat using high-power magnification. The tissue is transitional epithelium, which allows changes in its thickness from unstretched when the bladder is empty to stretched when the bladder distends with urine.
8. Prepare a labeled sketch of a segment of the urinary bladder wall in Part D of the laboratory report.

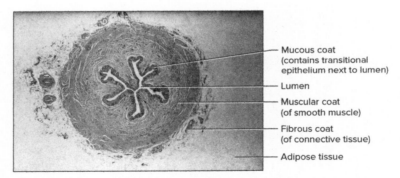

FIGURE 48.6   Micrograph of a cross section of a ureter (75×). **APR** (Biophoto Associates/Science Source)

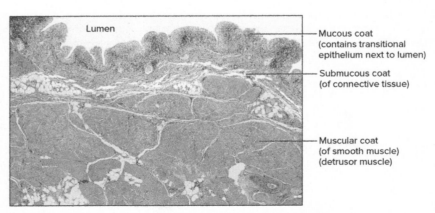

FIGURE 48.7   Micrograph of a segment of the human urinary bladder wall (6×). **APR** (Alvin Telser/McGraw-Hill Education)

LABORATORY EXERCISE

49

Urinalysis

MATERIALS NEEDED

Normal and abnormal simulated urine specimens can
  be substituted for collected urine.
Disposable urine-collecting container
Paper towel
Urinometer cylinder (jar)
Urinometer hydrometer
Laboratory thermometer
pH test paper
Reagent strips (individual or combination test strips
  such as Chemstrip or Multistix) to test for the
  presence of the following:
  Glucose
  Protein
  Ketones
  Bilirubin
  Hemoglobin/occult blood
Compound light microscope
Microscope slide
Coverslip
Centrifuge
Centrifuge tube
Graduated cylinder, 10 mL
Medicine dropper
Sedi-stain

⚠ SAFETY

- Consider using normal and abnormal simulated
  urine samples available from various laboratory
  supply houses.
- Use safety glasses, laboratory coats, and dispos-
  able gloves when working with body fluids.
- Work only with your own urine sample.
- Use an appropriate disinfectant to wash the labora-
  tory table before and after the procedures.
- Place glassware in a disinfectant when finished.
- Dispose of contaminated items as directed by your
  laboratory instructor.
- Wash your hands before leaving the laboratory.

PURPOSE OF THE EXERCISE

To perform the observations and tests commonly used to
analyze the characteristics and composition of urine.

LEARNING OUTCOMES

After completing this exercise, you should be able to

1. Evaluate the color, transparency, and specific gravity of
   a urine sample.
2. Measure the pH of a urine sample.
3. Test a urine sample for the presence of glucose, protein,
   ketones, bilirubin, and hemoglobin.
4. Perform a microscopic study of urine sediment.
5. Summarize the results of these observations and tests.

Urine is the product of three processes of the nephrons within the kidneys: glomerular filtration, tubular reabsorption, and tubular secretion. As a result of these
processes, various waste substances are removed from the
blood, and body fluid and electrolyte balance are main-
tained. Consequently, the composition of urine varies
considerably because of differences in dietary intake and
physical activity from day to day and person to person. The
normal urinary output ranges from 1.0 to 1.8 liters per day.
Typically, urine consists of 95% water and 5% solutes. The
volume of urine produced by the kidneys varies with such
factors as fluid intake, environmental temperature, relative
humidity, respiratory rate, and body temperature.

An analysis of urine composition and volume often
is used to evaluate the functions of the kidneys and other
organs. This procedure, called urinalysis, is a clinical assessment and a diagnostic tool for certain pathological conditions
and general overall health. A urinalysis and a complete blood
analysis complement each other for an evaluation of certain
diseases such as diabetes mellitus, and general health.

A urinalysis involves three aspects: physical characteristics, chemical analysis, and a microscopic examination. Physical characteristics of urine that are noted include volume, color,

transparency, and odor. The chemical analysis of solutes in urine addresses urea and other nitrogenous wastes; electrolytes; pigments; and possible glucose, protein, ketones, bilirubin, and hemoglobin. The specific gravity and pH of urine are greatly influenced by the components and amounts of solutes. An examination of microscopic solids, including cells, casts, and crystals, assists in the diagnosis of injury, various diseases, and urinary infections.

WARNING *While performing the following tests, you should wear disposable latex gloves so that skin contact with urine is avoided. Observe all safety procedures listed for this laboratory exercise. (Normal and abnormal simulated urine specimens can be used instead of real urine.)*

 PRACTICE

PROCEDURE A—Physical and Chemical Analysis

1. Proceed to the restroom with a clean, disposable container if collected urine is to be used for this exercise. The first small volume of urine should not be collected because it contains abnormally high levels of microorganisms from the urethra. Collect a midstream sample of about 50 mL of urine. The best collections are the first specimen in the morning or one taken 3 hours after a meal. Refrigerate samples if they are not used immediately.

2. Place a sample of urine in a clean, transparent container. Describe the *color* of the urine. Normal urine varies from light yellow to amber, depending on the presence of urochromes, end-product pigments produced during the decomposition of hemoglobin. Dark urine indicates a high concentration of pigments.

   Abnormal urine colors include yellow-brown or green, due to elevated concentrations of bile pigments, and red to dark brown, due to the presence of blood. Certain foods, such as beets or carrots, and various drug substances also can cause color changes in urine, but in such cases the colors have no clinical significance. Enter the results of this and the following tests in Part A of Laboratory Report 49.

3. Evaluate the *transparency* of the urine sample (judge whether the urine is clear, slightly cloudy, or very cloudy). Normal urine is clear enough to see through. You can read newsprint through slightly cloudy urine; you can no longer read newsprint through cloudy urine. Cloudy urine indicates the presence of various substances, including mucus, bacteria, epithelial cells, fat droplets, and inorganic salts.

4. Gently wave your hand over the urine sample toward your nose to detect the *odor*. Normal urine should have a slight ammonia-like odor due to the nitrogenous wastes. Some vegetables (such as asparagus), drugs (such as certain vitamins), and diseases (such as diabetes mellitus) will influence the odor of urine.

5. Determine the *specific gravity* of the urine sample, which indicates the solute concentration. Specific gravity is the ratio of the weight of something to the weight of an equal volume of pure water. For example, mercury (at 15°C) weighs 13.6 times as much as an equal volume of water; thus, it has a specific gravity of 13.6. Although urine is mostly water, it has substances dissolved in it and is slightly heavier than an equal volume of water. The specific gravity of pure (distilled) water is 1.000, while the specific gravity of urine is higher, varying from 1.003 to 1.035 under normal circumstances. If the specific gravity is too low, the urine contains few solutes and represents dilute urine, a likely result of excessive fluid intake or the use of diuretics. A specific gravity above the normal range represents a higher concentration of solutes, likely from a limited fluid intake. Concentrated urine over an extended time increases the risk of the formation of kidney stones (renal calculi).

   To determine the specific gravity of a urine sample, follow these steps:

   a. Pour enough urine into a clean urinometer cylinder to fill it about three-fourths full. Any foam that appears should be removed with a paper towel.

   b. Use a laboratory thermometer to measure the temperature of the urine.

   c. Gently place the urinometer hydrometer into the urine, and *make sure that the float is not touching the sides or the bottom of the cylinder* (fig. 49.1).

   d. Position your eye at the level of the urine surface. Determine which line on the stem of the hydrometer

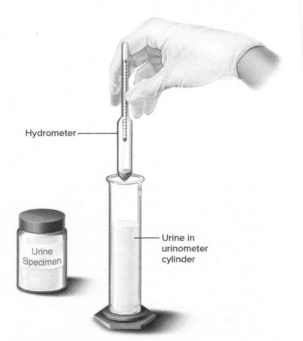

Hydrometer

Urine Specimen

Urine in urinometer cylinder

FIGURE 49.1 Float the hydrometer in the urine, making sure that it does not touch the sides or the bottom of the cylinder.

intersects the lowest level of the concave surface (meniscus) of the urine.

e. Liquids tend to contract and become denser as they are cooled, or to expand and become less dense as they are heated, so it may be necessary to make a temperature correction to obtain an accurate specific gravity measurement. To do this, add 0.001 to the hydrometer reading for each 3 degrees of urine temperature above 25°C or subtract 0.001 for each 3 degrees below 25°C. Enter this calculated value in the table in Part A of the laboratory report as the test result.

6. Individual or combination reagent strips can be used to perform a variety of urine tests (fig. 49.2). In each case, directions for using the strips are found on the strip container. *Be sure to read them.*

   To perform each test, follow these steps:

   a. Obtain a urine sample and the proper reagent test strip.

   b. Read the directions on the strip container.

   c. Dip the strip in the urine sample as directed on the container.

   d. Remove the strip at an angle and let it touch the inside rim of the urine container to remove any excess liquid.

   e. Wait for the length of time indicated by the directions on the container before you compare the color of the test strip with the standard color scale provided with the container. The value or amount represented by the matching color should be used as the test result for the following tests performed and recorded in Part A of the laboratory report.

7. Perform the *pH test.* The pH of normal urine varies from 4.6 to 8.0, but most commonly, it is near 6.0 (slightly acidic). The pH of urine may decrease as a result of a diet high in protein, or it may increase with a vegetarian diet. Significant daily variations within the broad normal range are results of concentrations of excesses from variable diets.

8. Perform the *glucose test.* Normally, there is no glucose in urine. However, glucose may appear in the urine temporarily following a meal high in carbohydrates. Glucose also may appear in the urine as a result of uncontrolled diabetes mellitus.

9. Perform the *protein test.* Normally, proteins of large molecular size are not present in urine. However, those of small molecular sizes, such as albumins, may appear in trace amounts, particularly following strenuous exercise. Increased amounts of proteins also may appear as a result of kidney diseases in which the glomeruli are damaged or as a result of high blood pressure.

10. Perform the *ketone test.* Ketones are products of fat metabolism. Usually, they are not present in urine. However, they may appear in the urine if the diet fails to provide adequate carbohydrate, as in the case of prolonged fasting or starvation, or as a result of insulin deficiency (diabetes mellitus).

11. Perform the *bilirubin test.* Bilirubin, which results from hemoglobin decomposition in the liver, normally is absent in urine. It may appear, however, as a result of liver disorders that cause obstructions of the biliary tract. Urochrome, a normal yellow component of urine, is a result of additional breakdown of bilirubin.

12. Perform the *hemoglobin/occult blood test.* Hemoglobin occurs in the red blood cells, and because such cells normally do not pass into the renal tubules, hemoglobin is not found in normal urine. Its presence in urine usually indicates a disease process, a transfusion reaction, an injury to the urinary organs, or menstrual blood.

13. Perform the *leukocytes test.* White blood cells in the urine are indicative of an infection of the urinary tract, urinary bladder, or kidney; normally, they are not present in urine.

14. Complete Part A of the laboratory report.

 PRACTICE

PROCEDURE B—Microscopic Sediment Analysis

1. A urinalysis usually includes an analysis of urine sediment—the microscopic solids present in a urine sample. This sediment normally includes mucus; certain crystals; and a variety of cells, such as the epithelial cells that line the urinary tubes and an occasional white blood cell. Other types of solids, such as casts or red blood cells, may indicate a disease or injury if they are present in excess. (Casts are cylindrical masses of cells or other substances that form in the renal tubules and are flushed out by the flow of urine.)

   To observe urine sediment, follow these steps:

   a. Thoroughly stir or shake your urine sample to suspend the sediment, which tends to settle to the bottom of the container.

   b. Pour 10 mL of urine into a clean centrifuge tube and centrifuge it for 5 minutes at slow speed (1,500 rpm). Be sure to balance the centrifuge with an even number of tubes filled to the same levels.

   c. Carefully decant 9 mL (leave 1 mL) of the liquid from the sediment in the bottom of the centrifuge tube, as directed by your laboratory instructor. Resuspend the 1 mL of sediment.

   d. Use a medicine dropper to remove some of the sediment and place it on a clean microscope slide.

FIGURE 49.2 An example of a reagent test strip dipped into urine to determine a variety of urine components. (McGraw Hill Education)

e. Add a drop of Sedi-stain to the sample, and add a coverslip.

f. Examine the sediment with low-power (reduce the light when using low power) and high-power magnifications.

g. With the aid of figure 49.3, identify the types of solids present.

2. In Part B of the laboratory report, make a sketch of each type of sediment that you observed.

3. Complete Part B of the laboratory report.

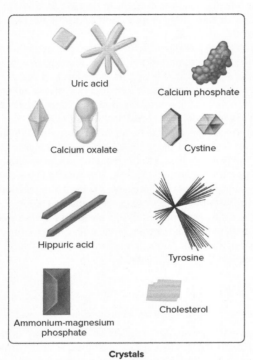

Crystals

Uric acid

Calcium phosphate

Calcium oxalate

Cystine

Hippuric acid

Tyrosine

Ammonium-magnesium phosphate

Cholesterol

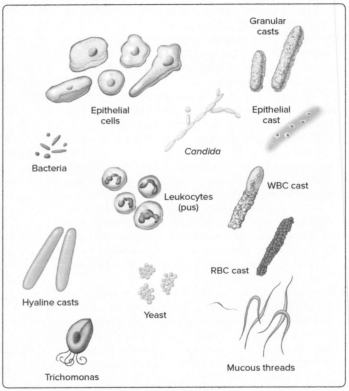

Cells and casts

Epithelial cells

Granular casts

Candida

Epithelial cast

Bacteria

Leukocytes (pus)

WBC cast

Hyaline casts

RBC cast

Yeast

Trichomonas

Mucous threads

FIGURE 49.3 Types of urine sediment. Healthy individuals lack many of these sediments and possess only occasional to trace amounts of others. *(Note*: Shades of white to purple sediments are most characteristic when using Sedi-stain.)

Name _Grace Bryant_____

Date _5/8/23_____

Section _BISC 228 001_____

The 🄐 corresponds to the indicated Learning Outcome(s) found at the beginning of the laboratory exercise.

Urinalysis

PART A ASSESSMENTS

1. Enter your observations, test results, and evaluations in the following table: 🄐 🄐 🄐

Urine Characteristics	Observations and Test Results	Normal Values	Evaluations
Color		Light yellow to amber	
Transparency		Clear	
Odor		Slight ammonia-like	
Specific gravity (corrected for temperature)	1.003	1.003–1.035	
pH	6	4.6–8.0	
Glucose	normal	Negative (0)	
Protein	trace	Negative to trace	
Ketones	negative	Negative (0)	
Bilirubin		Negative (0)	
Hemoglobin/occult blood	negative	Negative (0)	
Leukocytes	negative	Negative to trace	
(Other)			

2. Summarize the results of the physical and chemical analyses of urine. 🄐 _____

_____

_____

_____

_____

_____

ASSESS

CRITICAL THINKING

Why do you think it is important to refrigerate a urine sample if an analysis cannot be performed immediately after collecting it?

_____
_____
_____
_____

PART B ASSESSMENTS

1. Make a sketch for each type of sediment you observed. Label any from those shown in figure 49.3. **A**

2. Summarize the results of the microscopic sediment analysis of urine. **5** _____

_____
_____
_____
_____
_____
_____

Reproductive System and Development

LABORATORY EXERCISE

50

Male Reproductive System

MATERIALS NEEDED

Textbook
Human torso model
Model of the male reproductive system
Anatomical chart of the male reproductive system
Compound light microscope
Prepared microscope slides of the following:
    Testis section
    Epididymis, cross section
    Ductus deferens, cross section
    Penis, cross section

PURPOSE OF THE EXERCISE APR

To review the structures and functions of the male reproductive organs and to examine some of their features.

 ## LEARNING OUTCOMES APR

After completing this exercise, you should be able to

1. Locate and identify the organs of the male reproductive system.

2. Describe the functions of these organs.

3. Sketch and label the major features from microscopic sections of the testis, epididymis, ductus deferens, and penis.

The organs of the male reproductive system are specialized to produce and maintain the male sex cells, to transport these cells together with supporting fluids to the female reproductive tract, and to produce and secrete male sex hormones.

    These organs include the testes, in which sperm and male sex hormones are produced, and sets of internal and external accessory structures. The internal structures include various tubes and glands, whereas the external structures are the scrotum and the penis.

 ## PRACTICE

PROCEDURE A—Male Reproductive Organs

1. Review the concept headings titled "Testes," "Male Internal Accessory Reproductive Organs," and "Male External Accessory Reproductive Organs" in section 22.2 in chapter 22 of the textbook.

2. As a review activity, label figures 50.1 and 50.2.

3. Observe the human torso model, the model of the male reproductive system, and the anatomical chart of the male reproductive system. Locate the following features:

testes
    seminiferous tubules
    interstitial cells (cells of Leydig)
inguinal canal
spermatic cord
epididymis
ductus deferens (vas deferens)
ejaculatory duct
seminal vesicles
prostate gland
bulbourethral glands
scrotum
penis
    corpora cavernosa
    corpus spongiosum
    tunica albuginea
    glans penis
    external urethral orifice
    prepuce (foreskin)

4. Complete Part A of Laboratory Report 50.

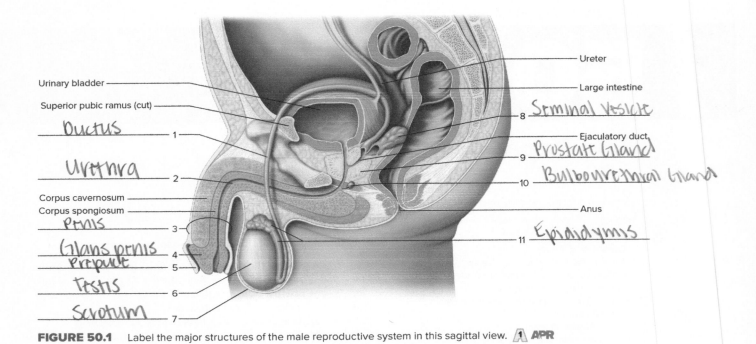

Ureter

Urinary bladder

Large intestine

Superior pubic ramus (cut)

Seminal vesicle — 8

Ductus — 1

Ejaculatory duct — 9
Prostate Gland

Urethra — 2

Bulbourethral Gland — 10

Corpus cavernosum

Corpus spongiosum

Anus

Penis — 3

Glans penis — 4
Prepuce — 5

Epididymis — 11

Testis — 6

Scrotum — 7

FIGURE 50.1   Label the major structures of the male reproductive system in this sagittal view. A APR

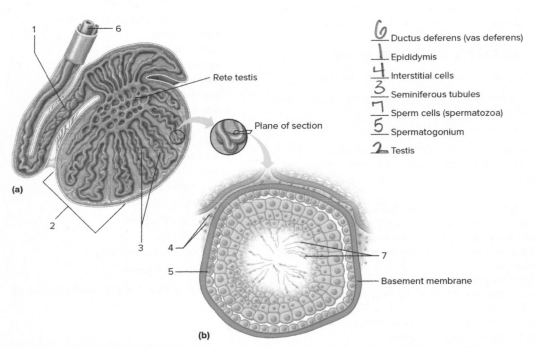

Rete testis

Plane of section

(a)

(b)

6 Ductus deferens (vas deferens)

1 Epididymis

4 Interstitial cells

3 Seminiferous tubules

7 Sperm cells (spermatozoa)

5 Spermatogonium

2 Testis

Basement membrane

FIGURE 50.2   Label the diagram of (a) the sagittal section of a testis and (b) a cross section of a seminiferous tubule by placing the correct numbers in the spaces provided. A APR

PRACTICE

PROCEDURE B—Microscopic Anatomy

1. Obtain a microscope slide of human testis section and examine it, using low-power magnification (fig. 50.3). Locate the thick *fibrous capsule* (tunica albuginea) on the surface and the numerous sections of *seminiferous tubules* inside.
2. Focus on some of the seminiferous tubules, using high-power magnification (fig. 50.4). Locate the *basement*

membrane and the layer of *spermatogonia* (undifferentiated spermatogenic cells) just beneath the basement membrane. Identify some *sustentacular cells* (supporting cells; Sertoli cells), which have pale, oval-shaped nuclei. Spermatogonia give rise to *spermatogenic cells* that are in various stages of spermatogenesis as they are forced toward the lumen. Near the lumen of the tube, find some darkly stained, elongated heads of developing sperm cells. In the spaces between adjacent seminiferous tubules, locate some isolated *interstitial cells* (cells of Leydig) of the endocrine system. Interstitial cells produce the hormone testosterone, transported by the blood.
3. Prepare a labeled sketch of a representative section of the testis in Part B of the laboratory report.
4. Obtain a microscope slide of a cross section of *epididymis* (fig. 50.5). Examine its wall, using high-power magnification. Note the elongated, *pseudostratified columnar epithelial cells* that compose most of the inner lining. These cells have nonmotile stereocilia (elongated microvilli) on their free surfaces that absorb excess fluid secreted by the testes. Also note the thin layer of smooth muscle and connective tissue surrounding the tube.

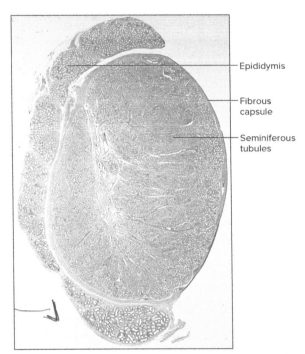

FIGURE 50.3 Micrograph of a human testis (1.7×). (Biophoto Associates/Science Source)

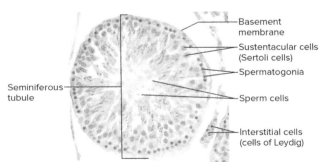

FIGURE 50.4 Micrograph of a seminiferous tubule (250×). **APR** (©Ed Reschke)

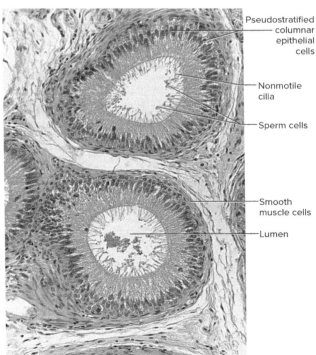

FIGURE 50.5 Micrograph of a cross section of a human epididymis (200×). (©Ed Reschke)

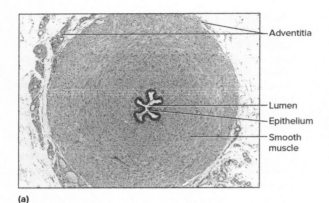

Adventitia

Lumen

Epithelium

Smooth muscle

(a)

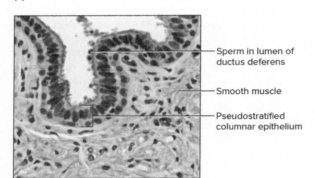

Sperm in lumen of ductus deferens

Smooth muscle

Pseudostratified columnar epithelium

(b)

FIGURE 50.6   Ductus deferens. (*a*) Micrograph of a cross section of the ductus deferens (40×). (*b*) Micrograph of the wall of the ductus deferens (400×). **APR**

((*a,b*) Alvin Telser/McGraw-Hill Education)

5. Prepare a labeled sketch of the epididymis wall in Part B of the laboratory report.
6. Obtain a microscope slide of a cross section of the *ductus deferens,* and examine it with scan and low-power magnification (fig. 50.6*a*). Note the small lumen with an epithelium and the thick *smooth muscle* layer composing most of the structure of the duct. Three layers of smooth muscle fibers are distinctive: a middle circular layer between two layers of longitudinal muscle fibers. The peristaltic contractions of the muscular layer transport sperm along the passageway toward the ejaculatory duct and the urethra. An outer, connective tissue covering of the ductus deferens is called the *adventitia.* Use high-power magnification and examine the epithelium layer composed of pseudostratified columnar epithelial cells (fig. 50.6*b*).
7. Prepare a labeled sketch of a representative section of a ductus deferens in Part B of the laboratory report.
8. Obtain a microscope slide of a *penis* cross section, and examine it with low-power magnification (fig. 50.7). Identify the following features:

   corpora cavernosa
   corpus spongiosum
   tunica albuginea
   urethra
   skin

9. Prepare a labeled sketch of a penis cross section in Part B of the laboratory report.
10. Complete Part B of the laboratory report.

Sperm cell
↳ little strings look like lines.

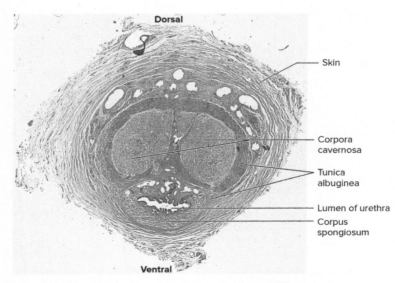

Dorsal

Skin

Corpora cavernosa

Tunica albuginea

Lumen of urethra

Corpus spongiosum

Ventral

FIGURE 50.7   Micrograph of a cross section of the body of the penis (5×). **APR** (©Ed Reschke)

LABORATORY EXERCISE

51

Female Reproductive System

MATERIALS NEEDED

Textbook
Human torso model
Model of the female reproductive system
Anatomical chart of the female reproductive system
Compound light microscope
Prepared microscope slides of the following:
    Ovary section with maturing follicles
    Uterine tube, cross section
    Uterine wall section
    Uterine wall, early proliferative phase
    Uterine wall, secretory phase
    Uterine wall, early menstrual phase

PURPOSE OF THE EXERCISE APR

To review the structures and functions of the female reproductive organs and to examine some of their features.

LEARNING OUTCOMES APR

After completing this exercise, you should be able to

(1) Locate and identify the organs of the female reproductive system.

(2) Describe the functions of these organs.

(3) Sketch and label the major features from microscopic sections of the ovary, uterine tube, and uterine wall.

The organs of the female reproductive system are specialized to produce and maintain the female sex cells, to transport these cells to the site of fertilization, to provide a favorable environment for a developing offspring, to move the offspring to the outside, and to produce female sex hormones.

These organs include the ovaries, which produce the oocytes (egg cells) and female sex hormones, and sets of internal and external accessory structures. The internal accessory structures include the uterine tubes, uterus, and vagina. The external accessory structures are the labia majora, labia minora, clitoris, and vestibular glands.

PRACTICE

PROCEDURE A—Female Reproductive Organs

1. Review the concept headings titled "Ovaries," "Female Internal Accessory Reproductive Organs," "Female External Accessory Reproductive Organs," in section 22.5 and section 22.7 titled "Mammary Glands" in chapter 22 of the textbook.
2. As a review activity, label figures 51.1–51.5.
3. Observe the human torso model, the model of the female reproductive system, and the anatomical chart of the female reproductive system. Locate the following features:

ovaries
  medulla
  cortex
ligaments
  broad ligament
  suspensory ligament
  ovarian ligament
  round ligament
uterine tubes (Fallopian tubes; oviducts)
  infundibulum
  fimbriae
uterus
  fundus
  body
  cervix
  cervical orifice
  uterine wall
    endometrium
    myometrium
    perimetrium
rectouterine pouch
vagina
  fornices
  vaginal orifice
  hymen
  mucosal layer

vulva (external accessory structures)
  mons pubis
  labium majus
  labium minus
  vestibule
  vestibular glands
  clitoris

breasts
  nipple
  areola
  alveolar glands (compose milk-producing parts of mammary glands)
  lactiferous ducts
  adipose tissue

4. Complete Part A of Laboratory Report 51.

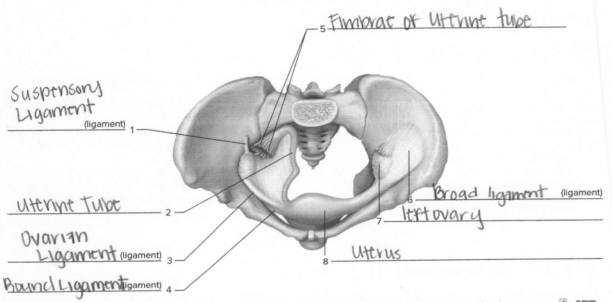

Suspensory Ligament _____ (ligament) 1

Uterine Tube _____ 2

Ovarian Ligament _____ (ligament) 3

Round Ligament _____ (ligament) 4

5 Fimbrae of Uterine tube _____

6 Broad ligament _____ (ligament)

7 left ovary _____

8 Uterus _____

FIGURE 51.1   Label the ligaments and other structures associated with the female internal reproductive organs. **1 APR**

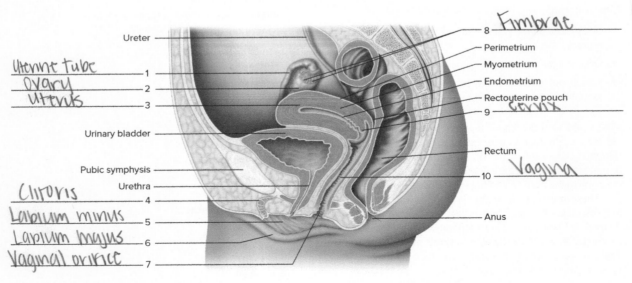

Uterine tube _____ 1
Ovary _____ 2
Uterus _____ 3

Ureter

Urinary bladder

Pubic symphysis

Clitoris _____ 4
Labium minus _____ 5
Labium majus _____ 6
Vaginal orifice _____ 7

Urethra

8 Fimbrae _____
Perimetrium
Myometrium
Endometrium
Rectouterine pouch
9 Cervix _____
Rectum
10 Vagina _____
Anus

FIGURE 51.2   Label the structures of the female reproductive system in this sagittal view. **1 APR**

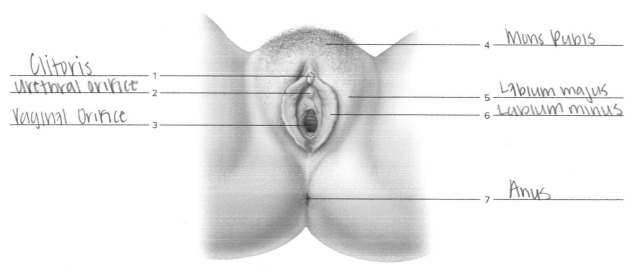

Clitoris — 1
Urethral orifice — 2
Vaginal orifice — 3

4 — Mons pubis
5 — Labium majus
6 — Labium minus
7 — Anus

FIGURE 51.3   Label the female external reproductive organs and associated structures. 🅰 APR

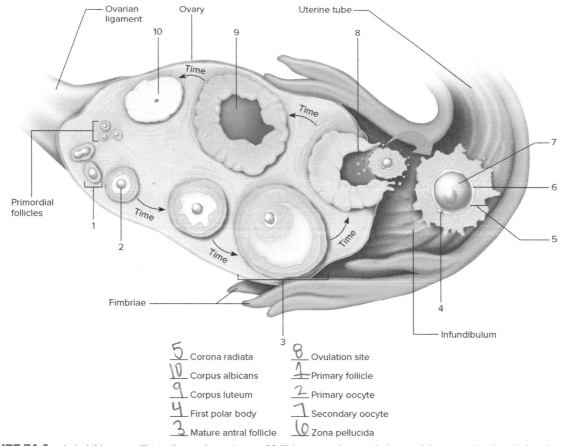

Ovarian ligament
Ovary
Uterine tube
10
9
8
Time
Time
Primordial follicles
1
Time
2
Time
Time
Fimbriae
3
7
6
5
4
Infundibulum

5 Corona radiata   8 Ovulation site
10 Corpus albicans   1 Primary follicle
9 Corpus luteum   2 Primary oocyte
4 First polar body   7 Secondary oocyte
3 Mature antral follicle 6 Zona pellucida

FIGURE 51.4   Label this ovary, illustrating various stages of follicle maturation, ovulation, and degeneration, by placing the correct numbers in the spaces provided. (*Note*: Although the various stages of ovarian development are depicted in a single illustration, the stages do not migrate around the ovary, but occur in a particular location during a particular ovarian cycle.) 🅰 APR

PRACTICE

PROCEDURE B—Microscopic Anatomy

1. Obtain a microscope slide of an ovary section with maturing follicles, and examine it with low-power magnification (fig. 51.6). Locate the outer layer, or *cortex*, composed of densely packed cells with developing follicles, and the inner layer, or *medulla,* which largely consists of loose connective tissue.

2. Focus on the cortex of the ovary, using high-power magnification (fig. 51.7). Note the thin layer of small cuboidal cells on the free surface. These cells constitute the

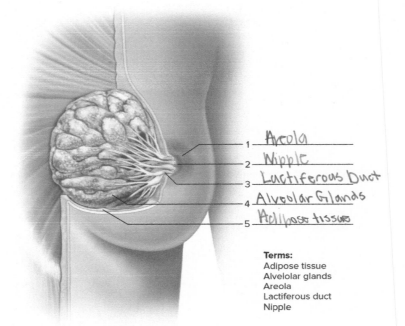

FIGURE 51.5   Label the structures of the breast (anterior view), using the terms provided. 🅐 **APR**

1 — Areola
2 — Nipple
3 — Lactiferous Duct
4 — Alveolar Glands
5 — Adipose tissues

Terms:
Adipose tissue
Alvelolar glands
Areola
Lactiferous duct
Nipple

wavy w/ dots

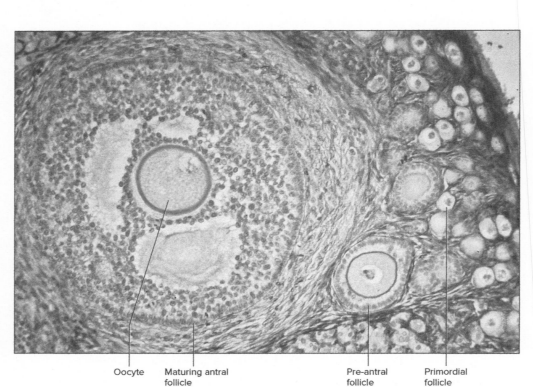

FIGURE 51.6
Micrograph of the ovary showing various stages of follicular development in the cortex (250×).
(Biophoto Associates/ Science Source)

Oocyte   Maturing antral follicle   Pre-antral follicle   Primordial follicle *germinal epithelium.* Also locate some *primordial follicles* just beneath the germinal epithelium. Each follicle consists of a single, relatively large *primary oocyte* with a prominent nucleus and a covering of *follicular cells.*

3. Use low-power magnification to search the ovarian cortex for maturing follicles in various stages of development (fig. 51.6). Locate and compare primordial follicles and *primary follicles.* A certain primary follicle can develop into a *pre-antral follicle* and then into a *mature antral follicle* just before ovulation (figs. 51.4 and 51.7).

4. Prepare two labeled sketches in Part B of the laboratory report to illustrate the changes that occur in a follicle as it matures.

5. Obtain a microscope slide of a cross section of a uterine tube. Examine it, using low-power magnification. The shape of the lumen is very irregular because of folds of the mucosa layer.

6. Focus on the inner lining of the uterine tube, using high-power magnification (fig. 51.8). The lining is composed of *simple columnar epithelium,* and some of the epithelial cells are ciliated on their free surfaces.

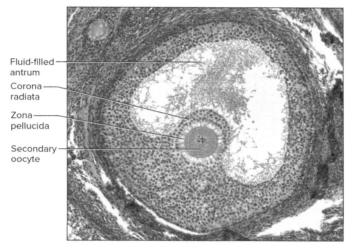

Fluid-filled antrum

Corona radiata

Zona pellucida

Secondary oocyte

FIGURE 51.7 Micrograph of a mature antral follicle (250×). **APR** (Alvin Telser/McGraw-Hill Education)

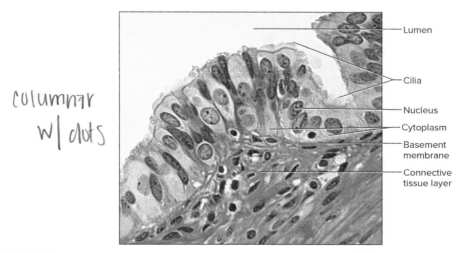

Lumen

Cilia

Nucleus

Cytoplasm

Basement membrane

Connective tissue layer

columnar
w/ dots

FIGURE 51.8 Micrograph of a cross section of the uterine tube (800×). **APR** (Ed Reschke/Getty Images)

7. Prepare a labeled sketch of a representative region of the wall of the uterine tube in Part B of the laboratory report.

8. Obtain a microscope slide of the uterine wall section (fig. 51.9). Examine it, using low-power magnification, and locate the following:

   endometrium (inner mucosal layer)
   myometrium (middle thick, muscular layer)
   perimetrium (outer serosal layer)

9. Prepare a labeled sketch of a representative section of the uterine wall in Part B of the laboratory report.

10. Complete Part B of the laboratory report.

LEARN: ACTIVITY

Observe the slides in the demonstration microscopes. Each slide contains a section of uterine mucosa taken during a different phase in the reproductive cycle. In the *early proliferative phase,* note the simple columnar epithelium on the free surface of the mucosa and the many sections of tubular uterine glands in the tissues beneath the epithelium. In the *secretory phase,* the endometrium is thicker and the uterine glands appear more extensive and they are coiled. In the *early menstrual phase,* the endometrium is thinner because its surface functional layer has been shed, while retaining the basal layer. Also, the uterine glands are less apparent, and the spaces between the glands contain many leukocytes. The basal layer of the endometrium renews the functional layer after menstruation ends.

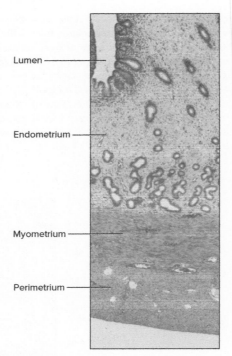

FIGURE 51.9   Micrograph of the uterine wall (10×).
APR (Carol D. Jacobson, Ph.D., Department of Veterinary Anatomy, Iowa State University/McGraw-Hill Education)

52

Fertilization and Early Development

MATERIALS NEEDED

Textbook
Sea urchin egg suspension*
Sea urchin sperm suspension*
Compound light microscope
Depression microscope slide
Coverslip
Medicine droppers
Prepared microscope slide of the following:
    Sea urchin embryos (cleavage, blastula, and
        gastrula stages)
Models of human development
Vaseline
Toothpick
Preserved mammalian embryos

*See the Instructor's Manual for a source of materials.

PURPOSE OF THE EXERCISE

To review the process of fertilization, to observe sea urchin eggs being fertilized, and to examine embryos in early stages of development.

 LEARNING OUTCOMES APR

After completing this exercise, you should be able to

1. Describe the process and structures of fertilization.

2. Describe and distinguish the early developmental stages of a human.

3. Identify and sketch the early developmental stages of a sea urchin.

Fertilization is the process by which the nuclei of a secondary oocyte and a sperm cell come together and combine their chromosomes, forming a single cell called a zygote.

Ordinarily, before fertilization can occur in a human female, an oocyte (egg cell) must be released from the ovary and must be carried into a uterine tube. Also, semen containing sperm cells must be deposited in the vagina; some of these sperm cells must travel through the uterus and into the uterine tubes. Although many sperm cells may reach an oocyte (egg cell), only one will participate in the fertilization of the secondary oocyte.

Shortly after fertilization, the zygote undergoes division (mitosis) to form two cells. These two cells become four, they in turn divide into eight, and so forth. The resulting mass of cells continues to grow and undergoes developmental changes and growth that give rise to an offspring.

 PRACTICE

PROCEDURE A—Fertilization

1. Review section 23.2 titled "Fertilization" and the concept heading titled "Period of Cleavage" in section 23.3 in chapter 23 of the textbook.

2. As a review activity, label figure 52.1.

3. Complete Part A of Laboratory Report 52.

4. Although it is difficult to observe fertilization in mammals since the process occurs internally, it is possible to view forms of external fertilization. For example, egg and sperm cells can be collected from sea urchins, and the process of fertilization can be observed microscopically. To make this observation, follow these steps:

   a. Place a drop of sea urchin egg-cell suspension in the chamber of a depression slide, and add a coverslip.

   b. Examine the egg cells, using low-power magnification.

   c. Focus on a single unfertilized egg cell with high-power magnification, and sketch the cell in Part B of the laboratory report.

   d. Remove the coverslip and add a drop of sea urchin sperm-cell suspension to the depression slide. Replace the coverslip, and observe the sperm cells with high-power magnification as they cluster around the egg cells. This attraction is stimulated by gamete secretions.

   e. Observe the egg cells with low-power magnification once again, and watch for the appearance of *fertilization membranes*. Such a membrane forms as soon as an egg cell is penetrated by a sperm cell; it looks like a clear halo surrounding the egg cell.

   f. Focus on a single fertilized egg cell, and sketch it in Part B of the laboratory report.

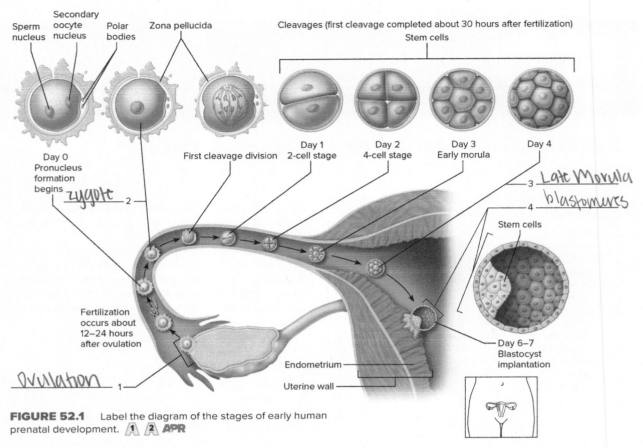

Sperm nucleus
Secondary oocyte nucleus
Polar bodies
Zona pellucida

Cleavages (first cleavage completed about 30 hours after fertilization)
Stem cells

Day 0
Pronucleus formation begins
zygote   2

First cleavage division

Day 1
2-cell stage

Day 2
4-cell stage

Day 3
Early morula

Day 4

3 — _Late Morula_
4 — _blastomeres_

Stem cells

Fertilization occurs about 12–24 hours after ovulation

Ovulation   1

Endometrium

Uterine wall

Day 6–7
Blastocyst implantation

FIGURE 52.1   Label the diagram of the stages of early human prenatal development. 1 2 **APR**

LEARN: LAB IN MOTION

Use a toothpick to draw a thin line of Vaseline around the chamber of the depression slide containing the fertilized sea urchin egg cells. Place a coverslip over the chamber, and gently press it into the Vaseline to seal the chamber and prevent the liquid inside from evaporating. Keep the slide in a cool place so that the temperature never exceeds 22°C (72°F). Using low-power magnification, examine the slide every 30 minutes, and look for the appearance of two-, four-, and eight-cell stages of developing sea urchin embryos.

2. Obtain a prepared microscope slide of developing sea urchin embryos. This slide contains embryos in various stages of cleavage. Search the slide, using low-power magnification, and locate embryos in two-, four-, and eight-cell and morula stages. Observe that cleavage results in an increase of cell numbers; however, the cells get progressively smaller.

3. Using low-power magnification, locate the blastula, and identify its blastocoel (hollow central cavity). Then locate the gastrula, and identify ectoderm (outermost germ layer) and endoderm (innermost germ layer).

4. Prepare a sketch of each stage in Part C of the laboratory report.

PRACTICE

PROCEDURE B—Sea Urchin Early Development

1. Review the concept headings titled "Period of Cleavage" and "Gastrulation and Organogenesis" in section 23.3 in chapter 23 of the textbook.

PRACTICE

PROCEDURE C—Human Early Development

1. Review the concept headings titled "Pre-embryonic Stage" and "Extraembryonic Membrane Formation and Placentation" in section 23.3 in chapter 23 of the textbook.

2. As a review activity, study and label figure 52.2.

3. Study figures 52.3 and 52.4.
4. Observe the models of human development, and identify the following features:

   blastocyst
       inner cell mass (embryoblast)
       trophoblast
       blastocyst cavity (blastocoel)
   primary germ layers
       ectoderm
       endoderm
       mesoderm
   chorion
   chorionic villi
   amnion
   amniotic fluid
   yolk sac
   allantois
   umbilical cord
       umbilical arteries (2)
       umbilical vein (1)
   placenta

5. Complete Part D of the laboratory report.

LEARN: ACTIVITY

Observe the preserved mammalian embryos that are on display. In addition to observing the developing external body structures, identify such features as the chorion, chorionic villi, amnion, yolk sac, umbilical cord, and placenta. What special features provide clues to the types of mammals these embryos represent?

_____

_____

_____

_____

_____

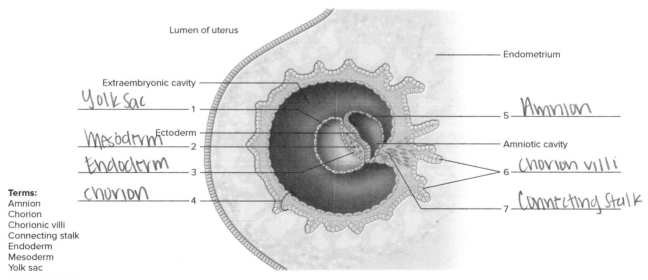

Lumen of uterus

Extraembryonic cavity

Yolk sac   1

Mesoderm   Ectoderm   2

Endoderm   3

Terms: Chorion   4
Amnion
Chorion
Chorionic villi
Connecting stalk
Endoderm
Mesoderm
Yolk sac

Endometrium

5 Amnion

Amniotic cavity

6 Chorion villi

7 Connecting Stalk

FIGURE 52.2   Label the major features of the early embryo and the structures associated with it, using the terms provided.

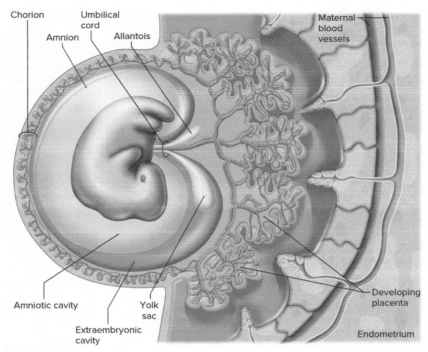

FIGURE 52.3   Human embryo with extraembryonic membranes and developing umbilical cord and placenta.

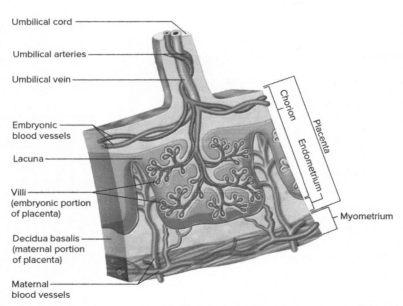

FIGURE 52.4   The umbilical cord, containing two oxygen-poor arteries and a single oxygen-rich vein, is connected to the placenta. The placenta, composed of an embryonic and maternal portion, serves as an exchange site of nutrients, gases, and wastes between embryonic and maternal blood.

53

Genetics

MATERIALS NEEDED

Textbook
Pennies (or other coins)
Dice
PTC paper
Astigmatism chart
Ichikawa's or Ishihara's color plates for
    colorblindness test

PURPOSE OF THE EXERCISE

To observe some selected human traits, to use pennies and dice to demonstrate laws of probability, and to solve some genetic problems using a Punnett square.

 ## LEARNING OUTCOMES

After completing this exercise, you should be able to

1. Examine and record twelve genotypes and phenotypes of selected human traits.

2. Demonstrate the laws of probability using tossed pennies and dice and interpret the results.

3. Predict genotypes and phenotypes of complete dominance, codominance, and sex-linked problems using Punnett squares.

Many of human genetics in this laboratory exercise are basic, external features to observe. Some traits are based upon simple Mendelian genetics. As our knowledge of genetics continues to develop, traits such as tongue roller and free earlobe may no longer be considered examples of simple Mendelian genetics. We may need to continue to abandon some simple Mendelian models. New evidence may involve polygenic inheritance, effects of other genes, or environmental factors as more appropriate explanations. Therefore, the analysis of family genetics is not always feasible or meaningful.

Genetics is the study of the inheritance of characteristics. The genes that transmit this information are coded in segments of DNA in chromosomes. Homologous chromosomes possess the same gene at the same *locus.* These genes may exist in variant forms, called *alleles.* If a person possesses two identical alleles, the person is said to be *homozygous* for that particular trait. If a person possesses two different alleles, the person is said to be *heterozygous* for that particular trait. The particular combination of these gene variants (alleles) represents the person's *genotype;* the appearance or physical manifestation of the individual that develops as a result of the way the genes are expressed represents the person's *phenotype.*

If one allele determines the phenotype by masking the expression of the other allele in a heterozygous individual, the allele is termed *dominant.* The allele whose expression is masked is termed *recessive.* If the heterozygous condition determines an intermediate phenotype, the inheritance represents *incomplete dominance.* However, different alleles are *codominant* if both are expressed in the heterozygous condition. Some characteristics inherited on the sex chromosomes result in phenotype frequencies that might be more prevalent in males or females. Such characteristics are called *sex-linked (X-linked or Y-linked) characteristics.*

As a result of meiosis during the formation of eggs and sperm, a mother and father each transmit an equal number of chromosomes (the haploid number 23) to form the zygote (diploid number 46). An offspring will receive one allele from each parent. These gametes combine randomly in the formation of each offspring. Hence, the *laws of probability* can be used to predict possible genotypes and phenotypes of offspring. A genetic tool called a *Punnett square* simulates all possible combinations (probabilities) in offspring genotypes and resulting phenotypes.

 PRACTICE

PROCEDURE A—Human Genotypes and Phenotypes

A complete set of genetic instructions in one human cell constitutes one's *genome.* The human genome contains about 2.9 billion base pairs representing approximately

TABLE 53.1   Examples of Some Common Human Phenotypes

Dominant Traits and Genotypes	Recessive Traits and Genotypes
Tongue roller (R___)	Nonroller (rr)
Freckles (F___)	No freckles (ff)
Widow's peak (W___)	Straight hairline (ww)
Dimples (D___)	No dimples (dd)
Free earlobe (E___)	Attached earlobe (ee)
Normal skin coloration (M___)	Albinism (mm)
Astigmatism (A___)	Normal vision (aa)
Curly hair (C___)	Straight hair (cc)
PTC taster (T___)	Nontaster (tt)
Blood type A (I^A___), B (I^B___), or AB ($I^A I^B$)	Blood type O (ii)
Normal color vision ($X^C X^C$), ($X^C X^c$), or ($X^C Y$)	Red-green colorblindness ($X^c X^c$) or ($X^c Y$)

20,500 protein-encoding genes. These instructions represent our genotypes and are expressed as phenotypes sometimes clearly observable on our bodies. Some of these traits are listed in table 53.1 and are discernible in figure 53.1.

A dominant trait might be homozygous or heterozygous, so only one capital letter is used along with a blank for the possible second dominant or recessive allele. For a recessive trait, two lowercase letters represent the homozygous recessive genotype for that characteristic. Dominant does not always correlate with the predominance of the allele in the gene pool; dominant means one allele will determine the appearance of the phenotype.

1. **Tongue roller/nonroller:** The dominant allele (R) determines the person's ability to roll the tongue into a U-shaped trough. The homozygous recessive condition (rr) prevents this tongue rolling (fig. 53.1). Record your results in the table in Part A of Laboratory Report 53.
2. **Freckles/no freckles:** The dominant allele (F) determines the appearance of freckles. The homozygous recessive condition (ff) does not produce freckles (fig. 53.1). Record your results in the table in Part A of the laboratory report.
3. **Widow's peak/straight hairline:** The dominant allele (W) determines the appearance of a hairline above the forehead that has a distinct downward point in the center, called a widow's peak. The homozygous recessive condition (ww) produces a straight hairline (fig. 53.1). A receding hairline would prevent this phenotype determination. Record your results in the table in Part A of the laboratory report.
4. **Dimples/no dimples:** The dominant allele (D) determines the appearance of a distinct dimple in one or both cheeks upon smiling. The homozygous recessive condition (dd) results in the absence of dimples

(fig. 53.1). Record your results in the table in Part A of the laboratory report.

5. **Free earlobe/attached earlobe:** The dominant allele (E) codes for the appearance of an inferior earlobe that hangs freely below the attachment to the head. The homozygous recessive condition (ee) determines the earlobe attaching directly to the head at its inferior border (fig. 53.1). Record your results in the table in Part A of the laboratory report.
6. **Normal skin coloration/albinism:** The dominant allele (M) determines the production of some melanin, producing normal skin coloration. The homozygous recessive condition (mm) determines albinism due to the inability to produce or use the enzyme tyrosinase in pigment cells. An albino does not produce melanin in the skin, hair, or the middle tunic (choroid coat, ciliary body, and iris) of the eye. The absence of melanin in the middle tunic allows the pupil to appear slightly red to nearly black. Remember that the pupil is an opening in the iris filled with transparent aqueous humor. An albino human has pale white skin, flax-white hair, and a pale blue iris. Record your results in the table in Part A of the laboratory report.
7. **Astigmatism/normal vision:** The dominant allele (A) results in an abnormal curvature to the cornea or the lens. As a consequence, some portions of the image projected on the retina are sharply focused, and other portions are blurred. The homozygous recessive condition (aa) generates normal cornea and lens shapes and normal vision. Use the astigmatism chart and directions to assess this possible defect described in Laboratory Exercise 35. Other eye defects, such as nearsightedness (myopia) and farsightedness (hyperopia), are different genetic traits due to genes at other locations. Record your results in the table in Part A of the laboratory report.

Dominant Traits

Recessive Traits

(a) Tongue roller

(b) Nonroller

(c) Freckles

(d) No freckles

(e) Widow's peak

(f) Straight hairline

FIGURE 53.1   Representative genetic traits comparing dominant and recessive phenotypes: (*a*) tongue roller; (*b*) nonroller; (*c*) freckles; (*d*) no freckles; (*e*) widow's peak; (*f*) straight hairline; (*g*) dimples; (*h*) no dimples; (*i*) free earlobe; (*j*) attached earlobe.

((*a–j*) ©J & J Photography)

Dominant Traits

Recessive Traits

(g) Dimples

(h) No dimples

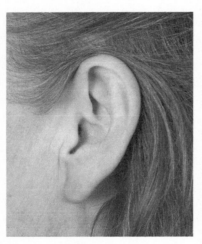

(i) Free earlobe

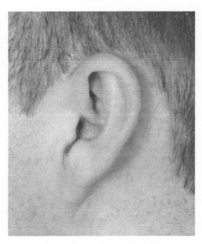

(j) Attached earlobe

FIGURE 53.1 *Continued.*

8. **Curly hair/straight hair:** The dominant allele (*C*) determines the appearance of curly hair. Curly hair is somewhat flattened in cross section, as the hair follicle of a similar shape served as a mold for the root of the hair during its formation. The homozygous recessive condition (*cc*) produces straight hair. Straight hair is nearly round in cross section from being molded into this shape in the hair follicle. In some populations (Caucasians), the heterozygous condition (*Cc*) expresses the intermediate wavy hair phenotype (incomplete dominance). This trait determination assumes no permanents or hair straightening procedures have been performed. Such hair alterations do not change the hair follicle shape, and future hair growth results in original genetic hair conditions. Record your unaltered hair appearance in the table in Part A of the laboratory report.

9. **PTC taster/nontaster:** The dominant allele (*T*) determines the ability to experience a bitter sensation when PTC paper is placed on the tongue. About 70% of people possess this dominant gene. The homozygous recessive condition (*tt*) makes a person unable to notice the substance. Place a piece of PTC (phenylthiocarbamide) paper on the upper tongue surface and chew it slightly to see if you notice a bitter sensation from this harmless chemical. The nontaster of the PTC paper does not detect any taste at all from this substance. Record your results in the table in Part A of the laboratory report.

10. **Blood type A, B, or AB/blood type O:** There are three alleles (I^A, I^B, and i) in the human population affecting RBC membrane structure. These alleles are located on a single pair of homologous chromosomes, so a person possesses either two of the three alleles or two of the same allele. All of the possible combinations of these alleles of genotypes and the resulting phenotypes are depicted in table 53.2. The expression of the blood type AB is a result of both codominant alleles located in the same individual. Possibly you have already determined your blood type in Laboratory Exercise 37 or have it recorded on a blood donor card. (If simulated blood-typing kits were used for Laboratory Exercise 37, those results would not be valid for your genetic factors.) Record your results in the table in Part A of the laboratory report.

11. **Sex determination:** A person with sex chromosomes XX displays a female phenotype. A person with sex chromosomes XY displays a male phenotype. Record your results in the table in Part A of the laboratory report.

12. **Normal color vision/red-green colorblindness:** This condition is a sex-linked (X-linked) characteristic. The alleles for color vision are also on the X chromosome, but they are absent on the Y chromosome. As a result, a female might possess both alleles (C and c), one on each of the X chromosomes. The dominant allele (C) determines normal color vision; the homozygous recessive condition (cc) results in red-green colorblindness. However, a male would possess only one of the two alleles for color vision because there is only a single X chromosome in a male. Hence, a male with even a single recessive gene for colorblindness possesses the defect. Note all the possible genotypes and phenotypes for this condition (table 53.1). Review the color vision test in Laboratory Exercise 35 using the color plate in figure 35.4 and color plates in Ichikawa's or Ishihara's book. Record your results in the table in Part A of the laboratory report.

13. Complete Part A of the laboratory report.

TABLE 53.2 Genotypes and Phenotypes (Blood Types)

Genotypes	Phenotypes (Blood Types)
$I^A I^A$ or $I^A i$	A
$I^B I^B$ or $I^B i$	B
$I^A I^B$ (codominant)	AB
ii	O

PRACTICE

PROCEDURE B—Laws of Probability

The laws of probability provide a mathematical way to determine the likelihood of events occurring by chance. This prediction is often expressed as a ratio of the number of results from experimental events to the number of results considered possible. For example, when tossing a coin there is an equal chance of the results displaying heads or tails. Hence, the probability is one-half of obtaining either a heads or a tails (there are two possibilities for each toss). When all of the probabilities of all possible outcomes are considered for the result, they will always add up to 1. To predict the probability of two or more events occurring in succession, multiply the probabilities of each individual event. For example, the probability of tossing a die and displaying a 4 twice in a row is $1/6 \times 1/6 = 1/36$ (there are six possibilities for each toss). Each toss in a sequence is an *independent event* (chance has no memory). The same laws apply when parents have multiple children (each fertilization is an independent event). Perform the following experiments to demonstrate the laws of probability:

1. Use a single penny (or other coin) and toss it 20 times. Predict the number of heads and tails that would occur from the 20 tosses. Record your prediction and the actual results observed in Part B of the laboratory report.

2. Use a single die (*pl.* dice) and toss it 24 times. Predict the number of times a number below 3 (numbers 1 and 2) would occur from the 24 tosses. Record your prediction and the actual results in Part B of the laboratory report.

3. Use two pennies and toss them simultaneously 32 times. Predict the number of times two heads, a heads and a tails, and two tails would occur. Record your prediction and the actual results in Part B of the laboratory report.

4. Use a pair of dice and toss them simultaneously 32 times. Predict the number of times for both dice coming up with odd numbers, one die an odd and the other an even number, and both dice coming up with even numbers. Record your prediction and the actual results in Part B of the laboratory report.

5. Obtain class totals for all of the coins and dice tossed by adding your individual results to a class tally location, as on the blackboard.

6. A Punnett square can be used to visually demonstrate the probable results for two pennies tossed simultaneously. For the purpose of a genetic comparison, an h (heads) will represent one "allele" on the coin; a t (tails) will represent a different "allele" on the coin.

Coin #1 Possibilities

	h	*t*
h	*hh*	*ht*
t	*ht*	*tt*

Coin #2 Possibilities

Possible Combinations of Two Tossed Pennies (in boxes)

7. Complete Part B of the laboratory report.

 PRACTICE

PROCEDURE C—Genetic Problems

1. A Punnett square can be constructed to demonstrate a visual display of the predicted offspring from parents with known genotypes. Recall that, in complete dominance, a dominant allele is expressed in the phenotype, as it can mask the other recessive allele on the homologous chromosome pair. During meiosis, the homologous chromosomes with their alleles separate (Mendel's Law of Segregation) into different gametes. An example of such a cross might be a homozygous dominant mother for dimples (*DD*) who has offspring with a father homozygous recessive (*dd*) for the same trait. The results of such a cross, according to the laws of probability, would be represented by the following Punnett square:

Female Gametes

	D	*D*
d	*Dd*	*Dd*
d	*Dd*	*Dd*

Male Gametes

Possible Genotypes of Offspring (in boxes)

Results: Genotypes: 100% *Dd* (all heterozygous)

Phenotypes: 100% dimples

In another example, assume that both parents are heterozygous (*Dd*) for dimples. The results of such a cross, according to the laws of probability, would be represented by the following Punnett square:

Female Gametes

	D	*d*
D	*DD*	*Dd*
d	*Dd*	*dd*

Male Gametes

Possible Genotypes of Offspring (in boxes)

Results: Genotypes: 25% *DD* (homozygous dominant); 50% Dd (heterozygous); 25% *dd* (homozygous recessive) (1:2:1 genotypic ratio)

Phenotypes: 75% dimples; 25% no dimples (3:1 phenotypic ratio)

2. Work genetic problems 1 and 2 in Part C of the laboratory report.

3. The ABO blood type inheritance represents an example of codominance. Review table 53.2 for the genotypes and phenotypes for the expression of this trait. A Punnett square can be constructed to predict the offspring of parents of known genotypes. In this example, assume the genotype of the mother is $I^A I^B$, and the father is *ii*. The results of such a cross would be represented by the following Punnett square:

Female Gametes

	I^A	I^B
i	$I^A i$	$I^B i$
i	$I^A i$	$I^B i$

Male Gametes

Possible Genotypes of Offspring (in boxes)

Results: Genotypes: 50% $I^A i$ (heterozygous for A); 50% $I^B i$ (heterozygous for B) (1:1 genotypic ratio)

Phenotypes: 50% blood type A; 50% blood type B (1:1 phenotypic ratio)

Note: In this cross, all of the children would have blood types unlike either parent.

4. Work genetic problems 3 and 4 in Part C of the laboratory report.

5. Review the inheritance of red-green colorblindness, an X-linked characteristic, in table 53.1. A Punnett square can be constructed to predict the offspring of parents of known genotypes. In this example, assume the genotype of the mother is heterozygous $X^C X^c$ (normal color vision but a carrier for the colorblindness defect), and the father is $X^C Y$ (normal color vision; no allele on the Y chromosome). The results of such a cross would be represented by the following Punnett square:

Female Gametes

	X^C	X^c
X^C	$X^C X^C$	$X^C X^c$
Y	$X^C Y$	$X^c Y$

Male Gametes

Possible Genotypes of Offspring (in boxes)

Results: Genotypes: 25% $X^C X^C$; 25% $X^C X^c$; 25% $X^C Y$; 25% $X^c Y$ (1:1:1:1 genotypic ratio)

Phenotypes for sex determination: 50% females; 50% males (1:1 phenotypic ratio)

Phenotypes for color vision: Females 100% normal color vision (however, 50% are heterozygous carriers for colorblindness)

Phenotypes for color vision: Males 50% normal; 50% with red-green colorblindness (1:1 phenotypic ratio)

Note: In X-linked inheritance, the colorblind males received the recessive gene from their mothers.

6. Complete Part C of the laboratory report.

Name _____

Date _____

Section _____

The Ⓐ corresponds to the indicated Learning Outcome(s) found at the beginning of the laboratory exercise.

Genetics

PART A ASSESSMENTS

1. Enter your test results for genotypes and phenotypes in the table. Circle your phenotype and genotype for each of the twelve traits. ⓵

Trait	Dominant Phenotype	Genotype	Recessive Phenotype	Genotype
Tongue movement	Roller	R___	Nonroller	rr
Freckles	Freckles	F___	No freckles	ff
Hairline	Widow's peak	W___	Straight	ww
Dimples	Dimples	D___	No dimples	dd
Earlobe	Free	E___	Attached	ee
Skin coloration	Normal (some melanin)	M___	Albinism	mm
Vision	Astigmatism	A___	Normal	aa
Hair shape*	Curly	C___	Straight	cc
Taste	PTC taster	T___	Nontaster for PTC	tt
Blood type	A, B, or AB	I^A___; I^B___; or $I^A I^B$	O	ii
Sex		XX or XY		
Color vision	Normal	$X^C X$___ or $X^C Y$	Red-green colorblindness	$X^c X^c$ or $X^c Y$

*In some populations (Caucasians) the heterozygous condition (Cc) results in the appearance of wavy hair, which actually represents an example of incomplete dominance for this trait.

2. Choose at least three dominant phenotypes that you circled in question 1, and analyze the genotypes for those traits. If it is feasible to observe your biological parents and siblings for any of these traits, are you able to determine if any of your dominant genotypes are homozygous dominant or heterozygous? _____ If so, which ones? _____ _____ Explain the rationale for your response.

_____

_____

_____

_____

_____

_____

_____

PART B ASSESSMENTS

1. Single penny tossed 20 times and counting heads and tails: [2]

   Probability (prediction): _____/20 heads _____/20 tails

   (Note: Traditionally, probabilities are converted to the lowest fractional representation.)

   Actual results: _____ heads _____ tails

   Class totals: _____ heads _____ tails

2. Single die tossed 24 times and counting the number of times a number below 3 occurs: [2]

   Probability: _____/24 number below 3 (numbers 1 and 2)

   Actual results: _____ number below 3

   Class totals: _____ number below 3 _____ total tosses by class members

3. Two pennies tossed simultaneously 32 times and counting the number of two heads, a heads and a tails, and two tails: [2]

   Probability: _____/32 of two heads _____/32 of a heads and a tails _____/32 of two tails

   Actual results: _____ two heads _____ heads and tails _____ two tails

   Class totals: _____ two heads _____ heads and tails _____ two tails

4. Two dice tossed simultaneously 32 times and counting the number of two odd numbers, an odd and an even number, and two even numbers: [2]

   Probability: _____/32 of two odd numbers _____/32 of an odd and an even number _____/32 of two even numbers

   Actual results: _____ two odd numbers _____ an odd and an even number _____ two even numbers

   Class totals: _____ two odd numbers _____ an odd and an even number _____ two even numbers

5. Use the example of the two dice tossed 32 times, and construct a Punnett square to represent the possible combinations that could be used to determine the probability (prediction) of odd and even numbers for the resulting tosses. Your construction should be similar to the Punnett square for the two coins tossed that is depicted in Procedure B of the laboratory exercise. [2]

6. Complete the following:

   a. Are the class totals closer to the predicted probabilities than your results? _____ Explain your response.

   _____

   _____

   b. Does the first toss of the penny or the first toss of the die have any influence on the next toss? _____
   Explain your response.

   _____

c. Assume a family has two boys or two girls. They wish to have one more child but hope for the child to be of the opposite sex from the two they already have. What is the probability that the third child will be of the opposite sex? _____ Explain your response.

_____

d. What is the probability (prediction) that a couple without children will eventually have four children, all girls? _____ Explain your response.

_____

_____

PART C ASSESSMENTS

For each of the genetic problems, (a) determine the parents' genotypes, (b) determine the possible gametes for each parent, (c) construct a Punnett square, and (d) record the resulting genotypes and phenotypes as ratios from the cross. Problems 1 and 2 involve examples of complete dominance; problems 3 and 4 are examples of codominance; problem 5 is an example of sex-linked (X-linked) inheritance.

1. Determine the results from a cross of a mother who is heterozygous (*Rr*) for tongue rolling with a father who is homozygous recessive (*rr*). ◢3

2. Determine the results from a cross of a mother and a father who are both heterozygous for freckles. ◢3

3. Determine the results from a mother who is heterozygous for blood type B and a father who is homozygous dominant for blood type A. ◢3

4. Determine the results from a mother who is heterozygous for blood type A and a father who is heterozygous for blood type B.

5. Colorblindness is an example of X-linked inheritance. Hemophilia is another example of X-linked inheritance, also from a recessive allele (*h*). The dominant allele (*H*) determines whether the person possesses normal blood clotting. A person with hemophilia has a permanent tendency for hemorrhaging due to a deficiency of one of the clotting factors (VIII—antihemophilic factor). Determine the offspring from a cross of a mother who is a carrier (heterozygous) for the disease and a father with normal blood coagulation.

ASSESS

CRITICAL THINKING

Assume that the genes for hairline and earlobes are on different pairs of homologous chromosomes. Determine the genotypes and phenotypes of the offspring from a cross if both parents are heterozygous for both traits. (1) First determine the genotypes for each parent. (2) Determine the gametes, but remember each gamete has one allele for each trait (gametes are haploid). (3) Construct a Punnett square, with 16 boxes, that has four different gametes from each parent along the top and the left edges. (This is an application to demonstrate Mendel's Law of Independent Assortment.) (4) List the results of genotypes and phenotypes as ratios.

Cat Dissection Laboratory Exercises

LABORATORY EXERCISE

54

Cat Dissection: Musculature

MATERIALS NEEDED

Textbook
Preserved cat (double injection)
Dissecting tray
Dissecting instruments
Large plastic bag
Identification tag
Disposable gloves
Bone shears
Human torso model
Human upper and lower limb models
Cat skeleton

⚠ SAFETY

- Wear disposable gloves when working on the cat dissection.
- Dispose of tissue remnants and gloves as instructed.
- Wash the dissecting tray and instruments as instructed.
- Wash your laboratory table.
- Wash your hands before leaving the laboratory.

PURPOSE OF THE EXERCISE

To examine and identify the musculature of the cat and to compare it with that of the human.

LEARNING OUTCOMES APR

After completing this exercise, you should be able to

(1) Name and locate the major skeletal muscles of the cat.

(2) Name the origins, insertions, and actions of the skeletal muscles designated by the laboratory instructor.

(3) Name and locate the corresponding skeletal muscles of the human.

Although the aim of this exercise is to become more familiar with the human musculature, human cadavers are not always available for dissection. Instead, preserved cats are often used for dissection because they are relatively small and can be purchased from biological suppliers. Because they are mammals, cats have many features in common with humans, including similar skeletal muscles (with similar names).

On the other hand, cats use four limbs for locomotion, whereas humans use only two limbs. Because the musculature of each type of organism is adapted to provide for its special needs, comparisons of the muscles of cats and humans may not be precise.

As you continue your dissection of various systems of the cat, many anatomical similarities between cats and humans will be observed. These fundamental similarities are called homologous structures. Although homologous structures have a similar structure and embryological origin, the functions are sometimes different.

MUSCLE DISSECTION TECHNIQUES

Dissect: to expose the entire length of the muscle from origin to insertion. Most of the procedures are accomplished with blunt probes used to separate various connective tissues that hold adjacent structures together. It does not mean to remove or to cut into the muscles or other organs.

Transect: to cut through the muscle near its midpoint. The cut is perpendicular to the length of the muscle fibers.

Reflect: to lift a transected muscle aside to expose deeper muscles or other organs.

PRACTICE

PROCEDURE A—Skin Removal

1. Cats usually are preserved with a solution that prevents microorganisms from causing the tissues to decompose. However, because fumes from this preserving fluid may

be annoying and may irritate skin, be sure to work in a well-ventilated room and wear disposable gloves to protect your hands.

2. Obtain a preserved cat, a dissecting tray, a set of dissecting instruments, a large plastic bag, and an identification tag.

3. If the cat is in a storage bag, dispose of any excess preserving fluid in the bag, as directed by the laboratory instructor. However, you may want to save some of the preserving fluid to keep the cat moist until you have completed your work later in the year.

4. If the cat is too wet with preserving fluid, blot the cat dry with paper towels.

5. Remove the skin from the cat. To do this, follow these steps:

   a. Place the cat in the dissecting tray with its ventral surface down.

   b. Use a sharp scalpel to make a short, shallow incision through the skin in the dorsal midline at the base of the cat's neck.

   c. Insert the pointed blade of a dissecting scissors through the opening in the skin and cut along the dorsal midline, following the tips of the vertebral spinous processes to the base of the tail (fig. 54.1).

   d. Use the scissors to make an incision encircling the region of the tail, anus, and genital organs. Do not remove the skin from this area.

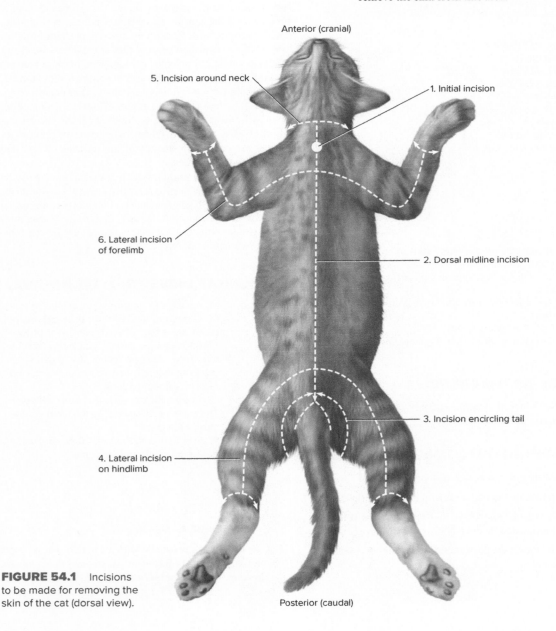

FIGURE 54.1 Incisions to be made for removing the skin of the cat (dorsal view).

e. Use bone shears or a bone saw to sever the tail at its base and discard it (optional procedure).

f. From the incision at the base of the tail, cut along the lateral surface of each hindlimb to the paw, and encircle the ankle. Also clip off the claws from each paw to prevent them from tearing the plastic storage bag and scratching your skin (optional procedure).

g. Make an incision from the initial cut at the base of the neck to encircle the neck.

h. From the incision around the neck, cut along the lateral surface of each forelimb to the paw and encircle the wrist (fig. 54.1).

i. Grasp the skin on either side of the dorsal midline incision, and carefully pull the skin laterally away from the body. At the same time, use your fingers or a blunt probe to help remove the skin from the underlying muscles by separating the loose connective tissue (superficial fascia). As you pull the skin away, you may note a thin sheet of muscle tissue attached to it. This muscle is called the *cutaneous maximus,* and it functions to move the cat's skin. Humans lack the cutaneous maximus, but a similar sheet of muscle (platysma) is present in the neck of a human.

j. As you remove the skin, work toward the ventral surface, then work toward the head, and finally work toward the tail. Pull the skin over each limb as if you were removing a glove.

k. If the cat is a female, note the mammary glands on the ventral surface of the thorax and abdomen (fig. 54.2). These glands will adhere to the skin and can be removed from it and discarded.

l. After the skin has been pulled away, carefully remove as much of the remaining connective tissue as possible to expose the underlying skeletal muscles. The muscles should appear light brown and fibrous.

6. After skinning the cat, follow these steps:

a. Discard the tissues you have removed as directed by the laboratory instructor.

b. Wrap the skin around the cat to help keep its body moist, and place it in a plastic storage bag.

c. Write your name in pencil on an identification tag, and tie the tag to the storage bag so that you can identify your specimen.

7. Observe the recommended safety procedures for the conclusion of a laboratory session.

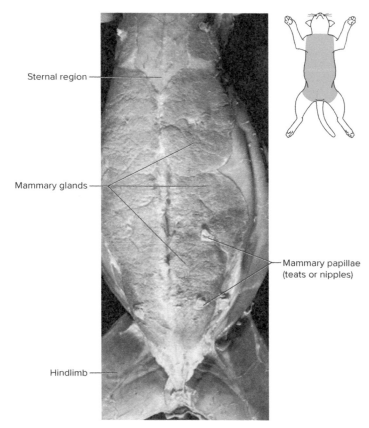

Sternal region

Mammary glands

Mammary papillae (teats or nipples)

Hindlimb

FIGURE 54.2   Mammary glands on the ventral surface of the cat's thorax and abdomen. (©Dr. Sheril Burton)

PRACTICE

PROCEDURE B—Skeletal Muscle Dissection

1. The purpose of a skeletal muscle dissection is to separate the individual muscles from any surrounding tissues and thus expose the muscles for observation. To do a muscle dissection, follow these steps:

   a. Use the appropriate figure as a guide and locate the muscle to be dissected in the specimen.

   b. Use a blunt probe or a finger to separate the muscle from the surrounding connective tissue along its natural borders. The muscle should separate smoothly. If the border appears ragged, you probably have torn the muscle fibers.

2. If it is necessary to cut through (transect) a superficial muscle to observe a deeper one, use scissors to transect the muscle about halfway between its origin and insertion. Then, lift aside (reflect) the cut ends, leaving their attachments intact.

3. The following procedures will instruct you to dissect some of the larger and more easily identified muscles of the cat. In each case, the procedure will include the names of the muscles to be dissected, figures illustrating their locations, and tables listing the origins, insertions, and actions of the muscles. Compare the muscle origins and insertions to the cat skeleton (fig. 54.3).

(More detailed dissection instructions can be obtained by consulting an additional guide or atlas for cat dissection.) As you dissect each muscle, study the figures and tables in chapter 9 of the textbook and identify any corresponding (homologous) muscles of the human body. Also locate these homologous muscles in the human torso model or models of the human upper and lower limbs.

PRACTICE

PROCEDURE C—Muscles of the Head and Neck

1. Place the cat in the dissecting tray with its ventral surface up.
2. Remove the skin and underlying connective tissues from each side of the neck, forward to the chin, and up to each ear.
3. Study figures 54.4 and 54.5, and then locate and dissect the following muscles:

   sternomastoid
   mylohyoid
   sternohyoid
   masseter
   digastric

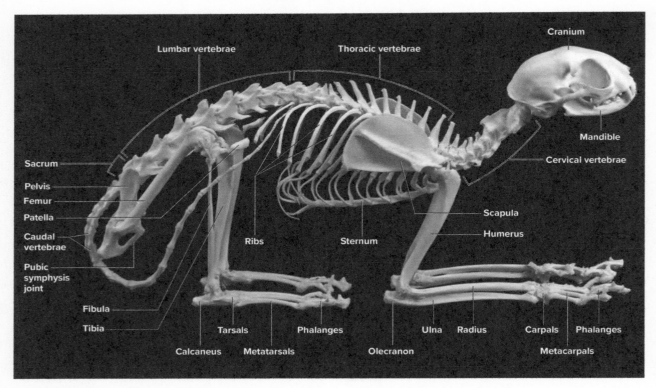

FIGURE 54.3   Skeleton of a cat, lateral view.   (McGraw-Hill Education/Photo and Dissection by Christine Eckel)

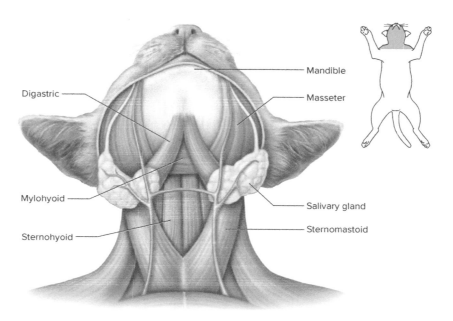

FIGURE 54.4   Muscles on the ventral surface of the cat's head and neck. **APR** (©McGraw-Hill Education/APR)

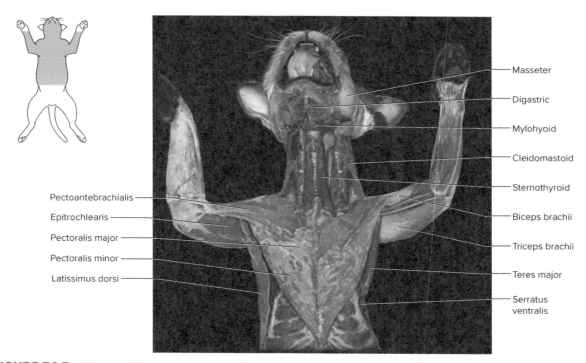

FIGURE 54.5   Muscles of the ventral surface of the cat's head, neck, thorax, and forelimb. (McGraw-Hill Education/APR)

4. To find the deep muscles of the cat's neck and chin, carefully remove the sternomastoid, sternohyoid, and mylohyoid muscles from the right side.

5. Study figure 54.6, and locate the following muscles:

hyoglossus
cleidomastoid
styloglossus
thyrohyoid
geniohyoid
sternothyroid

6. See table 54.1 for the origins, insertions, and actions of these head and neck muscles.

7. Complete Part A of Laboratory Report 54.

TABLE 54.1   Muscles of the Head and Neck

Muscle	Origin	Insertion	Action
Sternomastoid	Manubrium of sternum	Mastoid process of temporal bone	Turns and depresses head
Sternohyoid	Costal cartilage	Hyoid bone	Depresses hyoid bone
Digastric	Mastoid process and occipital bone	Ventral border of mandible	Depresses mandible
Mylohyoid	Inner surface of mandible	Hyoid bone	Raises floor of mouth
Masseter	Zygomatic arch	Coronoid fossa of mandible	Elevates mandible
Hyoglossus	Hyoid bone	Tongue	Pulls tongue back
Styloglossus	Styloid process	Tip of tongue	Pulls tongue back
Geniohyoid	Inner surface of mandible	Hyoid bone	Pulls hyoid bone forward
Cleidomastoid	Clavicle	Mastoid process	Turns head to side
Thyrohyoid	Thyroid cartilage	Hyoid bone	Raises larynx
Sternothyroid	Sternum	Thyroid cartilage	Pulls larynx back

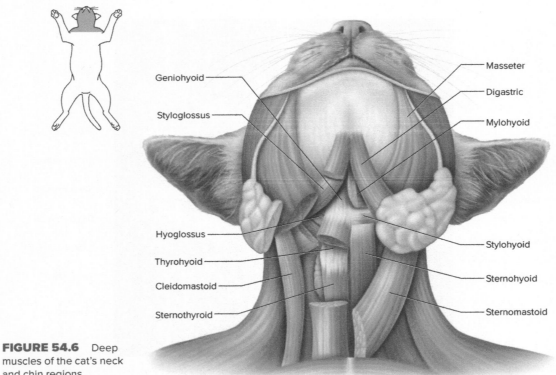

FIGURE 54.6   Deep muscles of the cat's neck and chin regions.

PRACTICE

PROCEDURE D—Muscles of the Thorax

1. Place the cat in the dissecting tray with its ventral surface up.
2. Remove any remaining fat and connective tissue to expose the muscles in the walls of the thorax and abdomen.
3. Study figures 54.5, 54.7, and 54.8. Transect the superficial pectoantebrachialis to expose the entire pectoralis major. Locate and dissect the following pectoral muscles:

   pectoantebrachialis (well developed in cats)
   pectoralis minor
   pectoralis major
   xiphihumeralis

4. To find the deep thoracic muscles, transect the pectoralis minor, pectoralis major, and pectoantebrachialis muscles, and reflect their cut edges to the sides. Remove the underlying fat and connective tissues to expose the following muscles, as shown in figure 54.9:

   scalenus
   levator scapulae
   serratus ventralis
   transversus costarum

5. See table 54.2 for the origins, insertions, and actions of these muscles.

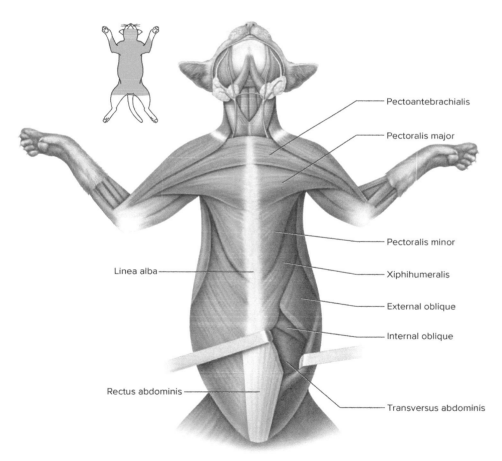

FIGURE 54.7   Muscles of the cat's thorax and abdomen. **APR**

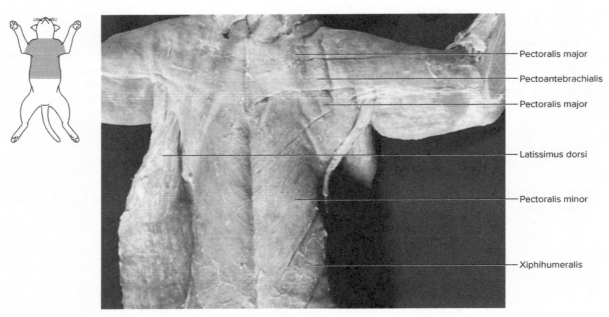

FIGURE 54.8   Superficial muscles of the cat's thorax.   (©Dr. Sheril Burton)

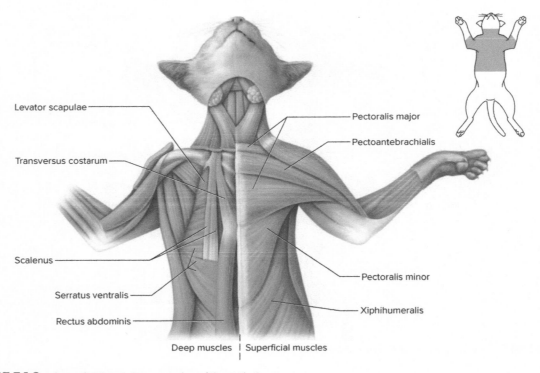

FIGURE 54.9   Superficial and deep muscles of the cat's thorax.

446

TABLE 54.2   Muscles of the Thorax

Muscle	Origin	Insertion	Action
Pectoantebrachialis	Manubrium of sternum	Fascia of forelimb	Adducts forelimb
Pectoralis major	Manubrium of sternum and costal cartilages	Greater tubercle of humerus	Adducts and rotates forelimb
Pectoralis minor	Sternum	Proximal end of humerus	Adducts and rotates forelimb
Xiphihumeralis	Xiphoid process of sternum	Proximal end of humerus	Adducts forelimb
Scalenus	Ribs	Cervical vertebrae	Flexes neck
Serratus ventralis	Ribs	Scapula	Pulls scapula ventrally and posteriorly
Levator scapulae	Cervical vertebrae	Scapula	Pulls scapula ventrally and anteriorly
Transversus costarum	Sternum	First rib	Pulls sternum toward head

 PRACTICE

PROCEDURE E—Muscles of the Abdominal Wall

1. Study figures 54.7 and 54.10.
2. Locate the *external oblique* muscle in the abdominal wall.
3. Make a shallow, longitudinal incision through the external oblique. Reflect the cut edge and expose the *internal oblique* muscle beneath (see fig. 54.7). The fibers of the internal oblique run at a right angle to those of the external oblique.
4. Make a longitudinal incision through the internal oblique. Reflect the cut edge and expose the *transversus abdominis* muscle.
5. Expose the *rectus abdominis* muscle on one side of the midventral line. This muscle lies beneath an aponeurosis.
6. See table 54.3 for the origins, insertions, and actions of these muscles.
7. Complete Part B of the laboratory report.

TABLE 54.3   Muscles of the Abdominal Wall

Muscle	Origin	Insertion	Action
External oblique	Ribs and fascia of back	Linea alba	Compresses abdominal wall
Internal oblique	Fascia of back	Linea alba	Compresses abdominal wall
Transversus abdominis	Lower ribs	Linea alba	Compresses abdominal wall
Rectus abdominis	Symphysis pubis	Sternum and costal cartilages	Compresses abdominal wall and flexes trunk

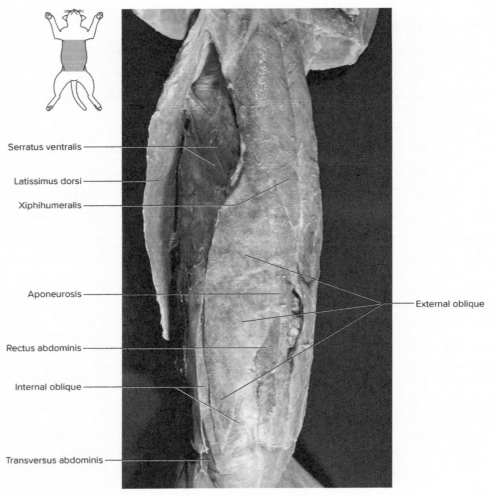

Serratus ventralis

Latissimus dorsi

Xiphihumeralis

Aponeurosis

External oblique

Rectus abdominis

Internal oblique

Transversus abdominis

FIGURE 54.10 Muscles of the cat's thorax and abdomen. **APR** (©Dr. Sheril Burton)

 PRACTICE

PROCEDURE F—Muscles of the Shoulder and Back

1. Place the cat in the dissecting tray with its dorsal surface up.
2. Remove any remaining fat and connective tissue to expose the muscles of the shoulder and back.
3. Study figures 54.11 and 54.12, and then locate and dissect the following superficial muscles:

clavotrapezius
acromiotrapezius
spinotrapezius
levator scapulae ventralis
clavodeltoid
acromiodeltoid
spinodeltoid
latissimus dorsi

Note: The clavodeltoid (clavobrachialis) appears as a continuation of the clavotrapezius.

4. Using scissors, transect the latissimus dorsi and the group of trapezius muscles. Reflect their cut edges and remove any underlying fat and connective tissue. Study figures 54.11 and 54.13, and then locate and dissect the following deep muscles of the shoulder and back:

supraspinatus
infraspinatus
teres major
rhomboid
rhomboid capitis
splenius

5. See table 54.4 for the origins, insertions, and actions of these muscles in the shoulder and back.
6. Review the locations, origins, insertions, and actions designated by the laboratory instructor without the aid of the figures and tables. **1** **2**
7. Complete Part C of the laboratory report.

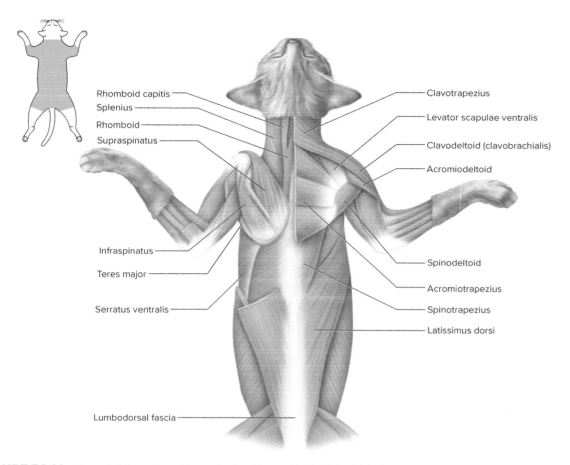

FIGURE 54.11 Superficial muscles of the cat's shoulder and back (*right side*); deep muscles of the shoulder and back (*left side*). **APR**

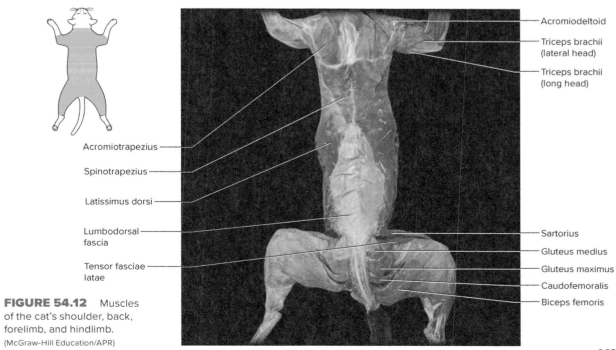

FIGURE 54.12 Muscles of the cat's shoulder, back, forelimb, and hindlimb.
(McGraw-Hill Education/APR)

449

TABLE 54.4   Muscles of the Shoulder and Back

Muscle	Origin	Insertion	Action
Clavotrapezius	Occipital bone of skull and spines of cervical vertebrae	Clavicle	Pulls clavicle upward
Acromiotrapezius	Spines of cervical and thoracic vertebrae	Spine and acromion process of scapula	Pulls scapula upward
Spinotrapezius	Spines of thoracic vertebrae	Spine of scapula	Pulls scapula upward and back
Levator scapulae ventralis	Occipital bone of skull	Ventral border of scapula	Pulls scapula toward head
Clavodeltoid	Clavicle	Ulna near semilunar notch	Flexes forelimb
Acromiodeltoid	Acromion process of scapula	Proximal end of humerus	Flexes and rotates forelimb
Spinodeltoid	Spine of scapula	Proximal end of humerus	Flexes and rotates forelimb
Latissimus dorsi	Thoracic and lumbar vertebrae	Proximal end of humerus	Pulls forelimb upward and back
Supraspinatus	Fossa above spine of scapula	Greater tubercle of humerus	Extends forelimb
Infraspinatus	Fossa below spine of scapula	Greater tubercle of humerus	Rotates forelimb
Teres major	Posterior border of scapula	Proximal end of humerus	Rotates forelimb
Rhomboid	Spines of cervical and thoracic vertebrae	Medial border of scapula	Pulls scapula back
Rhomboid capitis	Occipital bone of skull	Medial border of scapula	Pulls scapula forward
Splenius	Fascia of neck	Occipital bone	Turns and raises head

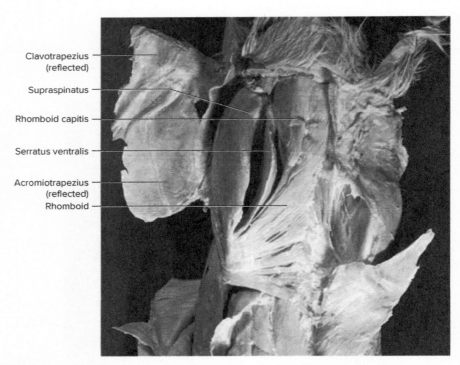

FIGURE 54.13   Deep muscles of the cat's back. (©Dr. Sheril Burton)

PRACTICE

PROCEDURE G—Muscles of the Forelimb

1. Place the cat in the dissecting tray with its ventral surface up.
2. Remove any remaining fat and connective tissue from a forelimb to expose the muscles.
3. Study figures 54.5 and 54.14a, and then locate and dissect the following muscles from the medial surface of the upper forelimb:

   epitrochlearis (well developed in cats)
   triceps brachii

4. Transect the pectoral muscles (pectoantebrachialis, pectoralis major, and pectoralis minor) and reflect their cut edges.

5. Study figures 54.14b and 54.15, and then locate and dissect the following muscles:

   coracobrachialis
   biceps brachii
   long head of the triceps brachii
   medial head of the triceps brachii

6. Study figures 54.16a and 54.17, which show the lateral surface of the upper forelimb.
7. Locate the lateral head of the triceps muscle, transect it, and reflect its cut edges.
8. Study figure 54.16b. Locate and dissect the following deeper muscles:

   anconeus
   brachialis

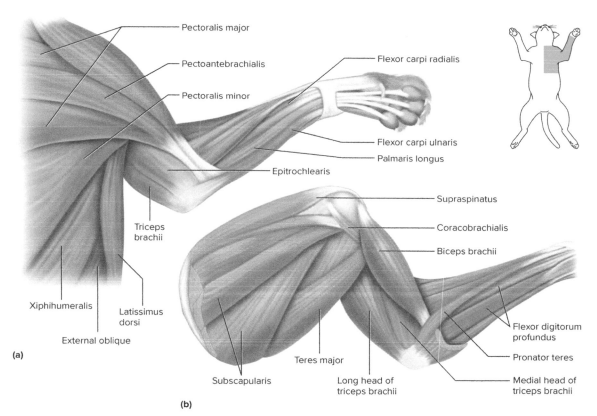

FIGURE 54.14   (a) Superficial muscles of the cat's left medial shoulder and forelimb. (b) Deep muscles of the left medial shoulder and forelimb. **APR**

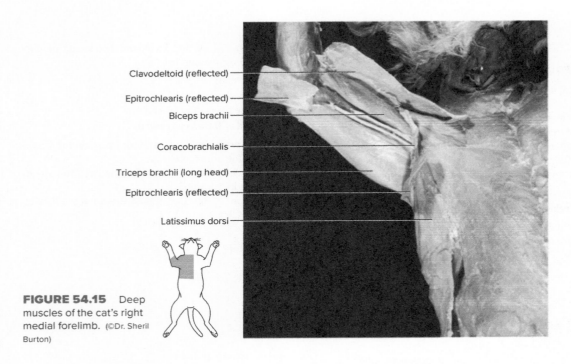

FIGURE 54.15 Deep muscles of the cat's right medial forelimb. (©Dr. Sheril Burton)

Clavodeltoid (reflected)
Epitrochlearis (reflected)
Biceps brachii
Coracobrachialis
Triceps brachii (long head)
Epitrochlearis (reflected)
Latissimus dorsi

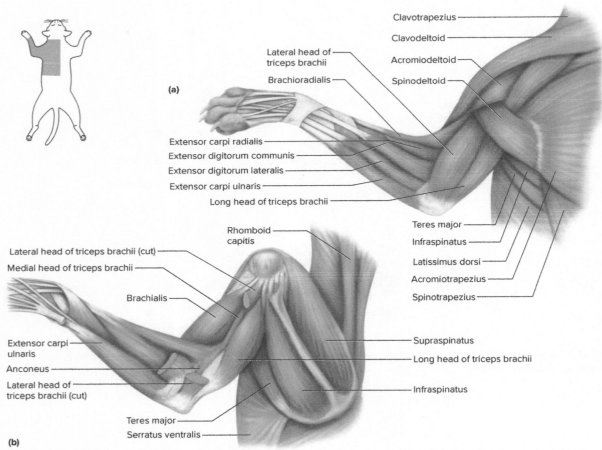

Clavotrapezius
Clavodeltoid
Lateral head of triceps brachii
Acromiodeltoid
Brachioradialis
Spinodeltoid

(a)

Extensor carpi radialis
Extensor digitorum communis
Extensor digitorum lateralis
Extensor carpi ulnaris
Long head of triceps brachii

Teres major
Infraspinatus
Latissimus dorsi
Acromiotrapezius
Spinotrapezius

Rhomboid capitis

Lateral head of triceps brachii (cut)
Medial head of triceps brachii

Brachialis

Extensor carpi ulnaris
Anconeus
Lateral head of triceps brachii (cut)

Supraspinatus
Long head of triceps brachii

Infraspinatus

Teres major
Serratus ventralis

(b)

FIGURE 54.16 (a) Superficial muscles of the cat's left lateral shoulder and forelimb. (b) Deep muscles of the left lateral shoulder and forelimb.

9. Separate the muscles in the lower forelimb into an *extensor group* (fig. 54.16*a*) and a *flexor group* (see fig. 54.14*a*). Generally, the extensors are located on the posterior surface of the lower forelimb, and the flexors are located on the anterior surface. A thick, tough layer of fascia (antebrachial fascia) surrounds the forelimb. Remove this fascia to expose and dissect the following muscles:

brachioradialis
extensor carpi radialis
extensor digitorum communis
extensor digitorum lateralis

extensor carpi ulnaris
flexor carpi ulnaris
palmaris longus
flexor carpi radialis
pronator teres

10. Transect the flexor carpi ulnaris, palmaris longus, and flexor carpi radialis to expose the flexor digitorum profundus.
11. See table 54.5 for the origins, insertions, and actions of these muscles of the forelimb.
12. Complete Part D of the laboratory report.

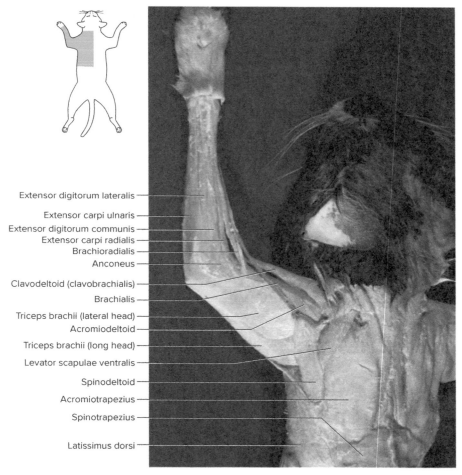

Extensor digitorum lateralis
Extensor carpi ulnaris
Extensor digitorum communis
Extensor carpi radialis
Brachioradialis
Anconeus
Clavodeltoid (clavobrachialis)
Brachialis
Triceps brachii (lateral head)
Acromiodeltoid
Triceps brachii (long head)
Levator scapulae ventralis
Spinodeltoid
Acromiotrapezius
Spinotrapezius
Latissimus dorsi

FIGURE 54.17   Superficial muscles of the cat's left lateral shoulder and forelimb, dorsal view. (©Dr. Sheril Burton)

TABLE 54.5   Muscles of the Forelimb

Muscle	Origin	Insertion	Action
Coracobrachialis	Coracoid process of scapula	Proximal end of humerus	Adducts forelimb
Epitrochlearis	Fascia of latissimus dorsi	Olecranon process of ulna	Extends forelimb
Biceps brachii	Border of glenoid cavity of scapula	Radial tuberosity of radius	Flexes forelimb
Triceps brachii			
Lateral head	Deltoid tuberosity of humerus	Olecranon process of ulna	Extends forelimb
Long head	Border of glenoid cavity of scapula	Olecranon process of ulna	Extends forelimb
Medial head	Shaft of humerus	Olecranon process of ulna	Extends forelimb
Anconeus	Lateral epicondyle of humerus	Olecranon process of ulna	Extends forelimb
Brachialis	Lateral surface of humerus	Proximal end of ulna	Flexes forelimb
Brachioradialis	Shaft of humerus	Distal end of radius	Supinates lower forelimb
Extensor carpi radialis	Lateral surface of humerus	Second and third metacarpals	Extends wrist
Extensor digitorum communis	Lateral surface of humerus	Digits two to five	Extends digits
Extensor digitorum lateralis	Lateral surface of humerus	Digits two to five	Extends digits
Extensor carpi ulnaris	Lateral epicondyle of humerus and ulna	Fifth metacarpal	Extends wrist
Flexor carpi ulnaris	Medial epicondyle of humerus	Carpals	Flexes wrist
Palmaris longus	Medial epicondyle of humerus	Digits	Flexes digits
Flexor carpi radialis	Medial epicondyle of humerus	Second and third metacarpals	Flexes metacarpals
Pronator teres	Medial epicondyle of humerus	Radius	Pronates lower forelimb
Flexor digitorum profundus	Radius, ulna, and medial epicondyle of humerus	Distal phalanges	Flexes digits

 PRACTICE

PROCEDURE H—Muscles of the Hip and Hindlimb

1. Place the cat in the dissecting tray with its ventral surface up.
2. Remove any remaining fat and connective tissue from the hip and hindlimb to expose the muscles.
3. Study figures 54.18a and 54.19, and then locate and dissect the following muscles from the medial surface of the thigh:

   sartorius
   gracilis

4. Using scissors, transect the sartorius and gracilis, and reflect their cut edges to observe the deeper muscles of the thigh.

5. Study figures 54.18b and 54.20, and then locate and dissect the following muscles:

   rectus femoris
   vastus medialis
   adductor longus
   adductor femoris
   semimembranosus
   semitendinosus
   tensor fasciae latae

6. Transect the tensor fasciae latae and rectus femoris muscles and turn their ends aside. Locate the *vastus intermedius* and *vastus lateralis* muscles beneath (fig. 54.21). (*Note*: In some specimens, the vastus intermedius, vastus lateralis, and vastus medialis muscles are closely united by connective tissue and are difficult to separate.)

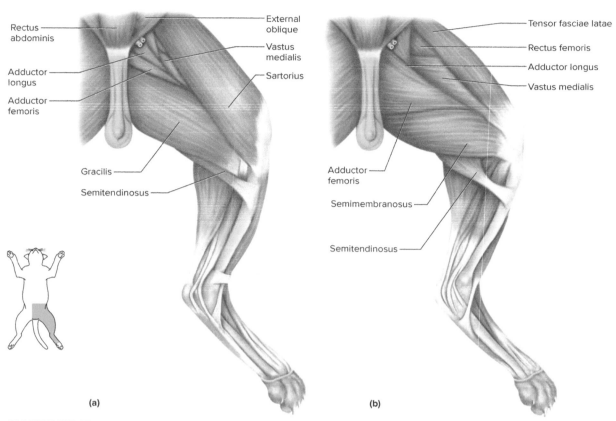

Rectus abdominis
Adductor longus
Adductor femoris
Gracilis
Semitendinosus
External oblique
Vastus medialis
Sartorius

Tensor fasciae latae
Rectus femoris
Adductor longus
Vastus medialis
Adductor femoris
Semimembranosus
Semitendinosus

(a)   (b)

FIGURE 54.18 (a) Superficial muscles of the cat's left medial hindlimb. (b) Deep muscles of the left medial hindlimb.
APR

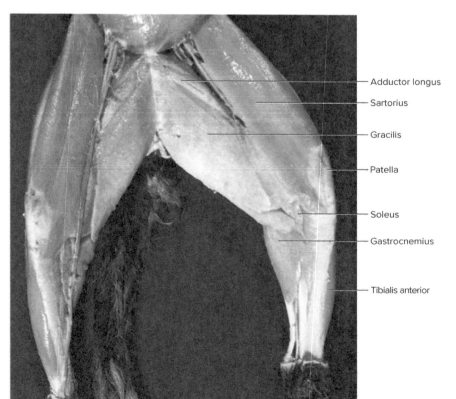

Adductor longus
Sartorius
Gracilis
Patella
Soleus
Gastrocnemius
Tibialis anterior

FIGURE 54.19
Superficial muscles of the cat's hindlimb, medial view. (©Dr. Sheril Burton)

455

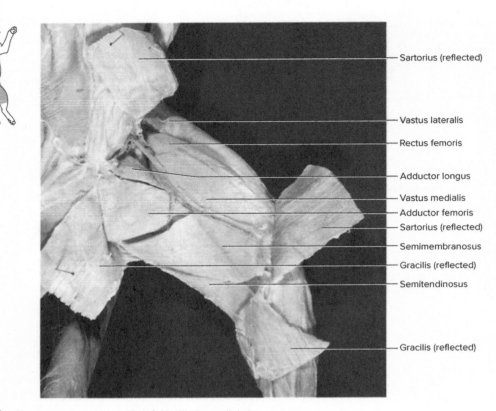

Sartorius (reflected)

Vastus lateralis

Rectus femoris

Adductor longus

Vastus medialis

Adductor femoris

Sartorius (reflected)

Semimembranosus

Gracilis (reflected)

Semitendinosus

Gracilis (reflected)

FIGURE 54.20   Deep muscles of the cat's left hindlimb, medial view.   (©Dr. Sheril Burton)

7. Study figures 54.12, 54.22a, and 54.23, and then locate and dissect the following muscles from the lateral surface of the hip and thigh:

biceps femoris
caudofemoralis
gluteus maximus (small in cats)

8. Using scissors, transect the tensor fasciae latae and biceps femoris, and reflect their cut edges to observe the deeper muscles of the thigh.

9. Locate and dissect the following muscles:

tenuissimus
gluteus medius

(*Note*: The tenuissimus muscle and the sciatic nerve may adhere tightly to the dorsal surface of the biceps femoris muscle. Take care to avoid cutting these structures.)

10. On the lateral surface of the lower hindlimb (figs. 54.22 and 54.24), locate and dissect the following muscles:

gastrocnemius
soleus
tibialis anterior
fibularis (peroneus) group

11. See table 54.6 for the origins, insertions, and actions of these muscles of the hip and hindlimb.

12. Review the locations, origins, insertions, and actions designated by the laboratory instructor without the aid of the figures and tables. ⓐ ⓑ

13. Complete Part E of the laboratory report.

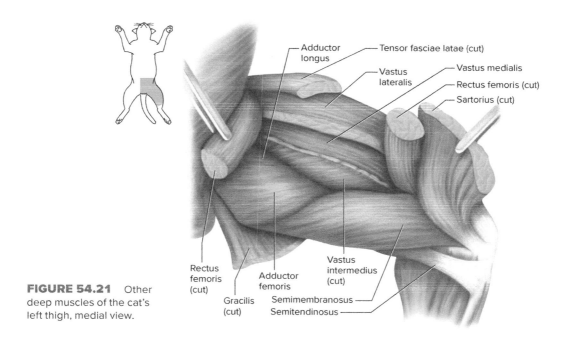

FIGURE 54.21 Other deep muscles of the cat's left thigh, medial view.

Adductor longus

Tensor fasciae latae (cut)

Vastus lateralis

Vastus medialis

Rectus femoris (cut)

Sartorius (cut)

Rectus femoris (cut)

Adductor femoris

Gracilis (cut)

Vastus intermedius (cut)

Semimembranosus

Semitendinosus

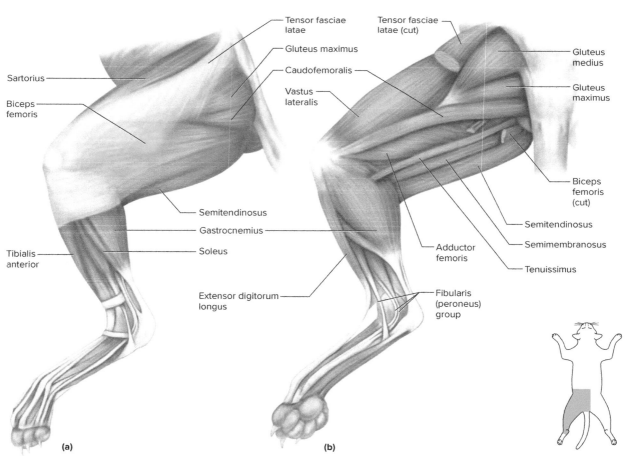

Tensor fasciae latae

Tensor fasciae latae (cut)

Gluteus maximus

Gluteus medius

Caudofemoralis

Sartorius

Vastus lateralis

Gluteus maximus

Biceps femoris

Biceps femoris (cut)

Semitendinosus

Semitendinosus

Gastrocnemius

Semimembranosus

Tibialis anterior

Soleus

Adductor femoris

Tenuissimus

Extensor digitorum longus

Fibularis (peroneus) group

(a)

(b)

FIGURE 54.22 (a) Superficial muscles of the cat's left lateral hip and hindlimb. (b) Deep muscles of the left lateral hip and hindlimb. **APR**

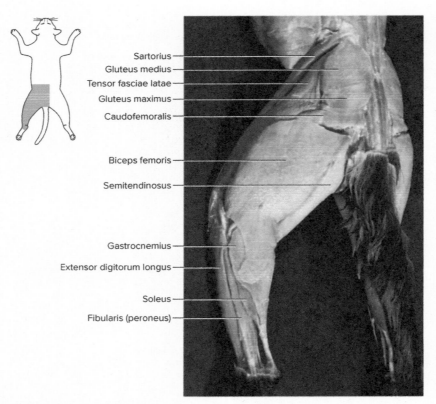

Sartorius
Gluteus medius
Tensor fasciae latae
Gluteus maximus
Caudofemoralis

Biceps femoris

Semitendinosus

Gastrocnemius

Extensor digitorum longus

Soleus

Fibularis (peroneus)

FIGURE 54.23   Muscles of the cat's left hip and hindlimb, lateral view. (©Dr. Sheril Burton)

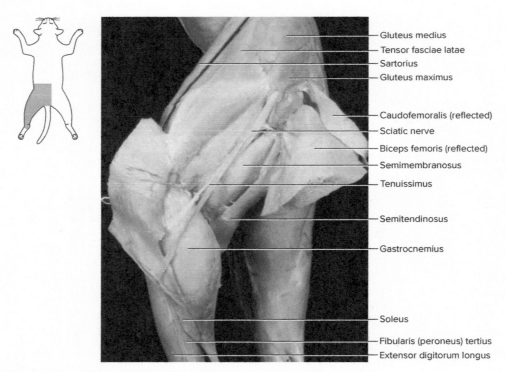

Gluteus medius
Tensor fasciae latae
Sartorius
Gluteus maximus

Caudofemoralis (reflected)
Sciatic nerve
Biceps femoris (reflected)
Semimembranosus
Tenuissimus

Semitendinosus

Gastrocnemius

Soleus

Fibularis (peroneus) tertius
Extensor digitorum longus

FIGURE 54.24   Deep muscles of the cat's left hindlimb, lateral view. (McGraw-Hill Education/APR)

TABLE 54.6   Muscles of the Hip and Hindlimb

Muscle	Origin	Insertion	Action
Sartorius	Crest of ilium	Fascia of knee and proximal end of tibia	Rotates and extends hindlimb
Gracilis	Symphysis pubis and ischium	Proximal end of tibia	Adducts hindlimb
Quadriceps femoris			
Vastus lateralis	Shaft of femur and greater trochanter	Patella	Extends hindlimb
Vastus intermedius	Shaft of femur	Patella	Extends hindlimb
Rectus femoris	Ilium	Patella	Extends hindlimb
Vastus medialis	Shaft of femur	Patella	Extends hindlimb
Adductor longus	Pubis	Proximal end of femur	Adducts hindlimb
Adductor femoris	Pubis and ischium	Shaft of femur	Adducts hindlimb
Semimembranosus	Ischial tuberosity	Medial epicondyle of femur	Extends thigh
Tensor fasciae latae	Ilium	Fascia of thigh	Extends thigh
Biceps femoris	Ischial tuberosity	Patella and tibia	Abducts thigh and flexes lower hindlimb
Tenuissimus	Second caudal vertebra	Tibia and fascia of biceps femoris	Abducts thigh and flexes lower hindlimb
Semitendinosus	Ischial tuberosity	Crest of tibia	Flexes lower hindlimb
Caudofemoralis	Caudal vertebrae	Patella	Abducts thigh and extends lower hindlimb
Gluteus maximus	Sacral and caudal vertebrae	Greater trochanter of femur	Abducts thigh
Gluteus medius	Ilium, and sacral and caudal vertebrae	Greater trochanter of femur	Abducts thigh
Gastrocnemius	Lateral and medial epicondyles of femur	Calcaneus	Extends foot
Soleus	Proximal end of fibula	Calcaneus	Extends foot
Tibialis anterior	Proximal end of tibia and fibula	First metatarsal	Flexes foot
Fibularis (peroneus) group	Shaft of fibula	Metatarsals	Flexes foot

Notes

Name _____

Date _____

Section _____

LABORATORY
REPORT

The Ⓐ corresponds to the indicated Learning Outcome(s) found at the beginning of the laboratory exercise.

Cat Dissection: Musculature

54

PART A ASSESSMENTS

Complete the following statements:

1. The _____ muscle of the human is homologous to the sternomastoid muscle of the cat. ③

2. The _____ muscle elevates the mandible in the human and in the cat. ② ③

3. Two muscles of the cat that are inserted on the hyoid bone are the _____ and the

   _____ . ① ②

PART B ASSESSMENTS

Complete the following:

1. Name two muscles that are found in the thoracic wall of the cat but are absent in the human. ① ③

   _____   _____

2. Name two muscles that are found in the thoracic wall of the cat and the human. ③

   _____   _____

3. Name four muscles that are found in the abdominal wall of the cat and the human. ③

   _____   _____

   _____   _____

PART C ASSESSMENTS

Complete the following:

1. Name three muscles of the cat that together correspond to the trapezius muscle in the human. ① ③

   _____

   _____

   _____

2. Name three muscles of the cat that together correspond to the deltoid muscle in the human. ① ③

   _____

   _____

   _____

3. Name the muscle in the cat and in the human that occupies the fossa above the spine of the scapula. ① ③

   _____

4. Name the muscle in the cat and in the human that occupies the fossa below the spine of the scapula.

_____

5. Name two muscles found in the cat and in the human that can rotate the forelimb.

_____   _____

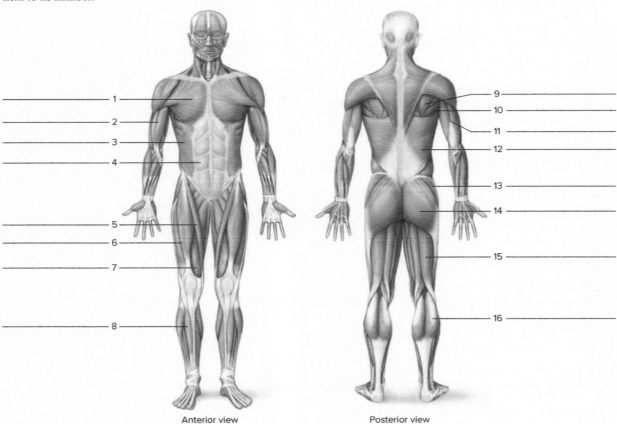 PART D ASSESSMENTS

Complete the following:

1. Name two muscles found in the cat and in the human that can flex the forelimb.

_____   _____

2. Name a muscle found in the cat but absent in the human that can extend the forelimb.

_____

3. Name a muscle that has three heads, can extend the forelimb, and is found in the cat and in the human.

_____

PART E ASSESSMENTS

Observe the muscles on the human torso model and the upper and lower limb models. Each of the muscles in figure 54.25 is found both in the cat and in the human. Identify each of the numbered muscles in the figure by placing its name in the space next to its number.

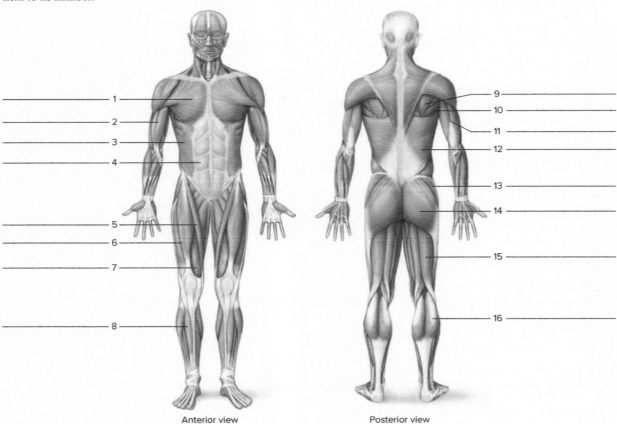

Anterior view   Posterior view

FIGURE 54.25   Identify the numbered muscles that occur in both the cat and the human.

LABORATORY EXERCISE

Cat Dissection: Cardiovascular System

MATERIALS NEEDED

Preserved cat
Dissecting tray
Dissecting instruments
Disposable gloves
Human heart model
Human torso model

⚠ SAFETY

- Wear disposable gloves when working on the cat dissection.
- Dispose of tissue remnants and gloves as instructed.
- Wash the dissecting tray and instruments as instructed.
- Wash your laboratory table.
- Wash your hands before leaving the laboratory.

PURPOSE OF THE EXERCISE

To examine and identify the major organs of the cardiovascular system of the cat and to compare them with the corresponding organs of the human torso model.

LEARNING OUTCOMES APR

After completing this exercise, you should be able to

1. Locate and identify the major organs of the cardiovascular system of the cat.

2. Compare the features of the cardiovascular system of the cat with those of the human.

3. Identify the corresponding cardiovascular organs of the human.

In this laboratory exercise, you will dissect the major organs of the cardiovascular system of the cat. As before, while you are examining the organs of the cat, compare them with the corresponding organs of the human torso model.

If the cardiovascular system of the cat has been injected, the systemic arteries will be filled with red latex, and the systemic veins will be filled with blue latex. This will make it easier for you to trace the vessels as you dissect them.

PRACTICE

PROCEDURE A—The Arterial System

1. Place the preserved cat in the dissecting tray with its ventral surface up.

2. Open the thoracic cavity, and expose its contents. To do this, follow these steps:

   a. Make a longitudinal incision passing anteriorly from the diaphragm along one side of the sternum. Continue the incision through the neck muscles to the mandible. Try to avoid damaging the internal organs as you cut.

   b. Make a lateral cut on each side along the anterior surface of the diaphragm, and cut the diaphragm loose from the thoracic wall.

   c. Spread the sides of the thoracic wall outward, and use a scalpel to make a longitudinal cut along each side of the inner wall of the rib cage to weaken the ribs. Continue to spread the thoracic wall laterally to break the ribs so that the flaps of the wall will remain open (figs. 55.1 and 55.2).

3. Note the location of the heart and the large blood vessels associated with it. Slit the outer *parietal pericardium* that surrounds the heart by cutting with scissors along the midventral line. Note how this membrane is connected to the *visceral pericardium* that is attached to the surface of the heart. Locate the *pericardial cavity,* the space between the two layers of the pericardium.

4. Examine the heart and closely associated blood vessels (figs. 55.1–55.5). The arrangement of blood vessels coming off the aortic arch is different in the cat than in the human body. Locate the following:

right atrium
left atrium
right ventricle
left ventricle
pulmonary trunk
aorta
coronary arteries

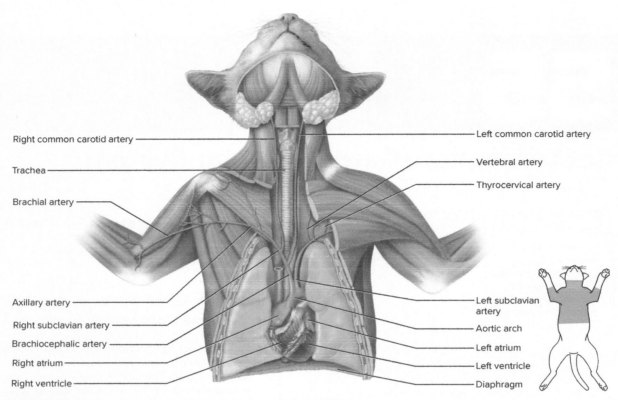

Right common carotid artery

Trachea

Brachial artery

Axillary artery

Right subclavian artery

Brachiocephalic artery

Right atrium

Right ventricle

Left common carotid artery

Vertebral artery

Thyrocervical artery

Left subclavian artery

Aortic arch

Left atrium

Left ventricle

Diaphragm

FIGURE 55.1   Arteries of the cat's thorax, neck, and forelimb. **APR**

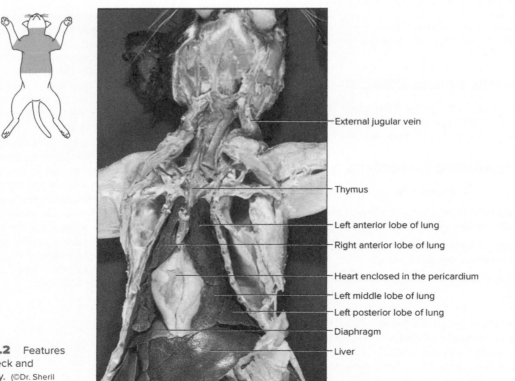

External jugular vein

Thymus

Left anterior lobe of lung

Right anterior lobe of lung

Heart enclosed in the pericardium

Left middle lobe of lung

Left posterior lobe of lung

Diaphragm

Liver

FIGURE 55.2   Features of the cat's neck and thoracic cavity. (©Dr. Sheril Burton)

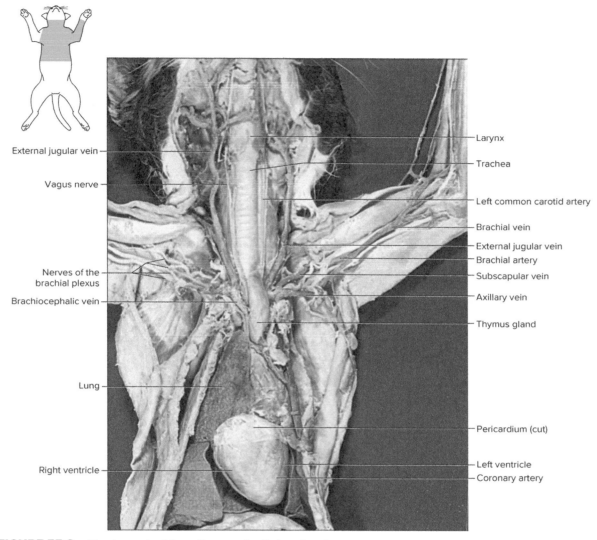

External jugular vein

Vagus nerve

Nerves of the
brachial plexus

Brachiocephalic vein

Lung

Right ventricle

Larynx

Trachea

Left common carotid artery

Brachial vein

External jugular vein

Brachial artery

Subscapular vein

Axillary vein

Thymus gland

Pericardium (cut)

Left ventricle

Coronary artery

FIGURE 55.3   Blood vessels of the cat's upper forelimb and neck. (©Dr. Sheril Burton)

5. Use a scalpel to open the heart chambers by making a cut along the frontal plane from its apex to its base. Remove any remaining latex from the chambers. Examine the valves between the chambers, and note the relative thicknesses of the chamber walls (see fig. 55.4). Do not remove the heart, as it is needed for future examination of the relationship between major blood vessels. Compare the heart structures of the cat with the model of the human heart.

6. Using figures 55.1, 55.3–55.5 as guides, locate and dissect the following arteries of the thorax and neck:

aortic arch

brachiocephalic artery (on right only; first branch of the aortic arch)

right subclavian artery

left subclavian artery
right common carotid artery
left common carotid artery

7. Trace the right subclavian artery into the forelimb, and locate the following arteries:

vertebral artery
thyrocervical artery
subscapular artery
axillary artery
brachial artery

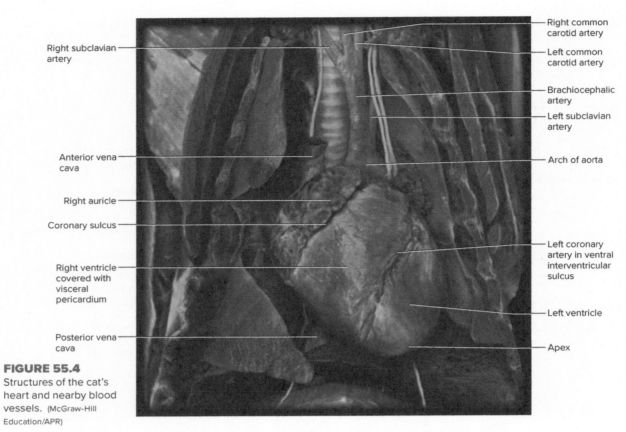

Right subclavian
artery

Anterior vena
cava

Right auricle

Coronary sulcus

Right ventricle
covered with
visceral
pericardium

Posterior vena
cava

Right common
carotid artery

Left common
carotid artery

Brachiocephalic
artery

Left subclavian
artery

Arch of aorta

Left coronary
artery in ventral
interventricular
sulcus

Left ventricle

Apex

FIGURE 55.4
Structures of the cat's
heart and nearby blood
vessels. (McGraw-Hill
Education/APR)

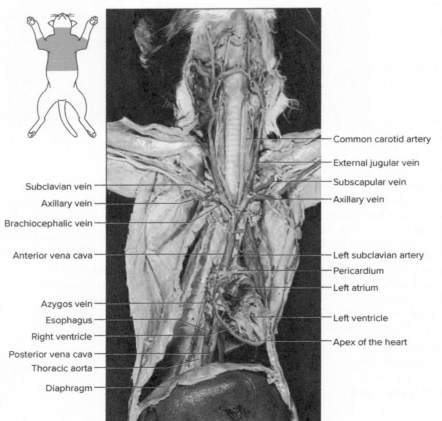

Common carotid artery

External jugular vein

Subscapular vein

Axillary vein

Left subclavian artery

Pericardium

Left atrium

Left ventricle

Apex of the heart

Subclavian vein

Axillary vein

Brachiocephalic vein

Anterior vena cava

Azygos vein

Esophagus

Right ventricle

Posterior vena cava

Thoracic aorta

Diaphragm

FIGURE 55.5   Blood
vessels of the cat's neck
and thorax. (©Dr. Sheril Burton)

8. Open the abdominal cavity. To do this, follow these steps:

   a. Use scissors to make an incision through the body wall along the midline from the symphysis pubis to the diaphragm.

   b. Make a lateral incision through the body wall along either side of the inferior border of the diaphragm and along the bases of the thighs.

   c. Reflect the flaps created in the body wall as you would open a book, and expose the contents of the abdominal cavity.

   d. Note the *parietal peritoneum* that forms the inner lining of the abdominal wall. Also note the *greater omentum*, a structure composed of a double layer of peritoneum that hangs from the border of the stomach and covers the lower abdominal organs like a fatty apron (fig. 55.6). Reflect the abdominal organs to one side to expose various blood vessels in the abdominopelvic cavity region.

9. As you expose and dissect blood vessels, try not to destroy other visceral organs needed for future studies.

Using figures 55.7–55.9 as guides, locate and dissect the following arteries of the abdomen:

abdominal aorta (unpaired)
celiac artery (unpaired)
hepatic artery (unpaired)
gastric artery (unpaired)
splenic artery (unpaired)
anterior mesenteric artery (corresponds to superior mesenteric artery; unpaired)
renal arteries (paired)
adrenolumbar arteries (paired)
gonadal arteries (paired)
posterior mesenteric artery (corresponds to inferior mesenteric artery; unpaired)
iliolumbar arteries (paired)
external iliac arteries (paired)
internal iliac arteries (paired)
caudal artery (unpaired)

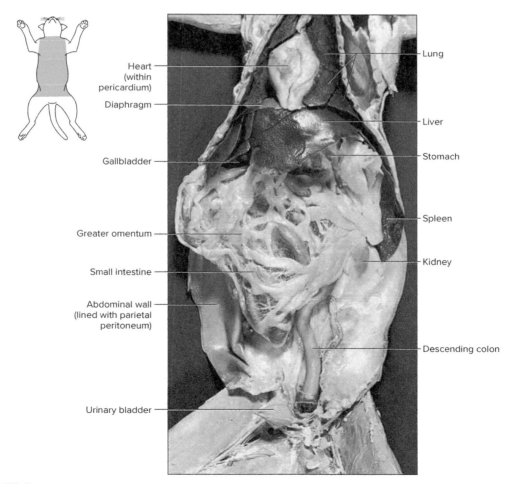

FIGURE 55.6 Organs of the cat's thoracic and abdominal cavities. (©Dr. Sheril Burton)

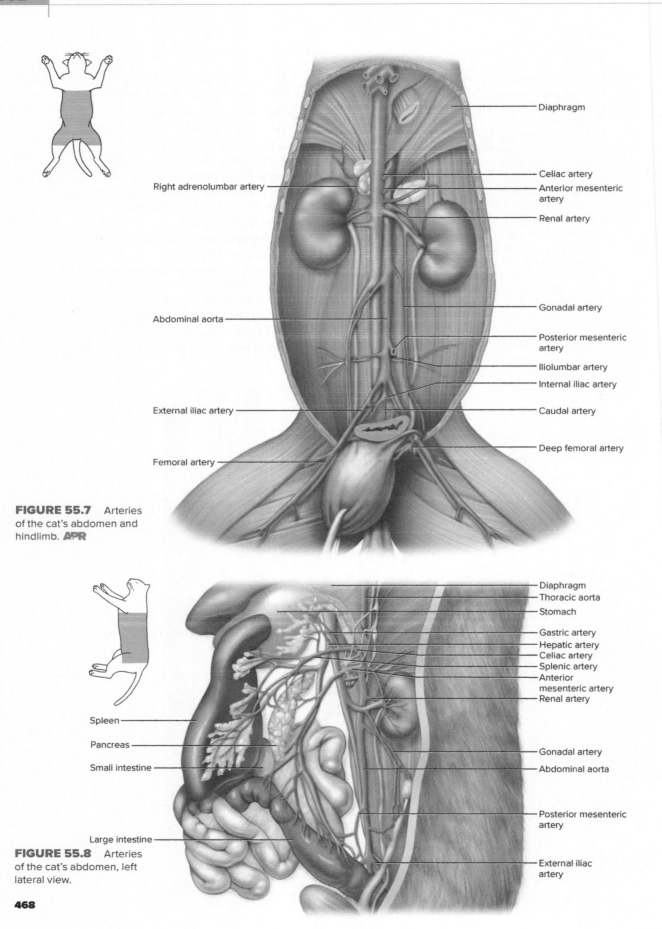

FIGURE 55.7 Arteries of the cat's abdomen and hindlimb. **APR**

Diaphragm

Right adrenolumbar artery

Celiac artery

Anterior mesenteric artery

Renal artery

Gonadal artery

Abdominal aorta

Posterior mesenteric artery

Iliolumbar artery

Internal iliac artery

External iliac artery

Caudal artery

Femoral artery

Deep femoral artery

FIGURE 55.8 Arteries of the cat's abdomen, left lateral view.

Diaphragm
Thoracic aorta
Stomach

Gastric artery
Hepatic artery
Celiac artery
Splenic artery
Anterior mesenteric artery
Renal artery

Spleen

Pancreas

Small intestine

Gonadal artery

Abdominal aorta

Posterior mesenteric artery

Large intestine

External iliac artery

468

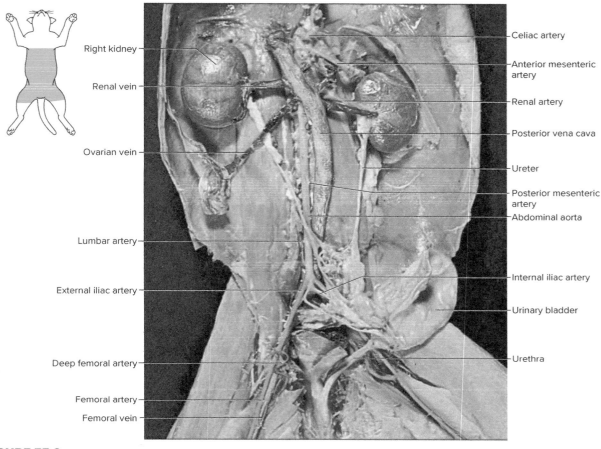

FIGURE 55.9   Blood vessels of the cat's abdominal cavity, with the digestive organs removed.   (©Dr. Sheril Burton)

10. Trace the external iliac artery into the right hindlimb (fig. 55.10), and locate the following:

    femoral artery
    deep femoral artery

11. Review the locations of the heart structures and arteries of the cat without the aid of the figures. ⒜

12. Examine the human torso model along with figures 55.1, 55.5, 55.7, and 55.9. Identify the arteries of the cat that correspond with those of the human. ⒊

13. Complete Part A of Laboratory Report 55.

 **PRACTICE**

PROCEDURE B—The Venous System

1. Examine the heart again, and locate the following veins:

    anterior vena cava (corresponds to superior
        vena cava)
    posterior vena cava (corresponds to inferior
        vena cava)
    pulmonary veins

2. Using figures 55.3, 55.5, and 55.11 as guides, locate and dissect the following veins in the thorax and neck:

    right brachiocephalic vein
    left brachiocephalic vein
    right subclavian vein
    left subclavian vein
    internal jugular vein
    external jugular vein

3. Trace the right subclavian vein into the forelimb, and locate the following veins:

    axillary vein
    subscapular vein
    brachial vein

4. Using figures 55.9, 55.12, and 55.13 as guides, locate and dissect the following veins in the abdomen:

    posterior vena cava
    adrenolumbar vein
    anterior mesenteric vein (corresponds to superior
        mesenteric vein)

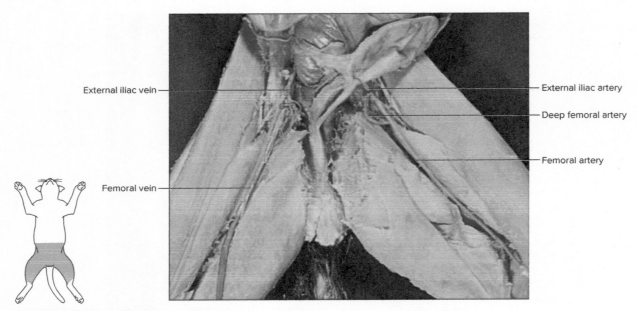

External iliac vein

Femoral vein

External iliac artery

Deep femoral artery

Femoral artery

FIGURE 55.10   Blood vessels of the cat's upper hindlimb.   (©Dr. Sheril Burton)

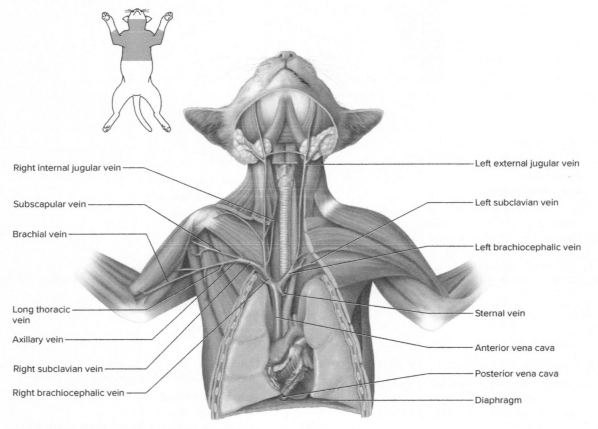

Right internal jugular vein

Subscapular vein

Brachial vein

Long thoracic vein

Axillary vein

Right subclavian vein

Right brachiocephalic vein

Left external jugular vein

Left subclavian vein

Left brachiocephalic vein

Sternal vein

Anterior vena cava

Posterior vena cava

Diaphragm

FIGURE 55.11   Veins of the cat's thorax, neck, and forelimb.   **APR**

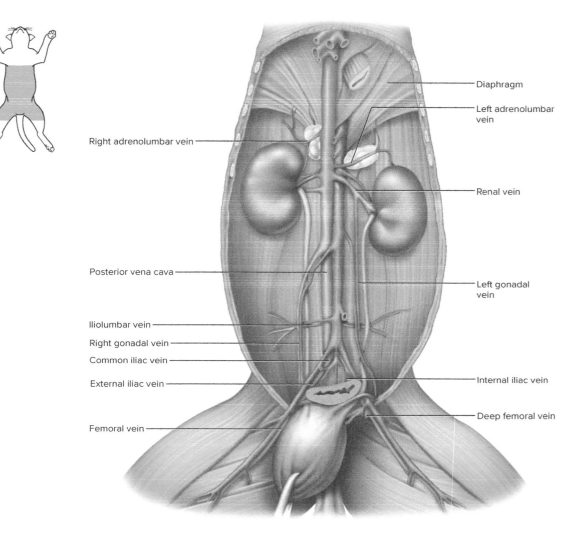

- Diaphragm
- Left adrenolumbar vein
- Right adrenolumbar vein
- Renal vein
- Posterior vena cava
- Left gonadal vein
- Iliolumbar vein
- Right gonadal vein
- Common iliac vein
- Internal iliac vein
- External iliac vein
- Deep femoral vein
- Femoral vein

FIGURE 55.12   Veins of the cat's abdomen and hindlimb. **APR**

posterior mesenteric vein (corresponds to inferior mesenteric vein)
hepatic portal vein
splenic vein
renal vein
gonadal vein
iliolumbar vein
common iliac vein
internal iliac vein
external iliac vein

5. Trace the external iliac vein into the hindlimb (figs. 55.10 and 55.12), and locate the following veins:

   femoral vein
   deep femoral vein

6. Review the locations of the veins of the cat without the aid of the figures. **1**

7. Examine the human torso model along with figures 55.5, 55.9, 55.11, and 55.12. Identify the veins of the cat that correspond with those of the human. **3**

8. Complete Part B of the laboratory report.

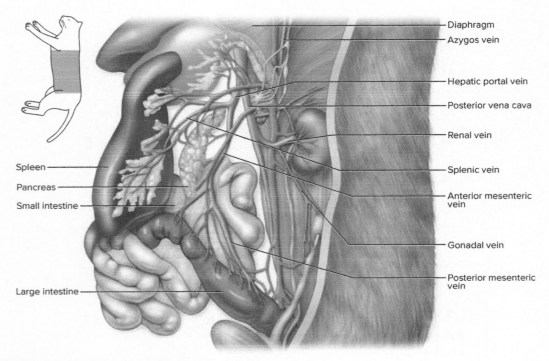

Diaphragm

Azygos vein

Hepatic portal vein

Posterior vena cava

Renal vein

Splenic vein

Anterior mesenteric
vein

Gonadal vein

Posterior mesenteric
vein

Spleen

Pancreas

Small intestine

Large intestine

FIGURE 55.13   Veins of the cat's abdomen and hindlimb.

Name _____

Date _____

Section _____

The Ⓐ corresponds to the indicated Learning Outcome(s) found at the beginning of the laboratory exercise.

Cat Dissection:
Cardiovascular System

PART A ASSESSMENTS

Complete the following:

1. Describe the position and attachments of the parietal pericardium of the heart of the cat. Ⓐ _____

_____

_____

_____

2. Describe the relative thicknesses of the walls of the heart chambers of the cat. Ⓐ _____

_____

_____

_____

3. Explain how the wall thicknesses are related to the functions of the chambers. Ⓐ _____

_____

_____

_____

_____

_____

4. Compare the origins of the common carotid arteries of the cat with those of the human. Ⓐ2 _____

_____

_____

_____

_____

_____

5. Compare the origins of the iliac arteries of the cat with those of the human. Ⓐ2 _____

_____

_____

_____

_____

PART B ASSESSMENTS

Complete the following:

1. Compare the origins of the brachiocephalic veins of the cat with those of the human. [2] _____

_____

_____

_____

2. Compare the relative sizes of the external and internal jugular veins of the cat with those of the human. [2] _____

_____

_____

_____

_____

_____

3. List twelve veins that cats and humans have in common. [2] [3] _____

_____

_____

_____

_____

_____

_____

LABORATORY EXERCISE

56

Cat Dissection: Digestive System

PURPOSE OF THE EXERCISE

To examine and identify the major digestive organs of the cat and to compare these organs with those of the human.

 LEARNING OUTCOMES APR

After completing this exercise, you should be able to

1. Locate and identify the major digestive organs of the cat.
2. Compare the digestive system of the cat with that of the human.
3. Identify the corresponding organs in the human torso model.

In this laboratory exercise, you will dissect the major digestive organs of the cat. As you observe these organs, compare them with those of the human by observing the parts of the human torso model; however, keep in mind that the cat is a carnivore (flesh-eating mammal) and that the organs of its digestive system are adapted to capturing, holding, eating, and digesting the bodies of other animals. In contrast, humans are omnivores (eat both plant and animal substances). Because humans are adapted to eating and digesting a greater variety of foods, comparisons of the cat and the human digestive organs may not be precise.

 PRACTICE

PROCEDURE—Digestive System Dissection

1. Place the preserved cat in the dissecting tray on its left side.
2. Locate the major salivary glands on one side of the head (fig. 56.1). To do this, follow these steps:
   a. Clear away any remaining fascia and other connective tissue from the region below the ear and near the joint of the mandible.
   b. Identify the *parotid gland,* a relatively large mass of glandular tissue just below the ear. Note the parotid duct (Stensen's duct) that passes over the surface of the masseter muscle and opens into the mouth.
   c. Look for the *submandibular gland* just below the parotid gland, near the angle of the jaw.
   d. Locate the *sublingual gland* that is adjacent and medial to the submandibular gland.
3. Open the oral cavity. To do this, follow these steps:
   a. Use scissors to cut through the soft tissues at the angle of the mouth.
   b. When you reach the bone of the jaw, use a bone cutter to cut through the bone, thus freeing the mandible.

c. Open the mouth wide, and locate the following features:

cheek

lip

vestibule

palate

    hard palate (has transverse ridges to help hold food)

    soft palate (lacks a uvula)

palatine tonsils (small, rounded masses of glandular tissue in the lateral wall of the soft palate)

tongue

    frenulum

    papillae (examine with a hand lens; spiny projections used to clean fur)

4. Examine the teeth of the upper jaw. The adult cat has 6 incisors, 2 canines, 8 premolars, and 2 molars. The lower jaw teeth are similar, except 4 premolars are present.

5. Reflect some of the structures of the neck and the thoracic cavity to one side. Look for the *esophagus* in the neck deep to the trachea, or deep to the heart in the thoracic cavity. This ventral view of the cat is also another way to observe the salivary glands (fig. 56.2).

6. Complete Part A of Laboratory Report 56.

7. Examine organs in the abdominal cavity with the cat positioned with its ventral surface up. Review the normal position and function of the greater omentum.

8. Examine the *liver,* which is located just beneath the diaphragm and is attached to the central portion of the diaphragm by the *falciform ligament.* Also, locate the *spleen,* which is posterior to the liver on the left side (fig. 56.3). The liver has five lobes—a right medial, right lateral (which is subdivided into two parts by a deep cleft), left medial, left lateral, and caudate lobe. The caudate lobe is the smallest lobe and is located in the median line, where it projects into the curvature of the stomach. The caudate lobe is covered by a sheet of mesentery, the *lesser omentum,* that connects the liver to the stomach (figs. 56.4 and 56.5). Find the greenish *gallbladder* on the inferior surface of the liver on the right side.

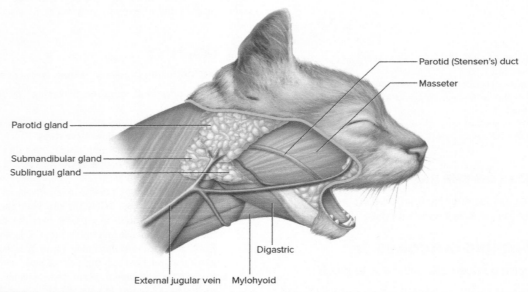

FIGURE 56.1   Major salivary glands of the cat, lateral view.

Also note the *cystic duct,* by which the gallbladder is attached to the *bile duct,* and the *hepatic duct,* which originates in the liver and attaches to the cystic duct. Trace the bile duct to its connection with the duodenum.

9. Locate the *stomach* on the left side of the abdominal cavity. At its anterior end, note the union with the esophagus, which passes through the diaphragm. Identify the *cardia, fundus, body,* and *pyloric part* (listed from entrance to exit) of the stomach. Use scissors to make an incision along the convex border of the stomach from the cardia to the pyloric part. The lining of the stomach

has numerous *gastric folds* (*rugae*). Examine the *pyloric sphincter,* which creates a constriction between the stomach and small intestine.

10. Locate the *pancreas.* It appears as a grayish, two-lobed, elongated mass of glandular tissue. One lobe lies dorsal to the stomach and extends across to the duodenum. The other lobe is enclosed by the *mesentery,* which supports the duodenum (figs. 56.4 and 56.5).

11. Trace the *small intestine,* beginning at the pyloric sphincter. The first portion, the *duodenum,* travels

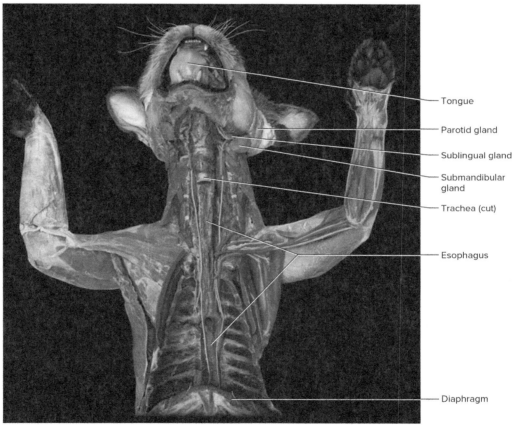

Tongue

Parotid gland

Sublingual gland

Submandibular gland

Trachea (cut)

Esophagus

Diaphragm

FIGURE 56.2 Ventral view of the cat's head, neck, and thoracic cavity with many structures removed in order to observe some deep digestive structures. (McGraw-Hill Education/APR)

posteriorly for several centimeters. Then it loops back around a lobe of the pancreas. The proximal half of the remaining portion of the small intestine is the *jejunum,* and the distal half is the *ileum.* Open the duodenum and note the velvety appearance of the villi. Note how the mesentery supports the small intestine from the dorsal body wall. The small intestine terminates on the right side, where it joins the large intestine.

12. Locate the *large intestine,* and identify the *cecum, ascending colon, transverse colon,* and *descending colon.*

Also locate the *rectum,* which extends through the pelvic cavity to the *anus.* Make an incision at the junction between the ileum and cecum, and look for the *ileocecal sphincter.* An appendix is absent in cats.

13. Review the locations of the digestive organs without the aid of the figures. ◢1◣

14. Examine the human torso model along with figures 56.1–56.5. For the human digestive organs, identify corresponding digestive organs of the cat. ◢3◣

15. Complete Part B of the laboratory report.

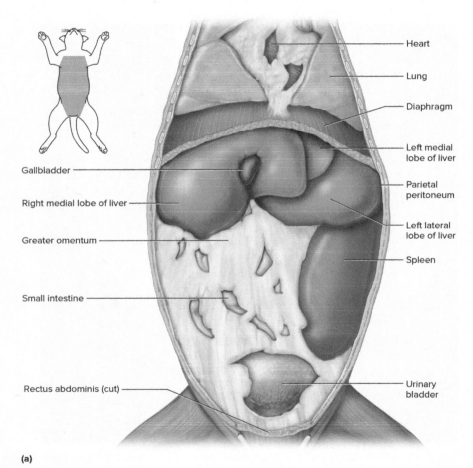

(a)

FIGURE 56.3   Ventral view of abdominal organs of the cat: (*a*) illustration; (*b*) dissected cat with greater omentum removed.
APR ((*b*) McGraw-Hill Education/APR)

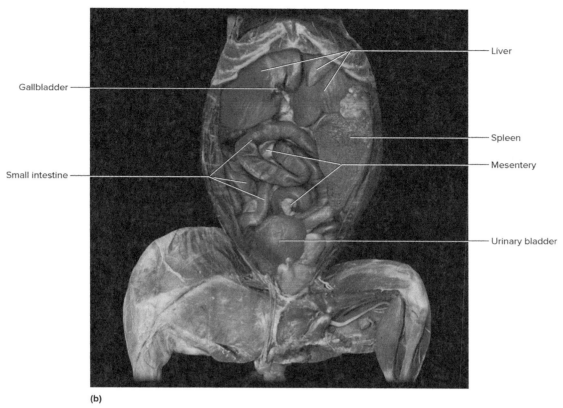

(b)

FIGURE 56.3 *Continued.*

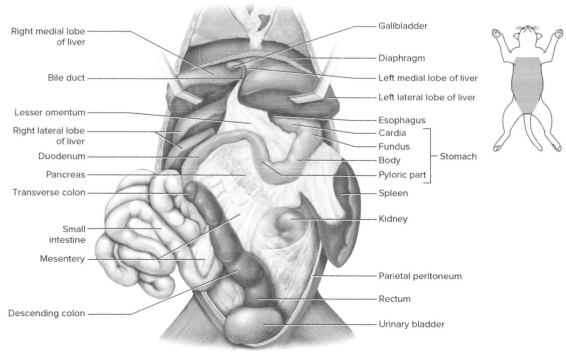

FIGURE 56.4 Ventral view of the cat's abdominal organs with intestines reflected to the right side.

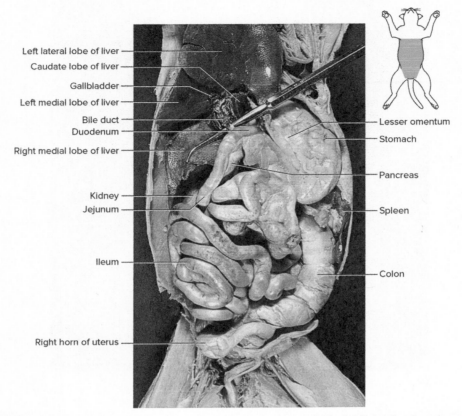

Left lateral lobe of liver
Caudate lobe of liver
Gallbladder
Left medial lobe of liver
Bile duct
Duodenum
Right medial lobe of liver

Kidney
Jejunum

Ileum

Right horn of uterus

Lesser omentum
Stomach

Pancreas

Spleen

Colon

FIGURE 56.5 Abdominal organs of the cat with the greater omentum removed. The left liver lobes have been reflected to the right side. (©Dr. Sheril Burton)

Name _____

Date _____

Section _____

The Ⓐ corresponds to the indicated Learning Outcome(s) found at the beginning of the laboratory exercise.

Cat Dissection: Digestive System

PART A ASSESSMENTS

Complete the following:

1. Compare the locations of the major salivary glands of the human with those of the cat. ② _____

   _____

   _____

   _____

   _____

2. Compare the types and numbers of teeth present in the cat's upper and lower jaws with those of the human. ② _____

   _____

   _____

   _____

   _____

3. In what ways do the cat's teeth seem to be adapted to a special diet? ① _____

   _____

   _____

   _____

   _____

4. What part of the human soft palate is lacking in the cat? ② _____

5. What do you think is the function of the transverse ridges (rugae) in the hard palate of the cat? ① _____

   _____

   _____

   _____

6. How do the papillae on the surface of the human tongue compare with those of the cat? ② _____

   _____

   _____

   _____

   _____

PART B ASSESSMENTS

Complete the following:

1. Describe how the peritoneum and mesenteries are associated with the organs in the abdominal cavity. **1** _____

_____

_____

_____

_____

_____

_____

_____

2. Describe the inner lining of the stomach. **1** _____

_____

_____

_____

_____

3. Compare the structure of the human liver with that of the cat. **2** _____

_____

_____

_____

_____

4. Compare the structure and location of the human pancreas with those of the cat. **2** _____

_____

_____

_____

_____

5. What feature of the human cecum is lacking in the cat? **2** _____

_____

57

Cat Dissection: Respiratory System

MATERIALS NEEDED

Preserved cat
Dissecting tray
Dissecting instruments
Disposable gloves
Human torso model

SAFETY

- Wear disposable gloves when working on the cat dissection.
- Dispose of tissue remnants and gloves as instructed.
- Wash the dissecting tray and instruments as instructed.
- Wash your laboratory table.
- Wash your hands before leaving the laboratory.

PURPOSE OF THE EXERCISE

To examine and identify the major respiratory organs of the cat and to compare these organs with those of the human.

LEARNING OUTCOMES APR

After completing this exercise, you should be able to

1. Locate and identify the major respiratory organs of the cat.
2. Compare the respiratory system of the human with that of the cat.
3. Identify the corresponding organs in the human torso model.

In this laboratory exercise, you will dissect the major respiratory organs of a cat. As you observe these structures in the cat, compare them with those of the human torso model.

PRACTICE

PROCEDURE—Respiratory System Dissection

1. Place the preserved cat in a dissecting tray with its ventral surface up.
2. Examine the *nostrils* (*external nares*) and *nasal septum*.
3. Open the oral cavity wide to expose the *hard palate* and the *soft palate* (fig. 57.1). Cut through the tissues of the soft palate, and observe the *nasopharynx* above it. Locate the small openings of the *auditory tubes* in the lateral walls of the nasopharynx. Insert a probe into an opening of an auditory tube.
4. Pull the tongue forward, and locate the *epiglottis* near its base. Also identify the *glottis*, an opening between the *vocal folds* (*vocal cords*) into the larynx. Locate the *esophagus*, which is dorsal to the larynx (figs. 57.1 and 57.2).
5. Open the thoracic cavity, and expose its contents.
6. Dissect the *trachea* in the neck, and expose the *larynx* at the anterior end near the base of the tongue. Note the *tracheal rings*, and locate the lobes of the *thyroid gland* on each side of the trachea, just posterior to the larynx. Also note the *thymus gland*, a rather diffuse mass of glandular tissue extending along the ventral surface of the trachea into the thorax (figs. 57.1 and 57.2).
7. Examine the larynx, and identify the *thyroid cartilage* and *cricoid cartilage* in its wall. Make a longitudinal incision through the ventral wall of the larynx, and locate the vocal folds (vocal cords) inside.
8. Trace the trachea posteriorly to where it divides into *main (primary) bronchi*, which pass into the lungs (fig. 57.3).
9. Examine the *lungs*, each of which is subdivided into three lobes—an *anterior (cranial)*, a *middle*, and a *posterior (caudal) lobe*. The right lung of the cat has a small fourth *accessory (mediastinal; cardiac) lobe* associated with the right posterior lobe. Notice the thin membrane, the *visceral pleura*, on the surface of each lung. Also

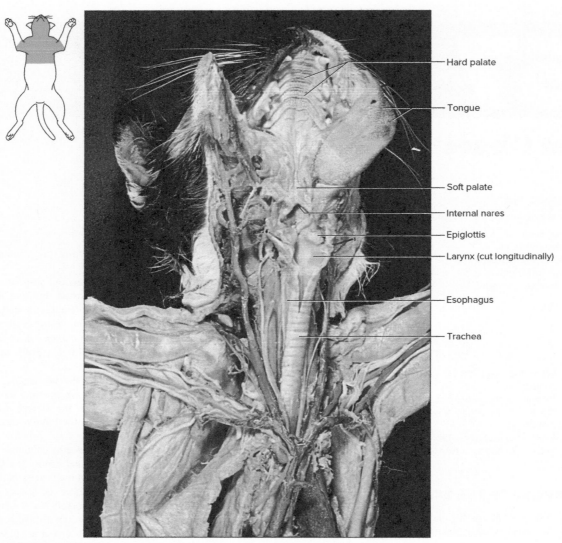

Hard palate

Tongue

Soft palate

Internal nares

Epiglottis

Larynx (cut longitudinally)

Esophagus

Trachea

FIGURE 57.1   Features of the cat's neck and oral cavity, ventral view.   **APR**   (©Dr. Sheril Burton)

notice the *parietal pleura*, which forms the inner lining of the thoracic wall, and locate the spaces of the *pleural cavities* (see fig. 57.2).

10. Make an incision through a lobe of a lung, and examine its interior. Note the branches of the smaller air passages. The main bronchi branch from the trachea and enter the lungs. Each main bronchus divides into *lobar bronchi;* lobar bronchi divide into even smaller *segmental bronchi* (fig. 57.3).

11. Examine the *diaphragm,* and note that its central portion is tendinous.

12. Locate the *phrenic nerve.* This nerve appears as a white thread passing along the side of the heart to the diaphragm.

13. Examine the organs located in the *mediastinum.* Note that the mediastinum separates the right and left lungs and pleural cavities within the thorax.

14. Review the locations of the respiratory organs without the aid of the figures. **1**

15. Examine the human torso model along with figures 57.1–57.3. For each human respiratory organ, identify the corresponding respiratory organ of the cat. **3**

16. Complete Laboratory Report 57.

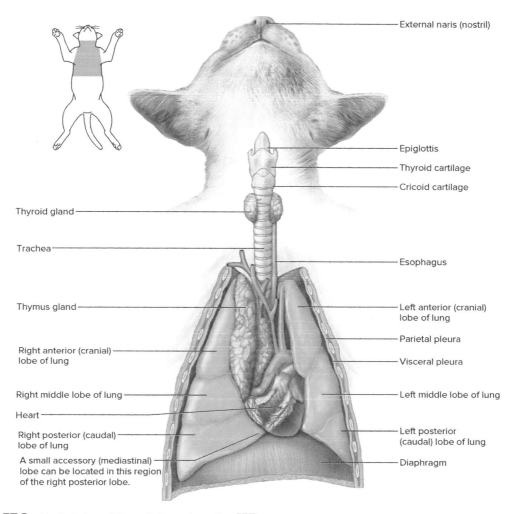

External naris (nostril)

Epiglottis

Thyroid cartilage

Cricoid cartilage

Thyroid gland

Trachea

Esophagus

Thymus gland

Left anterior (cranial) lobe of lung

Parietal pleura

Right anterior (cranial) lobe of lung

Visceral pleura

Right middle lobe of lung

Left middle lobe of lung

Heart

Right posterior (caudal) lobe of lung

Left posterior (caudal) lobe of lung

A small accessory (mediastinal) lobe can be located in this region of the right posterior lobe.

Diaphragm

FIGURE 57.2   Ventral view of the cat's thoracic cavity. **APR**

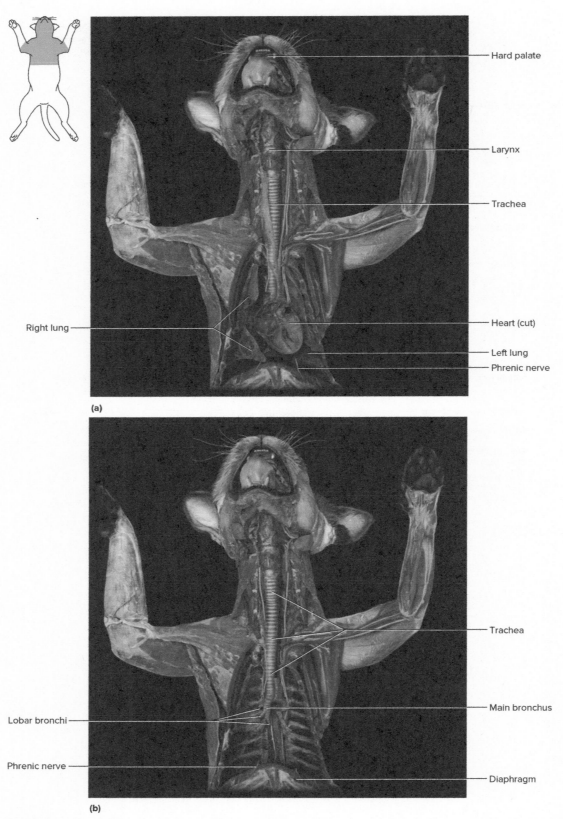

Hard palate

Larynx

Trachea

Right lung

Heart (cut)

Left lung

Phrenic nerve

(a)

Trachea

Lobar bronchi

Main bronchus

Phrenic nerve

Diaphragm

(b)

FIGURE 57.3   Ventral view of head, neck, and thoracic cavity of the cat: (*a*) lungs and heart included; (*b*) heart and lungs removed.   ((*a,b*) McGraw-Hill Education/APR)

Name _____

Date _____

Section _____

The ⚠ corresponds to the indicated Learning Outcome(s) found at the beginning of the laboratory exercise.

Cat Dissection:
Respiratory System

Complete the following:

1. Why do the auditory tubes open into the nasopharynx? ⚠1 _____

_____

_____

_____

2. Distinguish between the glottis and the epiglottis. ⚠1 _____

_____

_____

_____

3. Are the tracheal rings of the cat complete or incomplete? ⚠1 _____ How does this feature

   compare with that of the human? ⚠2 _____

_____

4. How does the structure of the primary bronchi compare with that of the trachea? ⚠1 _____

_____

_____

5. Compare the number of lobes in the human lungs with the number of lobes in the cat. ⚠2 _____

_____

_____

_____

_____

6. Describe the attachments of the diaphragm. ⚠1 _____

_____

_____

7. What major structures are located within the mediastinum of the cat? ⚠1 _____

_____

_____

8. Compare the organs located within the mediastinum of the cat with those within the human mediastinum. ⚠2 _____

_____

_____

58

Cat Dissection: Urinary System

MATERIALS NEEDED

Preserved cat
Dissecting tray
Dissecting instruments
Disposable gloves
Human torso model

SAFETY

- Wear disposable gloves when working on the cat dissection.
- Dispose of tissue remnants and gloves as instructed.
- Wash the dissecting tray and instruments as instructed.
- Wash your laboratory table.
- Wash your hands before leaving the laboratory.

PURPOSE OF THE EXERCISE

To examine and identify the urinary organs of the cat and to compare them with those of the human.

 ## LEARNING OUTCOMES APR

After completing this exercise, you should be able to

1. Locate and identify the urinary organs of the cat.
2. Compare the urinary organs of the cat with those of the human.
3. Identify the corresponding organs in the human torso model.

In this laboratory exercise, you will dissect the urinary organs of the cat. As you observe these structures, compare them with the corresponding human organs by observing the parts of the human torso model.

 ## PRACTICE

PROCEDURE—Urinary System Dissection

1. Place the preserved cat in a dissecting tray with its ventral surface up.
2. Open the abdominal cavity, and remove the liver, stomach, and spleen (figs. 58.1 and 58.2).
3. Push the intestines to one side, and locate the *kidneys* in the dorsal abdominal wall on either side of the vertebral column. The kidneys are located dorsal to the *parietal peritoneum* (retroperitoneal).
4. Carefully remove the parietal peritoneum and the adipose tissue surrounding the kidneys. Locate the following, using figures 58.1 and 58.2 as guides:

   ureters
   renal arteries
   renal veins

5. Locate the *adrenal glands,* which lie medially and anteriorly to the kidneys and are surrounded by connective tissues. The adrenal glands are attached to blood vessels and are separated by the abdominal aorta and posterior vena cava.
6. Expose the ureters by cleaning away the connective tissues along their lengths. They enter the *urinary bladder* on the dorsal surface.
7. Examine the urinary bladder, attached to the abdominal wall by folds of peritoneum that form ligaments. Locate the *medial ligament* on the ventral surface of the bladder that connects it with the linea alba and the *two lateral ligaments* on the dorsal surface.
8. Remove the connective tissue from around the urinary bladder. Use a sharp scalpel to open the bladder, and examine its interior. Locate the openings of the ureters and the urethra on the inside.
9. Expose the *urethra* at the posterior of the urinary bladder. The urethra terminates with the vagina in a urogenital sinus in the female cat; the urethra enters the penis in the male cat (figs. 58.1–58.3).

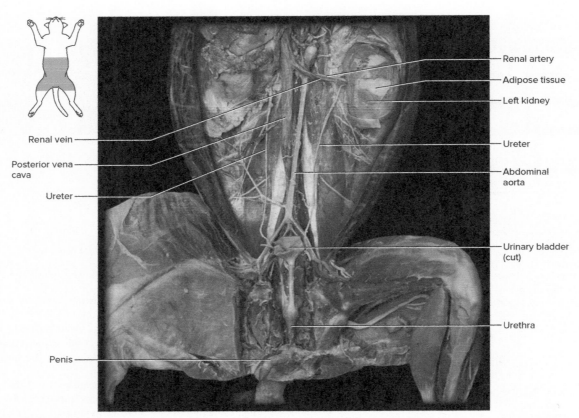

Renal artery
Adipose tissue
Left kidney
Ureter
Abdominal aorta
Urinary bladder (cut)
Urethra

Renal vein
Posterior vena cava
Ureter
Penis

FIGURE 58.1   Urinary system of the male cat's abdominal cavity, with most of the digestive and some reproductive organs removed. (McGraw-Hill Education/APR)

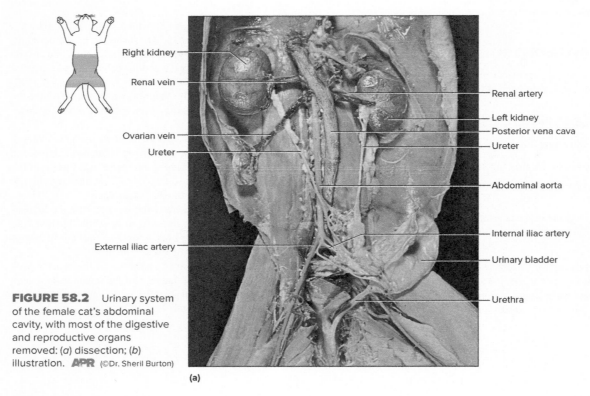

Right kidney
Renal vein
Ovarian vein
Ureter
External iliac artery

Renal artery
Left kidney
Posterior vena cava
Ureter
Abdominal aorta
Internal iliac artery
Urinary bladder
Urethra

FIGURE 58.2   Urinary system of the female cat's abdominal cavity, with most of the digestive and reproductive organs removed: (a) dissection; (b) illustration. **APR** (©Dr. Sheril Burton)

(a)

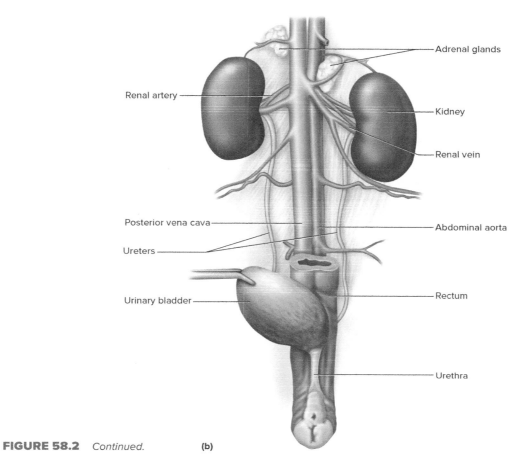

Adrenal glands

Renal artery

Kidney

Renal vein

Posterior vena cava

Abdominal aorta

Ureters

Urinary bladder

Rectum

Urethra

FIGURE 58.2 *Continued.* **(b)**

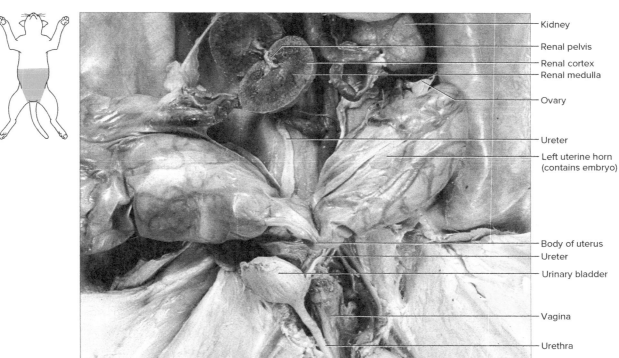

Kidney

Renal pelvis

Renal cortex

Renal medulla

Ovary

Ureter

Left uterine horn (contains embryo)

Body of uterus

Ureter

Urinary bladder

Vagina

Urethra

FIGURE 58.3 Female urinary and reproductive systems of a pregnant cat. **APR** (McGraw-Hill Education/Ralph W. Stevens III, Ph.d., photographer)

10. Remove one kidney, and section it longitudinally (figs. 58.3 and 58.4). Identify the following features:

fibrous capsule
renal cortex (superficial, lighter region)
renal medulla (deeper, darker region)
renal pyramid (single in cat)
renal papilla (single in cat)
renal sinus
renal pelvis
hilum of kidney

11. Review the locations of the urinary organs without the aid of the figures. ▲1

12. Discard the organs and tissues that were removed from the cat, as directed by the laboratory instructor.

13. Examine the human torso model along with figures 58.1–58.4. Identify urinary organs of the cat that correspond with those of the human. ▲3

14. Complete Laboratory Report 58.

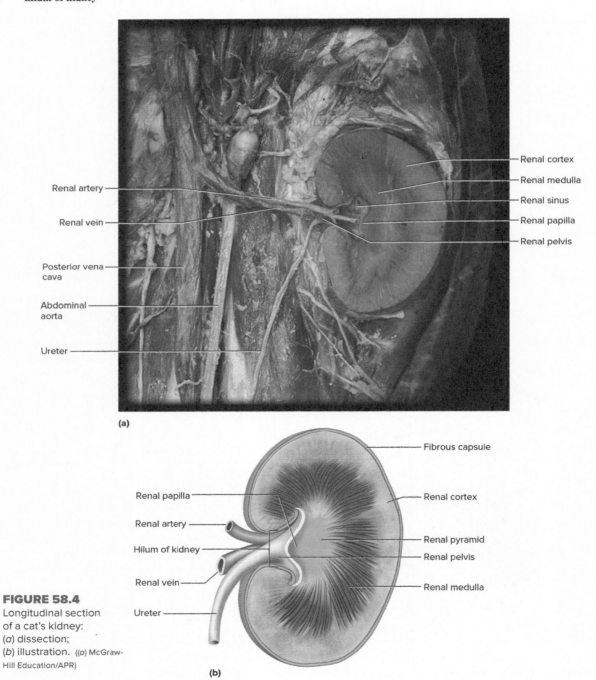

(a)

Renal artery
Renal vein
Posterior vena cava
Abdominal aorta
Ureter

Renal cortex
Renal medulla
Renal sinus
Renal papilla
Renal pelvis

Fibrous capsule

Renal papilla
Renal artery
Hilum of kidney
Renal vein
Ureter

Renal cortex
Renal pyramid
Renal pelvis
Renal medulla

FIGURE 58.4
Longitudinal section of a cat's kidney: (a) dissection; (b) illustration. ((a) McGraw-Hill Education/APR)

(b)

Name _____

Date _____

Section _____

The $\boxed{A}$ corresponds to the indicated Learning Outcome(s) found at the beginning of the laboratory exercise.

Cat Dissection:
Urinary System

Complete the following:

1. Compare the positions of the kidneys in the cat with those in the human. $\boxed{2}$ _____
   _____
   _____
   _____
   _____
   _____

2. Compare the locations of the adrenal glands in the cat with those in the human. $\boxed{2}$ _____
   _____
   _____
   _____

3. What structures of the cat urinary system are retroperitoneal? $\boxed{1}$ _____
   _____
   _____

4. Describe the wall of the urinary bladder of the cat. $\boxed{1}$ _____
   _____
   _____
   _____

5. Compare the renal pyramids and renal papillae of the human kidney with those of the cat. $\boxed{2}$ _____
   _____
   _____
   _____
   _____

59

Cat Dissection: Reproductive Systems

MATERIALS NEEDED

Preserved cat
Dissecting tray
Dissecting instruments
Bone cutters
Disposable gloves
Models of human reproductive systems

⚠ SAFETY

- Wear disposable gloves when working on the cat dissection.
- Dispose of tissue remnants and gloves as instructed.
- Wash the dissecting tray and instruments as instructed.
- Wash your laboratory table.
- Wash your hands before leaving the laboratory.

PURPOSE OF THE EXERCISE

To examine and identify the reproductive organs of the cat and to compare them with the corresponding organs of the human.

LEARNING OUTCOMES APR

After completing this exercise, you should be able to

1. Locate and identify the reproductive organs of a cat.

2. Compare the reproductive organs of the cat with those of the human.

3. Identify the corresponding organs in models of the human reproductive systems.

In this laboratory exercise, you will dissect the reproductive system of the cat. If you have a female cat, begin with Procedure A. If you have a male cat, begin with Procedure B. After completing the dissection, exchange cats with someone who has dissected one of the opposite sex, and examine its reproductive organs.

As you observe the cat reproductive organs, compare them with the corresponding human organs by examining the models of the human reproductive systems.

PRACTICE

PROCEDURE A—Female Reproductive System

1. Place the preserved female cat in a dissecting tray with its ventral surface up, and open its abdominal cavity.

2. Locate the small, oval *ovaries* just posterior to the kidneys (figs. 59.1 and 59.2).

3. Examine an ovary, suspended from the dorsal body wall by a fold of peritoneum called the *mesovarium.* Another attachment, the *ovarian ligament,* connects the ovary to the tubular *uterine horn* (see fig. 59.1).

4. Near the anterior end of the ovary, locate the funnel-shaped *infundibulum,* which is at the end of a very small *uterine tube* (*oviduct*). Note the tiny projections, or *fimbriae,* that create a fringe around the edge of the infundibulum. Trace the uterine tube around the ovary to its nearby connection with the uterine horn near the attachment of the ovarian ligament.

5. Examine the uterine horn. It is suspended from the body wall by a fold of peritoneum, the *broad ligament,* that is continuous with the mesovarium of the ovary. Also note the fibrous *round ligament,* which extends from the uterine horn laterally and posteriorly to the body wall (figs. 59.1 and 59.2).

6. To observe the remaining organs of the reproductive system more easily, remove the left hindlimb by severing it near the hip with bone cutters. Then use the bone cutters to remove the left anterior portion of the pelvic girdle. Also remove the necessary muscles and connective tissue to expose the structures shown in figure 59.3.

7. Trace the relatively long uterine horns posteriorly. They unite to form the *uterine body,* which is located between the urethra and rectum. This Y-shaped uterus allows ample space for several offspring to develop at the same time in the uterine horns (see fig. 59.2). The uterine body is continuous with the *vagina,* which leads to the outside.

8. Trace the *urethra* from the urinary bladder posteriorly. The urethra and the vagina open into a common chamber, called the *urogenital sinus*. The opening of this chamber, ventral to the anus, is called the *urogenital aperture*. Locate the *labia majora,* which are folds of skin on either side of the urogenital aperture.

9. Use scissors to open the vagina along its lateral wall, beginning at the urogenital aperture and continuing to the body of the uterus. Note the *urethral orifice* in the ventral wall of the urogenital sinus, and locate the small, rounded *cervix*

of the uterus, which projects into the vagina at its deep end. The *clitoris* is located in the ventral wall of the urogenital sinus near its opening to the outside.

10. Review the locations of the reproductive organs without the aid of the figures. ⓐ

11. Examine the models of the human reproductive systems along with figures 59.1–59.3. Identify the reproductive organs of the cat that correspond with those of the human. ⓐ

12. Complete Part A of Laboratory Report 59.

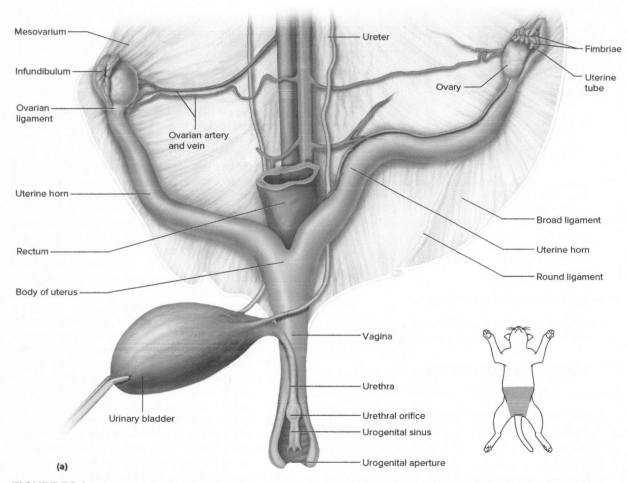

FIGURE 59.1 Ventral view of the female cat's reproductive system: (*a*) illustration; (*b*) dissection. (McGraw-Hill Education/APR)

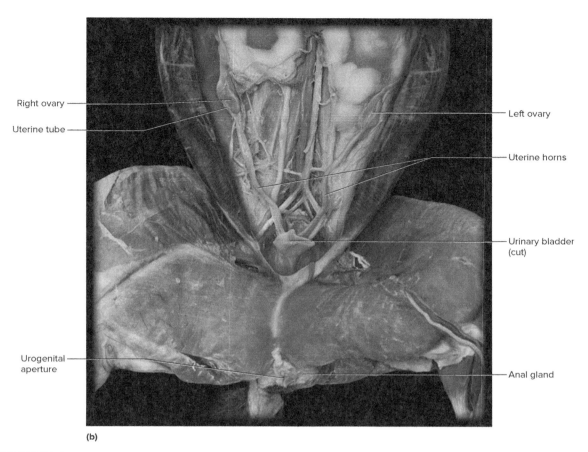

Right ovary

Uterine tube

Left ovary

Uterine horns

Urinary bladder (cut)

Urogenital aperture

Anal gland

(b)

FIGURE 59.1   *Continued.*

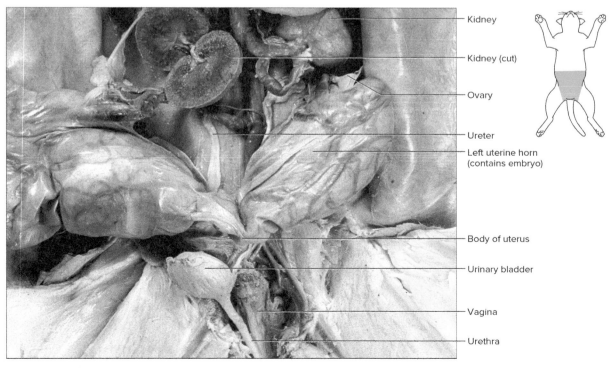

Kidney

Kidney (cut)

Ovary

Ureter

Left uterine horn (contains embryo)

Body of uterus

Urinary bladder

Vagina

Urethra

FIGURE 59.2   Female urinary and reproductive systems of a pregnant cat.   **APR**   (McGraw-Hill Education/Ralph W. Stevens III, Ph.d., photographer)

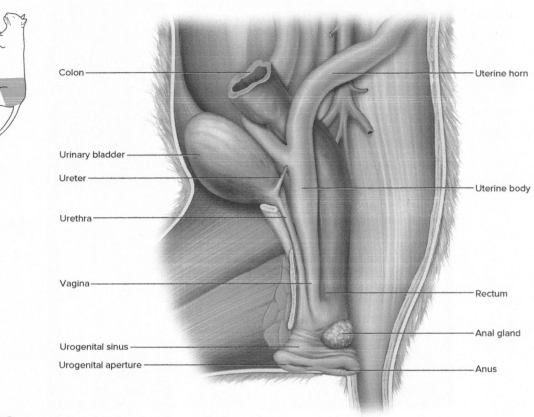

FIGURE 59.3   Lateral view of the female cat's reproductive system.

PRACTICE

PROCEDURE B—Male Reproductive System

1. Place the preserved male cat in a dissecting tray with its ventral surface up.
2. Locate the *scrotum* between the hindlimbs. Make an incision along the midline of the scrotum, and expose the *testes* inside (fig. 59.4). The testes are separated by a septum, and each testis is enclosed in a sheath of connective tissue.
3. Remove the sheath surrounding the testes, and locate the convoluted *epididymis* on the dorsal surface of each testis (fig. 59.5).
4. Locate the *spermatic cord* on the right side, leading away from the testis. This spermatic cord contains the *ductus (vas) deferens,* which is continuous with the epididymis, as well as with the nerves and blood vessels that supply the testis on that side. Trace the spermatic cord to the body wall, where its contents pass through the *inguinal canal* and enter the pelvic cavity (fig. 59.5).

5. Locate the *penis,* and identify the *prepuce,* which forms a sheath around the entire penis. Make an incision through the skin of the prepuce, and expose the *glans penis.* The glans penis has minute spines on its surface that assist copulation.
6. To observe the remaining structures of the reproductive system more easily, remove the left hindlimb by severing it near the hip with bone cutters. Then use the bone cutters to remove the left anterior portion of the pelvic girdle. Also remove the necessary muscles and connective tissues to expose the organs shown in figure 59.6.
7. Trace the ductus deferens from the inguinal canal to the penis. Note that the ductus deferens loops over the ureter within the pelvic cavity and passes downward behind the urinary bladder to join the urethra. Locate the *prostate gland,* which appears as an enlargement at the junction of the ductus deferens and urethra. The prostate gland is relatively small compared to that in humans and positioned along the urethra some distance from the urinary bladder.

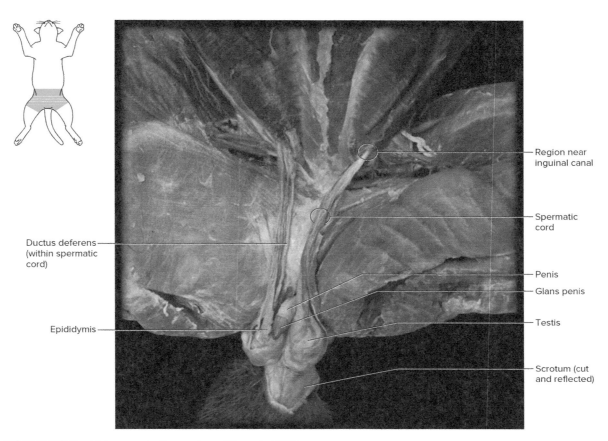

Region near inguinal canal

Spermatic cord

Ductus deferens (within spermatic cord)

Penis

Glans penis

Epididymis

Testis

Scrotum (cut and reflected)

FIGURE 59.4 Male reproductive system of the cat. **APR** (McGraw-Hill Education/APR)

8. Trace the urethra to the penis. Locate the *bulbourethral glands,* which form small swellings on either side at the proximal end of the penis. Seminal vesicles are absent in cats.

9. Use a sharp scalpel to cut a transverse section of the penis. Identify the *corpora cavernosa,* each of which contains many blood spaces surrounded by a sheath of connective tissue.

10. Review the locations of the reproductive organs without the aid of the figures. 1

11. Examine the models of the human reproductive systems along with figures 59.4–59.6. Identify the organs of the cat that correspond with those of the human. 3

12. Complete Part B of the laboratory report.

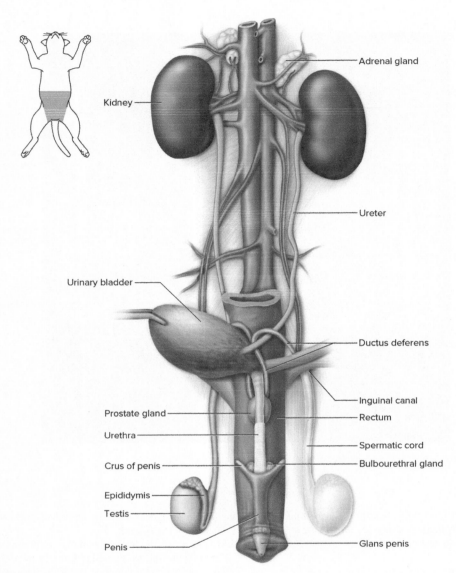

Kidney

Adrenal gland

Ureter

Urinary bladder

Ductus deferens

Inguinal canal

Prostate gland

Rectum

Urethra

Spermatic cord

Crus of penis

Bulbourethral gland

Epididymis

Testis

Penis

Glans penis

FIGURE 59.5 Ventral view of the male cat's reproductive system.

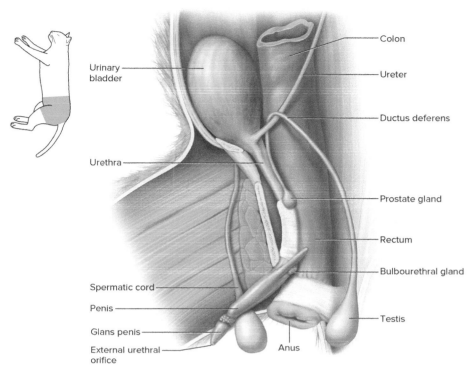

FIGURE 59.6 Lateral view of the male cat's reproductive system.

Notes

Name _____

Date _____

Section _____

The ⌂ corresponds to the indicated Learning Outcome(s) found at the beginning of the laboratory exercise.

Cat Dissection:
Reproductive Systems

PART A ASSESSMENTS

Complete the following:

1. Compare the relative lengths and paths of the uterine tubes (oviducts) of the cat and the human. ⌂2 _____

_____

_____

_____

_____

_____

2. How do the shape and structure of the uterus of the cat compare with that of the human? ⌂2 _____

_____

_____

_____

_____

_____

3. Assess the function of the uterine horns of the cat. ⌂1 _____

_____

_____

_____

_____

_____

4. Compare the relationship of the urethra and the vagina in the cat and in the human. ⌂2 _____

_____

_____

_____

_____

_____

PART B ASSESSMENTS

Complete the following:

1. Compare the glans penis in the cat and in the human. [2] _____
_____
_____
_____
_____
_____
_____

2. How do the location and the relative size of the prostate gland of the cat compare with that of the human? [2]
_____
_____
_____
_____
_____

3. What glands associated with the human ductus deferens are missing in the cat? [2] _____
_____
_____

4. Compare the prepuce in the cat and in the human. [2] _____
_____
_____
_____

Fetal Pig Dissection Laboratory Exercises

LABORATORY EXERCISE

60

Fetal Pig Dissection: Musculature

MATERIALS NEEDED

Textbook
Preserved fetal pig (double injection)
Dissecting tray
Dissecting instruments
Twine
Plastic bag
Identification tag
Disposable gloves
Bone shears
Human torso model
Human upper and lower limb models

⚠ SAFETY

- Wear disposable gloves when working on the fetal pig dissection.
- Dispose of tissue remnants and gloves as instructed.
- Wash the dissecting tray and instruments as instructed.
- Wash your laboratory table.
- Wash your hands before leaving the laboratory.

PURPOSE OF THE EXERCISE

To examine and identify the skeletal musculature of the fetal pig and to compare it with that of the human.

 LEARNING OUTCOMES APR

After completing this exercise, you should be able to

1. Name and locate the major skeletal muscles of the fetal pig.
2. Name the origins, insertions, and actions of the skeletal muscles designated by the laboratory instructor.
3. Identify the corresponding skeletal muscles of the human.

Although the aim of this exercise is to become more familiar with the human musculature, human cadavers are not always available for dissection. Instead, preserved fetal pigs are often used for dissection because they are relatively small and can be purchased from biological suppliers. Because they are mammals, pigs have many features in common with humans, including similar skeletal muscles (with similar names).

The gestational period of pigs is 112–115 days. At the time of birth, a fetal pig would be about 30 centimeters (12–13 inches) long measured from nose to base of tail. A fetal pig approaching 22 centimeters (9 inches) long would have developed about 100 days. The embryonic and fetal stages of development occur much faster than in humans, for whom the gestational period is closer to 40 weeks.

Despite their similarities, pigs use four limbs for locomotion, whereas humans use only two limbs. Because the musculature of each type of organism is adapted to provide for its special needs, comparisons of the muscles of pigs and humans may not be precise.

As you continue your dissection of various systems of the fetal pig, many anatomical similarities between pigs and humans will be observed. These fundamental similarities are called homologous structures. Although homologous structures have a similar structure and embryological origin, the functions are sometimes different.

 MUSCLE DISSECTION TECHNIQUES

Dissect: to expose the entire length of the muscle from origin to insertion. Most of the procedures are accomplished with blunt probes used to separate various connective tissues that hold adjacent structures together. It does not mean to remove or to cut into the muscles or other organs.

Transect: to cut through the muscle near its midpoint. The cut is perpendicular to the muscle fibers.

Reflect: to lift a transected muscle aside to expose deeper muscles or other organs.

PRACTICE

PROCEDURE A—External Features

1. Fetal pigs are usually preserved with a solution that pre-vents microorganisms from decomposing the tissues. However, because fumes from this preserving fluid may be annoying and may irritate skin, be sure to work in a well-ventilated room and wear disposable gloves to pro-tect your hands. If your fetal pig has an incision in the neck, it marks the location where colored latex (red into arteries and blue into veins) was injected into the blood vessels.

2. Obtain a preserved fetal pig, a dissecting tray, a set of dissecting instruments, a large plastic bag, twine, and an identification tag.

3. If the fetal pig is in a storage bag, dispose of any excess preserving fluid in the bag, as directed by the labora-tory instructor. However, you may want to save some of the preserving fluid to keep the fetal pig moist until you have completed your work later in the year.

4. Rinse the pig off before the dissection begins. Blot the pig dry with paper towels.

5. Study figure 60.1 and locate the following external ana-tomical features:

auricle
tongue
external nares
rostrum (snout)
eyelid
shoulder
arm
elbow
forearm
wrist
digits
hip
thigh
knee
leg
ankle
umbilical cord
tail
mammary papillae
anus

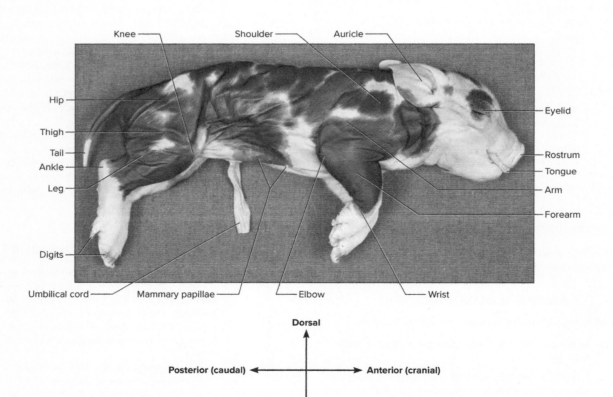

FIGURE 60.1   External features of a fetal pig, lateral view. (©J & J Photography)

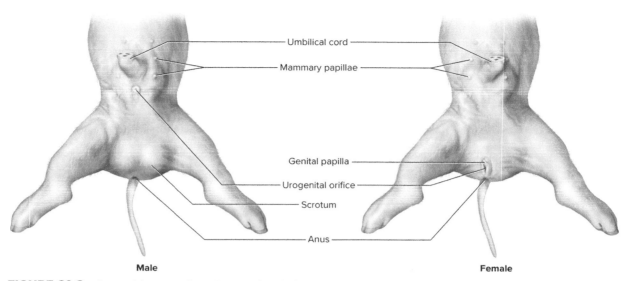

Male   Female

FIGURE 60.2   External features of a male and a female fetal pig, ventral view.

6. Study the external features of a male and a female pig (fig. 60.2) by comparing your pig to others in the laboratory. Identify the following:

male
   urogenital orifice
   scrotum
   mammary papillae

female
   genital papilla
   urogenital orifice
   mammary papillae

 **PRACTICE**

PROCEDURE B—Skin Removal

1. Remove the skin from the fetal pig. To do this, follow these steps:
   a. Place the fetal pig in the dissecting tray with its dorsal surface down.
   b. Use a sharp scalpel to make short, shallow incisions through the skin as illustrated in figure 60.3. The incisions are different for the male and female in the genital regions.
   c. Your incisions should encircle the neck, wrists, ankles, and urogenital regions. More skin is left attached on the male pig in the urogenital region.

   d. Use a blunt probe to separate the skin from the muscles. Observe the loose connective tissue (superficial fascia) that binds the skin to the muscle. As you pull the skin away, you may note a thin sheet of skeletal muscle attached to it. These are *cutaneous muscles,* and they function to move the skin to get rid of any irritants. Humans lack cutaneous muscles, but a similar sheet of muscle (platysma) is present in the neck of a human.
   e. As you remove the skin, work toward the dorsal surface, then work toward the head, and finally work toward the tail. Pull the skin over each limb as if you were removing a glove.
   f. After the skin has been pulled away, carefully remove as much of the remaining connective tissue as possible to expose the underlying skeletal muscles. This task only needs to be performed on the right side and ventrally so you can study the muscles. The muscles should appear light brown and fibrous.

2. After skinning the fetal pig, follow these steps:
   a. Discard the tissues you have removed as directed by the laboratory instructor.
   b. Wrap the skin around the pig to help keep its body moist, and place it in a plastic storage bag.
   c. Write your name in pencil on an identification tag, and tie the tag to the storage bag so that you can identify your specimen.

3. Observe the recommended safety procedures for the conclusion of a laboratory session.

External nares

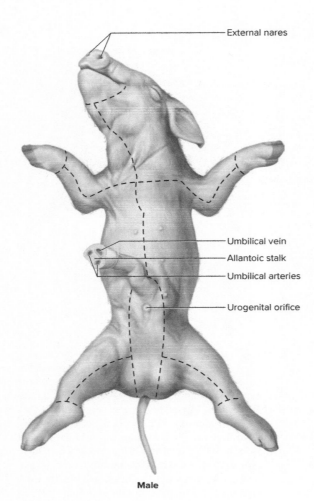

Umbilical vein

Allantoic stalk

Umbilical arteries

Urogenital orifice

Male

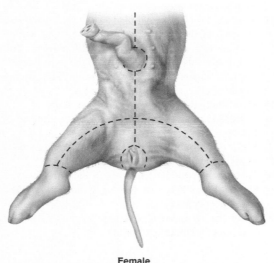

Female

FIGURE 60.3 The dotted lines indicate the shallow skin incisions on the male and female fetal pigs to view the underlying muscles.

 PRACTICE

PROCEDURE C—Skeletal Muscle Dissection

1. The purpose of a skeletal muscle dissection is to separate the individual muscles from any surrounding tissues in order to expose the muscles for observation. To do a muscle dissection, follow these steps:
   a. Use the appropriate figure as a guide and locate the muscle to be dissected in the specimen.
   b. Use a blunt probe to separate the muscle from the surrounding connective tissue along its natural borders. The muscle should separate smoothly. If the border appears ragged, you probably have torn the muscle fibers. Because the fetal pig's muscles are incompletely developed, they are easily damaged and hard to distinguish, especially in small to medium sizes.

2. If it is necessary to transect a superficial muscle to observe a deeper one, use scissors to transect the muscle about halfway between its origin and insertion. Then, reflect the cut ends, leaving their attachments intact.

3. The following procedures will instruct you to dissect some of the larger and more easily identified muscles of the fetal pig. In each case, the procedure will include the names of the muscles to be dissected, figures illustrating their locations, and tables listing the origins, insertions, and actions of the muscles. Compare the muscle origins and insertions to the fetal pig skeleton (fig. 60.4). (More detailed dissection instructions can be obtained by consulting an additional guide or an atlas for fetal pig dissection.) As you dissect each muscle, study the figures and tables in chapter 9 of the textbook and identify any corresponding (homologous) muscles of the human body. Also locate these homologous muscles in the human torso model or models of the human upper and lower limbs.

 PRACTICE

PROCEDURE D—Muscles of the Head and Neck

1. Place the fetal pig in the dissecting tray with its ventral surface up.
2. Remove the skin and underlying connective tissues from each side of the neck, forward to the chin, and up to each ear.
3. Study figures 60.5 and 60.6, and then locate and dissect the following muscles:

sternomastoid	**mylohyoid**
sternohyoid	**masseter**
digastric	

4. To find the deep *sternothyroid* muscle in the neck, transect and reflect the sternohyoid muscle from the right side.
5. See table 60.1 for the origins, insertions, and actions of these head and neck muscles.
6. Complete Part A of Laboratory Report 60.

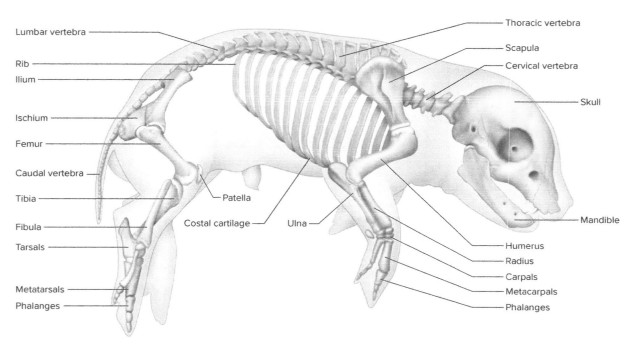

Lumbar vertebra

Rib

Ilium

Ischium

Femur

Caudal vertebra

Tibia

Fibula

Tarsals

Metatarsals

Phalanges

Thoracic vertebra

Scapula

Cervical vertebra

Skull

Mandible

Patella

Costal cartilage

Ulna

Humerus

Radius

Carpals

Metacarpals

Phalanges

FIGURE 60.4   Skeleton of a fetal pig, lateral view.

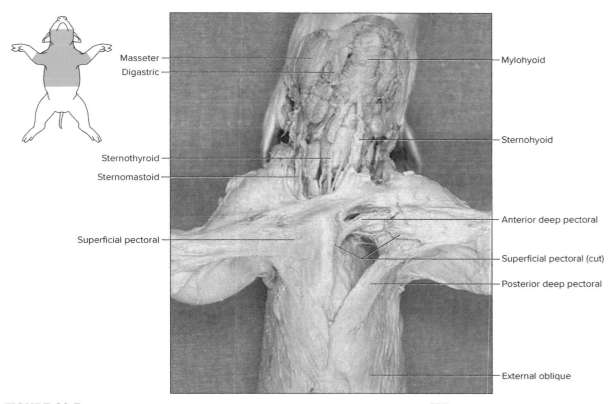

Masseter

Digastric

Sternothyroid

Sternomastoid

Superficial pectoral

Mylohyoid

Sternohyoid

Anterior deep pectoral

Superficial pectoral (cut)

Posterior deep pectoral

External oblique

FIGURE 60.5   Muscles of the head, neck, thorax, and abdominal wall, ventral view. **APR** (©J & J Photography)

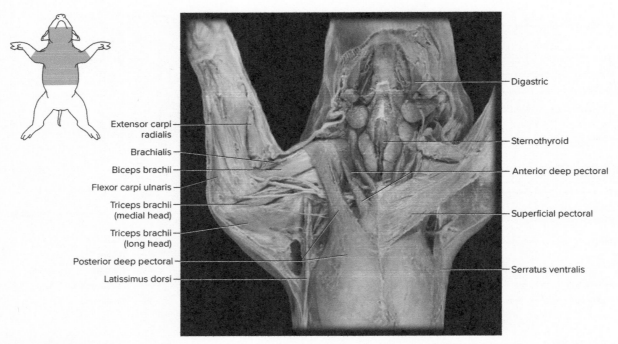

FIGURE 60.6   Muscles of the head, neck, thorax, and medial forelimb, ventral view.   **APR** (©McGraw-Hill Education/APR)

TABLE 60.1   Muscles of the Head and Neck

Muscle	Origin	Insertion	Action
Sternomastoid	Sternum	Mastoid process of temporal bone	Turns and depresses head
Sternohyoid	Costal cartilage	Hyoid bone	Depresses hyoid bone
Digastric	Mastoid process and occipital bone	Mandible	Depresses mandible
Mylohyoid	Mandible	Hyoid bone	Raises floor of mouth
Masseter	Zygomatic arch	Mandible	Elevates mandible
Sternothyroid	Sternum	Thyroid cartilage	Pulls larynx back

 PRACTICE

PROCEDURE E—Muscles of the Thorax

1. Place the fetal pig in the dissecting tray with its ventral surface up. Spread and secure all four limbs with twine.
2. Remove any remaining fat and connective tissue to expose the muscles in the walls of the thorax and abdomen.
3. Study figures 60.5 and 60.6. Locate and transect the superficial pectoral to expose the anterior deep pectoral. Locate and dissect the following pectoral muscles:

   superficial pectoral (pectoralis superficialis)— homologous to pectoralis major

   posterior deep pectoral (pectoralis profundus)— homologous to pectoralis minor

   anterior deep pectoral

4. To find the deep *serratus ventralis* thoracic muscle, transect the pectoral muscles, and reflect their cut edges to the sides. The fingerlike origins on the ribs can be located by pulling the forelimb away from the thorax.
5. See table 60.2 for the origins, insertions, and actions of these muscles.

 PRACTICE

PROCEDURE F—Muscles of the Abdominal Wall

1. Study figure 60.5.
2. Locate the *external oblique* muscle in the abdominal wall.
3. Make a shallow, longitudinal incision through the external oblique. Reflect the cut edge and expose the *internal*

oblique muscle beneath. The fibers of the internal oblique run at a right angle to those of the external oblique.

4. Make a longitudinal incision through the internal oblique. Reflect the cut edge and expose the *transversus abdominis* muscle.
5. Expose the *rectus abdominis* muscle on one side of the midventral line. This muscle lies beneath an aponeurosis.
6. See table 60.3 for the origins, insertions, and actions of these muscles.
7. Complete Part B of the laboratory report.

 PRACTICE

PROCEDURE G—Muscles of the Shoulder and Back

1. Place the fetal pig in the dissecting tray onto its lateral surface.
2. Remove any remaining fat and connective tissue to expose the muscles of the shoulder and back.
3. Study figure 60.7, and then locate and dissect the following superficial muscles:

   brachiocephalic

   acromiotrapezius

   spinotrapezius

   deltoid

   latissimus dorsi

4. Using scissors, transect the latissimus dorsi and the group of trapezius muscles. Reflect their cut edges and remove any underlying fat and connective tissue. Study

figure 60.8, and then locate and dissect the following deep muscles of the shoulder and back:

supraspinatus
infraspinatus
teres major
rhomboid
rhomboid capitis
splenius

5. See table 60.4 for the origins, insertions, and actions of these muscles in the shoulder and back.
6. Review the locations, origins, insertions, and actions designated by the laboratory instructor without the aid of the figures and tables.
7. Complete Part C of the laboratory report.

PRACTICE

PROCEDURE H—Muscles of the Forelimb

1. Place the fetal pig in the dissecting tray with its ventral surface up.
2. Remove any remaining fat and connective tissue from a forelimb to expose the muscles.
3. Transect the pectoral muscles and reflect their cut edges.
4. Study figures 60.6 and 60.9, and then locate and dissect the following muscles from the medial surface of the arm and forearm of the forelimb:

medial arm muscles
   biceps brachii
   triceps brachii
      long head
      medial head

TABLE 60.2   Muscles of the Thorax

Muscle	Origin	Insertion	Action
Superficial pectoral	Sternum	Humerus	Adducts arm
Posterior deep pectoral	Sternum and costal cartilages	Humerus	Adducts arm
Anterior deep pectoral	Sternum and costal cartilages	Scapula	Pulls scapula toward midline of body
Serratus ventralis	Ribs and cervical vertebrae	Vertebral border of scapula	Pulls scapula posteriorly and transfers weight from trunk to pectoral girdle

TABLE 60.3   Muscles of the Abdominal Wall

Muscle	Origin	Insertion	Action
External oblique	Ribs and fascia of back	Linea alba	Compresses abdominal wall
Internal oblique	Fascia of back	Linea alba	Compresses abdominal wall
Transversus abdominis	Lower ribs and fascia of back	Linea alba	Compresses abdominal wall
Rectus abdominis	Pubis	Sternum and costal cartilages	Compresses abdominal wall and flexes trunk

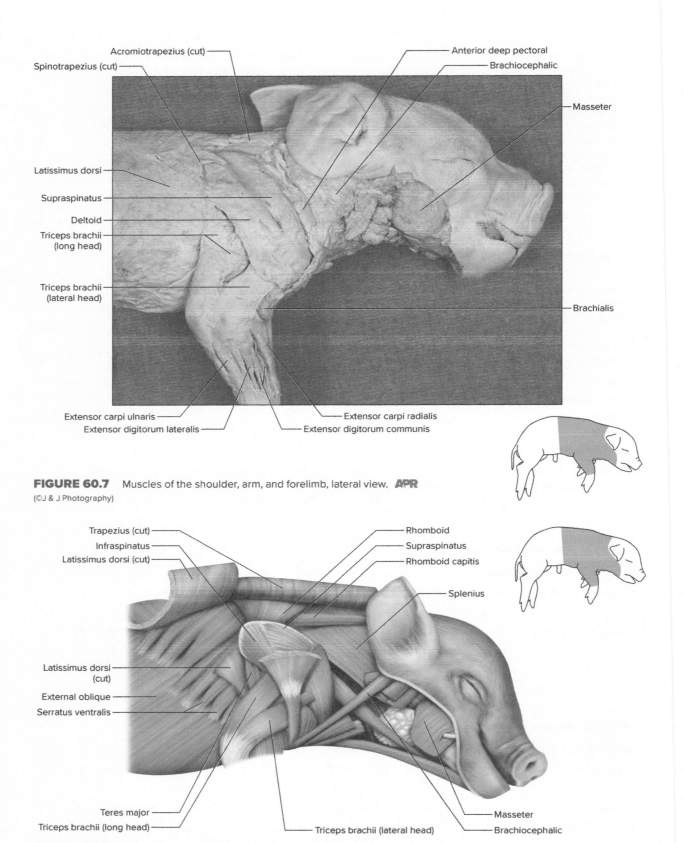

Spinotrapezius (cut)
Acromiotrapezius (cut)
Anterior deep pectoral
Brachiocephalic
Masseter
Latissimus dorsi
Supraspinatus
Deltoid
Triceps brachii
(long head)
Triceps brachii
(lateral head)
Brachialis
Extensor carpi ulnaris
Extensor digitorum lateralis
Extensor carpi radialis
Extensor digitorum communis

FIGURE 60.7   Muscles of the shoulder, arm, and forelimb, lateral view.   **APR**
(©J & J Photography)

Trapezius (cut)
Infraspinatus
Latissimus dorsi (cut)
Rhomboid
Supraspinatus
Rhomboid capitis
Splenius
Latissimus dorsi
(cut)
External oblique
Serratus ventralis
Teres major
Triceps brachii (long head)
Triceps brachii (lateral head)
Masseter
Brachiocephalic

FIGURE 60.8   Deep muscles of the shoulder and arm, lateral view.

TABLE 60.4   Muscles of the Shoulder and Back

Muscle	Origin	Insertion	Action
Acromiotrapezius	Spines of cervical and thoracic vertebrae	Spine and acromion process of scapula	Pulls scapula upward
Spinotrapezius	Spines of thoracic vertebrae	Spine of scapula	Pulls scapula upward and back
Brachiocephalic (clavotrapezius and clavobrachialis)	Occipital bone and mastoid process	Distal end of humerus	Flexes forelimb
Deltoid	Spine of scapula	Proximal end of humerus	Flexes forelimb
Latissimus dorsi	Thoracic and lumbar vertebrae	Proximal end of humerus	Pulls forelimb upward and back
Supraspinatus	Fossa above spine of scapula	Proximal end of humerus	Extends forelimb
Infraspinatus	Fossa below spine of scapula	Proximal end of humerus	Rotates and abducts forelimb
Teres major	Posterior border of scapula	Proximal end of humerus	Extends and adducts forelimb
Rhomboid	Spines of cervical and thoracic vertebrae	Medial border of scapula	Pulls scapula upward and forward
Rhomboid capitis	Occipital bone	Medial border of scapula	Pulls scapula forward
Splenius	Fascia of neck	Occipital bone	Raises head

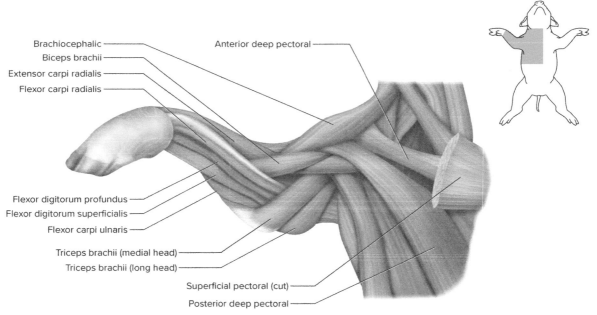

Brachiocephalic
Biceps brachii
Extensor carpi radialis
Flexor carpi radialis
Anterior deep pectoral
Flexor digitorum profundus
Flexor digitorum superficialis
Flexor carpi ulnaris
Triceps brachii (medial head)
Triceps brachii (long head)
Superficial pectoral (cut)
Posterior deep pectoral

FIGURE 60.9   Medial muscles of the right forelimb, ventral view.

medial forearm muscles

extensor carpi radialis

flexor carpi radialis

flexor digitorum profundus

flexor digitorum superficialis

flexor carpi ulnaris

5. Place your pig onto its lateral surface. Study figure 60.7, which shows the lateral surface of the forelimb. Locate and dissect the following muscles on the lateral surface of the arm and forearm of the forelimb:

lateral arm muscles

triceps brachii

long head

lateral head

brachialis

lateral forearm muscles

extensor carpi radialis

extensor digitorum communis

extensor digitorum lateralis

extensor carpi ulnaris

6. See table 60.5 for the origins, insertions, and actions of these muscles of the forelimb.

7. Complete Part D of the laboratory report.

 PRACTICE

PROCEDURE I—Muscles of the Hip and Hindlimb

1. Place the fetal pig in the dissecting tray with its ventral surface up.

2. Remove any remaining fat and connective tissue from the hip and hindlimb to expose the muscles.

3. Study figure 60.10, and then locate and dissect the following muscles from the medial surface of the thigh:

sartorius

gracilis

4. Using scissors, transect the sartorius and gracilis, and reflect their cut edges to observe the deeper muscles of the thigh.

5. Study figure 60.10, and then locate and dissect the following muscles:

tensor fasciae latae

rectus femoris

vastus medialis

adductor group

semimembranosus

semitendinosus

TABLE 60.5   Muscles of the Forelimb

Muscle	Origin	Insertion	Action
Triceps brachii			
Lateral head	Deltoid tuberosity of humerus	Olecranon process of ulna	Extends forelimb
Long head	Border of glenoid cavity of scapula	Olecranon process of ulna	Extends forelimb
Medial head	Shaft of humerus	Olecranon process of ulna	Extends forelimb
Brachialis	Lateral surface of humerus	Proximal end of ulna	Flexes forelimb
Biceps brachii	Scapula	Radius and ulna	Flexes forelimb
Extensor carpi ulnaris	Lateral epicondyle of humerus	Fifth metacarpal	Extends wrist
Extensor digitorum lateralis	Distal end of humerus	Digits	Extends digits
Extensor digitorum communis	Lateral surface of humerus	Digits	Extends digits
Extensor carpi radialis	Distal end of humerus	Distal end of radius	Rotates foot
Flexor carpi ulnaris	Medial epicondyle of humerus	Carpals	Flexes wrist
Flexor digitorum superficialis	Medial epicondyle of humerus	Distal phalanges	Flexes digits
Flexor digitorum profundus	Medial epicondyle of humerus	Distal phalanges	Flexes digits
Flexor carpi radialis	Medial epicondyle of humerus	Metacarpals	Flexes metacarpals

6. Transect the tensor fasciae latae and rectus femoris muscles and reflect their cut edges. Locate the *vastus intermedius* and *vastus lateralis* muscles beneath. (*Note:* In some specimens, the vastus intermedius, vastus lateralis, and vastus medialis muscles are closely united by connective tissue and are difficult to separate.)

7. Study figure 60.11, and then locate and dissect the following muscles from the lateral surface of the hip and thigh:

   biceps femoris
   gluteus superficialis (gluteus maximus)
   gluteus medius

8. Using scissors, transect the tensor fasciae latae and biceps femoris, and reflect their cut edges to observe the deeper muscles of the thigh.

9. On the lateral surface of the leg (fig. 60.11), locate and dissect the following muscles:

   gastrocnemius
   soleus
   extensor digitorum longus
   tibialis anterior
   fibularis (peroneus) group

10. See table 60.6 for the origins, insertions, and actions of these muscles of the hip and hindlimb.

11. Review the locations, origins, insertions, and actions designated by the laboratory instructor without the aid of the figures and tables. ⚠ ⚠

12. Complete Part E of the laboratory report.

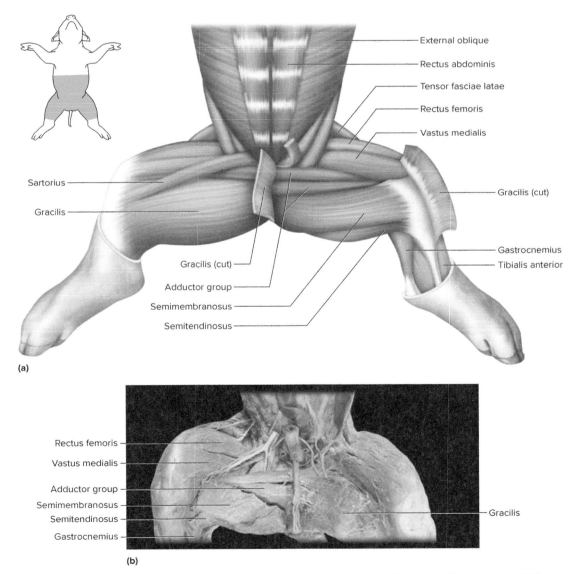

External oblique
Rectus abdominis
Tensor fasciae latae
Rectus femoris
Vastus medialis

Sartorius
Gracilis
Gracilis (cut)

Gracilis (cut)
Adductor group
Semimembranosus
Semitendinosus

Gastrocnemius
Tibialis anterior

(a)

Rectus femoris
Vastus medialis
Adductor group
Semimembranosus
Semitendinosus
Gastrocnemius

Gracilis

(b)

FIGURE 60.10   Medial superficial and deep muscles of the hindlimb, ventral view: (*a*) illustration; (*b*) dissection. **APR**
((*b*)McGraw-Hill Education/APR)

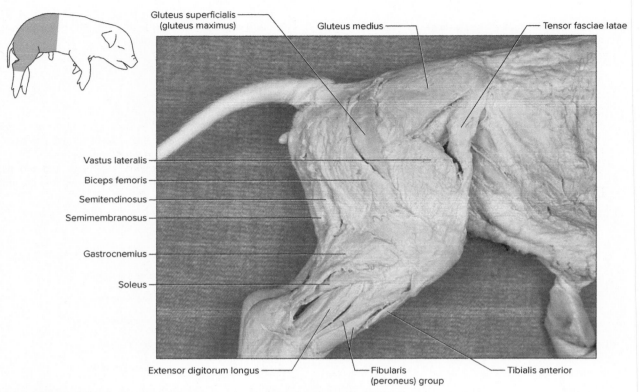

FIGURE 60.11   Muscles of the hip, thigh, and leg, lateral view. **APR** (©J & J Photography)

TABLE 60.6   Muscles of the Hip and Hindlimb

Muscle	Origin	Insertion	Action
Tensor fasciae latae	Iliac crest	Fascia lata of thigh	Tighten fascia lata
Gluteus superficialis	Sacral and caudal vertebrae	Greater trochanter of femur	Abducts thigh
Gluteus medius	Ilium	Greater trochanter of femur	Abducts thigh
Quadriceps femoris			
Vastus lateralis	Shaft of femur and greater trochanter	Patella	Extends hindlimb
Vastus intermedius	Shaft of femur	Patella	Extends hindlimb
Rectus femoris	Ilium	Patella	Extends hindlimb
Vastus medialis	Shaft of femur	Patella	Extends hindlimb
Hamstring muscles			
Biceps femoris	Ischium	Tibia	Abducts thigh and flexes lower hindlimb
Semimembranosus	Ischium	Tibia	Flexes lower hindlimb
Semitendinosus	Ischium	Tibia	Flexes lower hindlimb
Sartorius	Crest of ilium	Proximal end of tibia	Adducts thigh
Gracilis	Pubis	Proximal end of tibia	Adducts hindlimb
Adductor group	Pubis	Femur	Adducts hindlimb
Gastrocnemius	Lateral and medial epicondyles of femur	Calcaneus	Extends foot
Soleus	Proximal end of fibula	Calcaneus	Extends foot
Tibialis anterior	Proximal end of tibia	Second metatarsal	Flexes foot
Fibularis (peroneus) group	Shaft of tibia and fibula	Metatarsals	Flexes foot
Extensor digitorum longus	Proximal end of tibia and fibula	Digits	Extends digits

Name _____

Date _____

Section _____

The A corresponds to the indicated Learning Outcome(s) found at the beginning of the laboratory exercise.

Fetal Pig Dissection: Musculature

PART A ASSESSMENTS

Complete the following statements:

1. The _____ muscle of the human is homologous to the sternomastoid muscle of the pig. A3

2. The _____ muscle elevates the mandible in the human and in the pig. A2 A3

3. Two muscles of the pig that are inserted on the hyoid bone are the _____ and the

   _____ . A2

PART B ASSESSMENTS

Complete the following:

1. Name two pectoral muscles that are found in the thoracic wall of the human. A3

   _____   _____

2. Name three pectoral muscles that are found in the thoracic wall of the pig. A1

   _____   _____   _____

3. Name four muscles that are found in the abdominal wall of the pig and the human. A1 A3

   _____   _____

   _____   _____

PART C ASSESSMENTS

Complete the following:

1. Name three muscles of the pig that together correspond to the trapezius muscle in the human. A3

   _____

   _____

   _____

2. Name the muscle in the pig and in the human that occupies the fossa above the spine of the scapula. A3

   _____

3. Name the muscle in the pig and in the human that occupies the fossa below the spine of the scapula. A3

   _____

PART D ASSESSMENTS

Complete the following:

1. Name two muscles found in the pig and in the human that can flex the forelimb. 2 3

   _____

   _____

2. Name a muscle that has three heads, can extend the forelimb, and is found in the pig and in the human. 2 3

   _____

 PART E ASSESSMENTS

Observe the muscles on the human torso model and the upper and lower limb models. Each of the muscles in figure 60.12 is found both in the pig and in the human. Identify each of the numbered muscles in the figure by placing its name in the space next to its number.

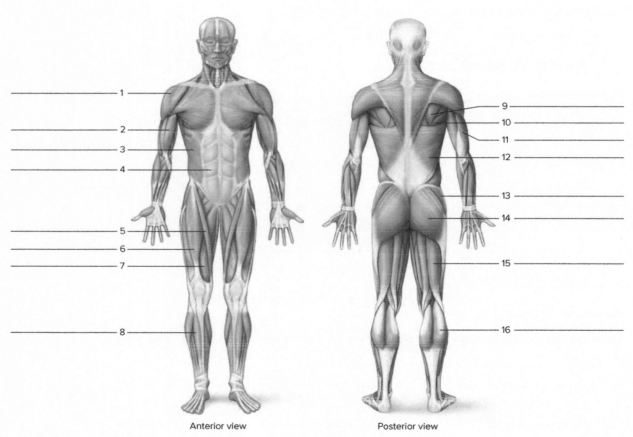

FIGURE 60.12   Identify the numbered muscles that occur in both the pig and the human. 3

Fetal Pig Dissection: Cardiovascular System

MATERIALS NEEDED

Preserved fetal pig
Dissecting tray
Dissecting instruments
Twine
Disposable gloves
Human torso model

⚠ SAFETY

- Wear disposable gloves when working on the fetal pig dissection.
- Dispose of tissue remnants and gloves as instructed.
- Wash the dissecting tray and instruments as instructed.
- Wash your laboratory table.
- Wash your hands before leaving the laboratory.

PURPOSE OF THE EXERCISE

To examine and identify the major organs of the cardiovascular system of the fetal pig and to compare them with the corresponding organs of the human torso model.

LEARNING OUTCOMES APR

After completing this exercise, you should be able to

1. Locate and identify the major organs of the cardiovascular system of the fetal pig.
2. Locate and identify fetal circulatory features of the fetal pig and contrast fetal and adult circulation.
3. Compare the features and organs of the cardiovascular system of the fetal pig with those of the human.
4. Identify the corresponding organs in the human torso model.

In this laboratory exercise, you will dissect the major organs of the cardiovascular system of the fetal pig. As before, while

you are examining the organs of the fetal pig, compare them with the corresponding organs of the human torso model.

If the cardiovascular system of the fetal pig has been injected, the systemic arteries will be filled with red latex (large arteries may not appear very red through a thick wall), and the systemic veins will be filled with blue latex. This will make it easier for you to trace the vessels as you dissect them.

PRACTICE

PROCEDURE A—The Arterial System

1. Place the fetal pig in the dissecting tray with its ventral surface up. Spread and secure all four limbs with twine.
2. Open the thoracic cavity, and expose its contents. To do this, follow these steps:
   a. Make a longitudinal incision passing anteriorly from the diaphragm along one side of the sternum. Continue the incision through the neck muscles to the mandible. Try to avoid damaging the internal organs as you cut. Make a lateral cut on each side of the sternum into each forelimb (fig. 61.1).
   b. Make a lateral cut on each side along the anterior surface of the diaphragm, and cut the diaphragm loose from the thoracic wall.
   c. Spread the sides of the thoracic wall outward, and use a scalpel to make a longitudinal cut along each side of the inner wall of the rib cage to weaken the ribs. Continue to spread the thoracic wall laterally to break the ribs so that the flaps of the wall will remain open. It is often necessary to rinse the body cavities to remove any coagulated blood or excess latex that has leaked out during injection.
3. Note the location of the heart and the large blood vessels associated with it. Slit the outer *parietal pericardium* that surrounds the heart by cutting with scissors along the midventral line. Note how this membrane is connected to the *visceral pericardium* that is attached to the surface of the heart. Locate the *pericardial cavity,* the space between the two layers of the pericardium.

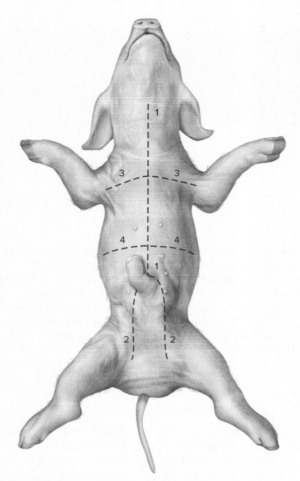

FIGURE 61.1 Incision lines indicate the locations for opening the ventral body cavity.

4. Remove the thymus gland to expose blood vessels anterior to the heart. Examine the heart (figs. 61.2 and 61.3). The arrangement of blood vessels coming off the aortic arch is different in the pig than in a human body. Locate the following:

right atrium
left atrium
right ventricle
left ventricle
pulmonary trunk
aorta
coronary arteries

5. Trace the pulmonary trunk and locate the short vessel, the *ductus arteriosus,* that is connected to the aorta (see fig. 61.2). This connection, found in fetal circulation, shunts blood from the pulmonary artery to the aorta as a partial bypass of nonfunctional, developing fetal lungs. This shunt becomes occluded after birth (forming the *ligamentum arteriosum*), allowing a fully functional pulmonary circuit.

6. Use a scalpel to open the heart chambers by making a cut along the frontal plane from its apex to its base. Remove any remaining latex from the chambers. Examine the valves between the chambers, and note the relative thicknesses of the chamber walls. Do not remove the heart, as it is needed for future study of the relationship of major blood vessels.

7. Using figure 61.3 as a guide, locate and dissect the following arteries of the thorax and neck:

aortic arch
brachiocephalic trunk (artery) (first branch of the aortic arch)
right subclavian artery
left subclavian artery
right common carotid artery
left common carotid artery

8. Trace the right subclavian artery into the forelimb and locate the following arteries:

axillary artery
brachial artery
radial artery
ulnar artery

9. Open the abdominal cavity. To do this, follow these steps:
   a. Use scissors to make two longitudinal incisions from the pubic bones and continuing around the umbilical cord to meet at the midline. Extend the incision anteriorly to the diaphragm.
   b. Make a lateral incision through the body wall along either side of the inferior border of the diaphragm and along the bases of the thighs.
   c. Reflect the flaps created in the body wall as you would open a book, and expose the contents of the abdominal cavity. Sever the *umbilical vein* that extends from the umbilical cord to the liver. Reflect the umbilical cord along with the midventral strip of tissues to expose the two *umbilical arteries* parallel to a central urinary bladder. The umbilical arteries, although injected with red latex, actually transport oxygen-poor fetal blood to the placenta. Flush the abdominal cavity of any coagulated blood or latex that leaked out during injection.
   d. Note the *parietal peritoneum* that forms the inner lining of the abdominal wall. Also note the *visceral peritoneum* that adheres to the surface of organs within the abdominal cavity.

10. As you expose and dissect blood vessels, try not to destroy other visceral organs needed for future studies. At times it will be necessary to displace some abdominal organs to locate some deep blood vessels.

   Using figure 61.4 as a guide, locate and dissect the following arteries of the abdomen:

abdominal aorta (unpaired)
celiac trunk (unpaired)

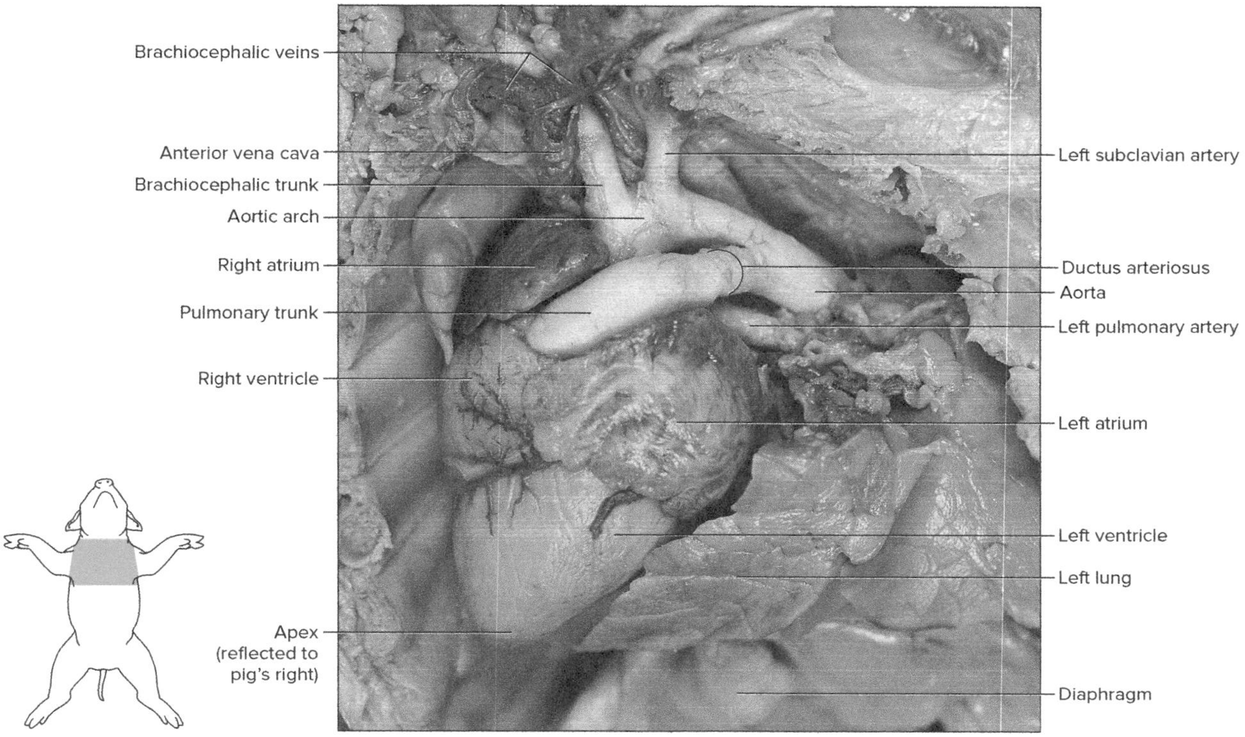
Brachiocephalic veins
Anterior vena cava
Brachiocephalic trunk
Aortic arch
Right atrium
Pulmonary trunk
Right ventricle
Apex
(reflected to
pig's right)
Left subclavian artery
Ductus arteriosus
Aorta
Left pulmonary artery
Left atrium
Left ventricle
Left lung
Diaphragm

FIGURE 61.2 Heart and associated arteries and veins of fetal circulation, ventral view. The anterior portion of the left lung has been removed. The apex is reflected to the pig's right to clearly show the arteries. (©J & J Photography)

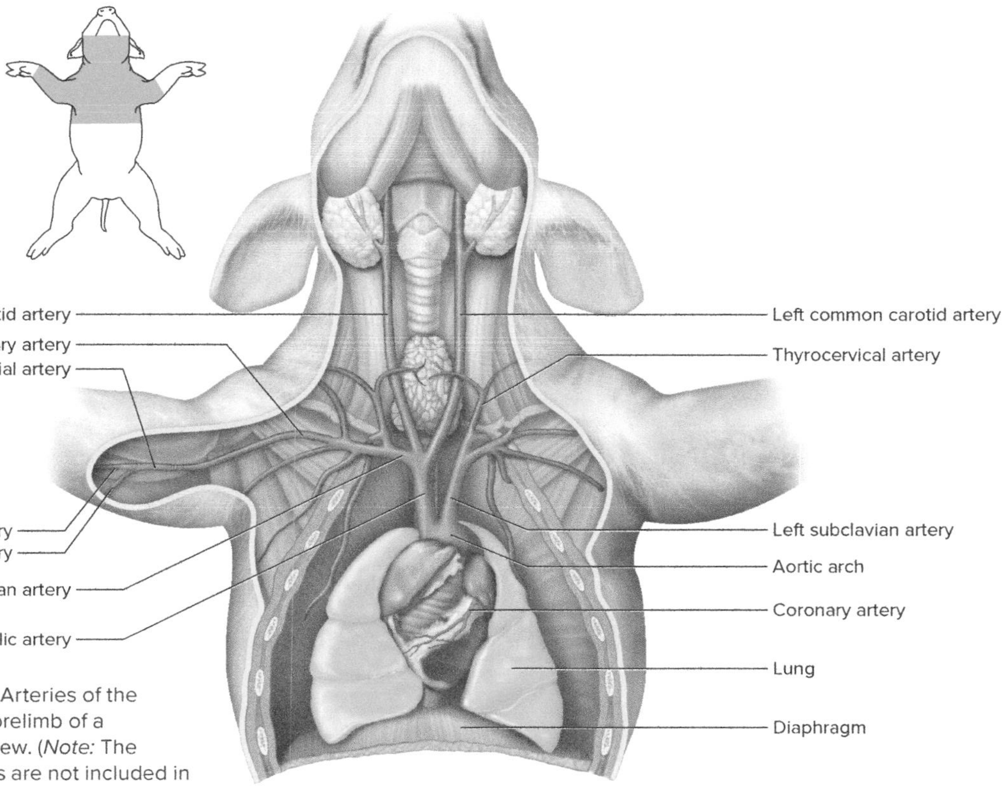
Right common carotid artery
Axillary artery
Brachial artery
Radial artery
Ulnar artery
Right subclavian artery
Brachiocephalic artery
Left common carotid artery
Thyrocervical artery
Left subclavian artery
Aortic arch
Coronary artery
Lung
Diaphragm

FIGURE 61.3 Arteries of the trunk, neck, and forelimb of a fetal pig, ventral view. (*Note:* The pulmonary arteries are not included in this illustration.) APR

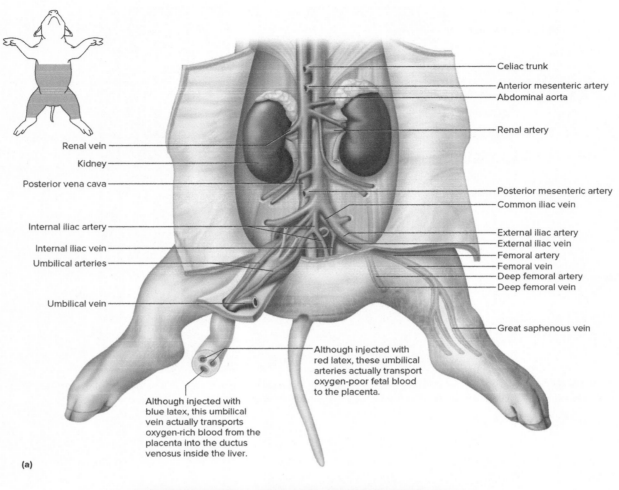

Celiac trunk
Anterior mesenteric artery
Abdominal aorta
Renal artery

Renal vein
Kidney
Posterior vena cava

Posterior mesenteric artery
Common iliac vein

Internal iliac artery
Internal iliac vein
Umbilical arteries

External iliac artery
External iliac vein
Femoral artery
Femoral vein
Deep femoral artery
Deep femoral vein

Umbilical vein

Great saphenous vein

Although injected with red latex, these umbilical arteries actually transport oxygen-poor fetal blood to the placenta.

Although injected with blue latex, this umbilical vein actually transports oxygen-rich blood from the placenta into the ductus venosus inside the liver.

(a)

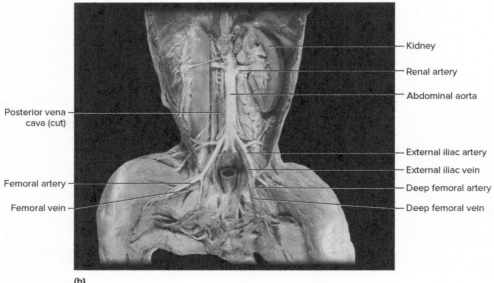

Kidney
Renal artery
Abdominal aorta

Posterior vena cava (cut)

External iliac artery
External iliac vein

Femoral artery
Femoral vein

Deep femoral artery
Deep femoral vein

(b)

FIGURE 61.4 Arteries and veins of the abdominal cavity of a fetal pig, ventral view: (*a*) illustration; (*b*) dissection. **APR**

((*b*) McGraw-Hill Education/APR)

anterior mesenteric artery (corresponds to superior mesenteric artery) (unpaired)

renal arteries (paired)

posterior mesenteric artery (corresponds to inferior mesenteric artery) (unpaired)

external iliac arteries (paired)

internal iliac arteries (paired)

umbilical arteries (paired)

11. Trace the external iliac artery into the left hindlimb and locate the following:

    femoral artery

    deep femoral artery (deep artery of thigh)

12. Review the locations of the heart structures and arteries without the aid of the figures.

13. Examine the human torso model along with figures 61.2–61.4. Identify arteries of the pig that correspond with those of the human.

14. Complete Part A of Laboratory Report 61.

PRACTICE

PROCEDURE B—The Venous System

1. Examine the heart again and locate the following veins:

   anterior vena cava (corresponds to superior vena cava)

   posterior vena cava (corresponds to inferior vena cava)

2. Using figure 61.5 as a guide, locate and dissect the following veins in the thorax and neck:

   right brachiocephalic vein
   left brachiocephalic vein
   right subclavian vein
   left subclavian vein
   internal jugular vein
   external jugular vein

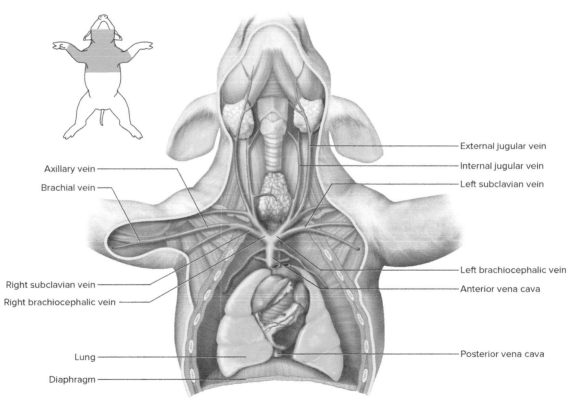

FIGURE 61.5 Veins of the trunk, neck, and forelimb of a fetal pig, ventral view. **APR**

3. Trace the right subclavian vein into the forelimb and locate the following veins:

   axillary vein
   brachial vein

4. Using figure 61.4 as a guide, locate and dissect the following veins in the abdomen:

   posterior vena cava
   renal veins
   common iliac vein
   internal iliac vein
   external iliac vein

5. Using figure 61.6 as a guide, locate and dissect the following veins (branches of the hepatic portal system):

   hepatic portal vein
   gastrosplenic vein
   mesenteric vein
   anterior mesenteric vein (corresponds to superior mesenteric vein)

   posterior mesenteric vein (corresponds to inferior mesenteric vein)

6. Locate and trace the umbilical vein (that was cut) from the umbilical cord into the liver. Inside the liver it becomes the *ductus venosus,* which shunts oxygen-rich fetal blood into the posterior vena cava. The umbilical vein and the ductus venosus degenerate after birth.

7. Trace the external iliac vein into the left hindlimb (see fig. 61.4) and locate the following veins:

   femoral vein
   deep femoral vein (deep vein of thigh)
   great saphenous vein

8. Review the locations of the veins without the aid of the figures. ⓐ1 ⓐ2

9. Examine the human torso model along with figures 61.4–61.6. Identify the veins of the pig that correspond with those of the human. Ⓐ4

10. Complete Part B of the laboratory report.

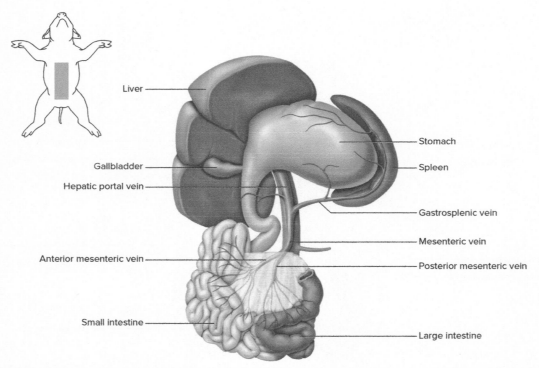

Liver — Stomach — Spleen — Gastrosplenic vein — Mesenteric vein — Gallbladder — Hepatic portal vein — Anterior mesenteric vein — Posterior mesenteric vein — Small intestine — Large intestine

FIGURE 61.6 Hepatic portal vein and its tributaries of a fetal pig, ventral view. The liver and stomach are retracted to obtain a clear view.

Name _____

Date _____

Section _____

The A corresponds to the indicated Learning Outcome(s) found at the beginning of the laboratory exercise.

Fetal Pig Dissection: Cardiovascular System

PART A ASSESSMENTS

Complete the following:

1. Describe the position and attachments of the parietal pericardium of the heart of the fetal pig. A1 _____

   _____

   _____

   _____

2. Describe the relative thicknesses of the walls of the heart chambers of the fetal pig. A1 _____

   _____

   _____

   _____

3. Explain how the wall thicknesses are related to the functions of the chambers. A1 _____

   _____

   _____

   _____

   _____

   _____

4. Compare the origins of the common carotid arteries of the fetal pig with those of the human. A3 _____

   _____

   _____

   _____

   _____

5. Compare the origins of the external and internal iliac arteries of the fetal pig with those of the human. A3

   _____

   _____

   _____

   _____

PART B ASSESSMENTS

Complete the following:

1. Compare the relative sizes of the external and internal jugular veins of the fetal pig with those of the human.   3

_____

_____

_____

_____

_____

2. List twelve veins that pigs and humans have in common.   3   4

_____

_____

_____

_____

_____

_____

ASSESS

CRITICAL THINKING

Explain the oxygen-rich blood in the umbilical veins of a human and a pig fetus.   2

_____

_____

_____

_____

LABORATORY EXERCISE

62

Fetal Pig Dissection: Digestive System

MATERIALS NEEDED

Preserved fetal pig
Dissecting tray
Dissecting instruments
Bone cutter
Disposable gloves
Hand lens
Human torso model

SAFETY

▪ Wear disposable gloves when working on the fetal pig dissection.
▪ Dispose of tissue remnants and gloves as instructed.
▪ Wash the dissecting tray and instruments as instructed.
▪ Wash your laboratory table.
▪ Wash your hands before leaving the laboratory.

PURPOSE OF THE EXERCISE

To examine and identify the major digestive organs of the fetal pig and to compare these organs with those of the human.

LEARNING OUTCOMES APR

After completing this exercise, you should be able to

1. Locate and identify the major digestive organs of the fetal pig.

2. Compare the digestive system of the fetal pig with that of the human.

3. Identify the corresponding organs in the human torso model.

In this laboratory exercise, you will dissect the major digestive organs of a fetal pig. As you observe these organs, compare them with those of the human by observing the parts of the human torso model. Because humans are adapted to eating and digesting a greater variety of foods, comparisons of the pig and human digestive organs may not be precise.

PRACTICE

PROCEDURE—Digestive System Dissection

1. Place the fetal pig in the dissecting tray on its left side.
2. Locate the major salivary glands on one side of the head (fig. 62.1). To do this, follow these steps:
   a. Clear away any remaining fascia and other connective tissue from the region below the ear and near the joint of the mandible.
   b. Identify the *parotid gland,* a relatively large, triangular mass of glandular tissue just below the ear. Although this gland covers a large area, it is poorly developed in the fetal stage of development.
   c. Look for the compact *submandibular gland* just below the parotid gland, near the angle of the jaw.
   d. Locate the small *sublingual gland* that is adjacent, anterior and medial to the submandibular gland.
3. Open the oral cavity. To do this, follow these steps:
   a. Use scissors to cut through the soft tissues at the angle of the mouth.
   b. When you reach the bone of the jaw, use a bone cutter to cut through the bone, thus freeing the mandible (fig. 62.2).
   c. Open the mouth wide and locate the following features:

   cheek
   lip
   palate
      hard palate (has transverse ridges to help hold food)
      soft palate (lacks a uvula)
   tongue
      papillae (examine with a hand lens)

4. Examine any erupted teeth of the maxilla. The primary (deciduous) teeth of a young pig would include six incisors, two canines, eight premolars, and zero molars on each jaw. Cut into some of the gum tissue to locate any developing teeth that have not erupted.
5. Complete Part A of Laboratory Report 62.
6. Examine organs in the abdominal cavity with the fetal pig positioned with its ventral surface up. You might

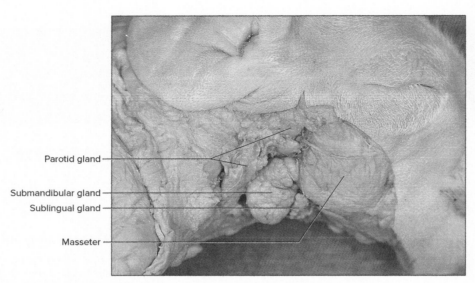

FIGURE 62.1   Salivary glands of a fetal pig, lateral view. **APR** (©J & J Photography)

Parotid gland

Submandibular gland

Sublingual gland

Masseter

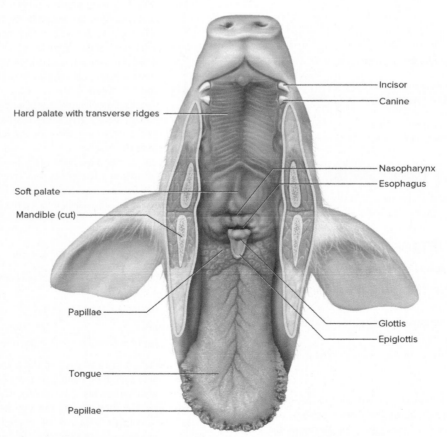

Incisor

Canine

Hard palate with transverse ridges

Nasopharynx

Esophagus

Soft palate

Mandible (cut)

Papillae

Glottis

Epiglottis

Tongue

Papillae

FIGURE 62.2   Oral cavity of the fetal pig with the lower jaw and tongue retracted. **APR**

wish to remove some of the side walls of the body cavity to make observations easier.

7. Examine the large *liver,* which is located just beneath the diaphragm and is attached to the central portion of the diaphragm and the ventral body wall by the *falciform ligament.* Also, locate the elongated *spleen,* which is lateral and ventral to the stomach on the left side (fig. 62.3). Locate the five lobes of the liver. A *greater omentum* extends from the spleen to the stomach, and the *lesser omentum* connects the liver to the stomach. Lift the liver to find the greenish to nearly colorless *gallbladder* embedded in the underside of the liver on the right side. Also note the *cystic duct,* by which the gallbladder is attached to the *bile duct,* and the *hepatic duct,* which originates in the liver and attaches to the cystic duct. Trace the bile duct to its connection with the duodenum (fig. 62.4).

8. Locate the *stomach* in the left side of the abdominal cavity. At its anterior end, note the union with the *esophagus,* which passes through the diaphragm. Identify the *cardia, fundus, body,* and *pyloric part* (listed from entrance to exit) of the stomach. Use scissors to make an incision along the convex border of the stomach from the cardia to the pyloric part. The greenish substance found in the stomach and the rest of the digestive tract is called *meconium.* Meconium found in a fetal digestive tract is a combination of sloughed-off epithelial cells, amniotic fluid residues that were swallowed, and bile-stained mucus. It will be the first substance of bowel movements after birth. The lining of the stomach has numerous *gastric folds* (*rugae*). Examine the *pyloric sphincter,* which creates a constriction between the stomach and small intestine.

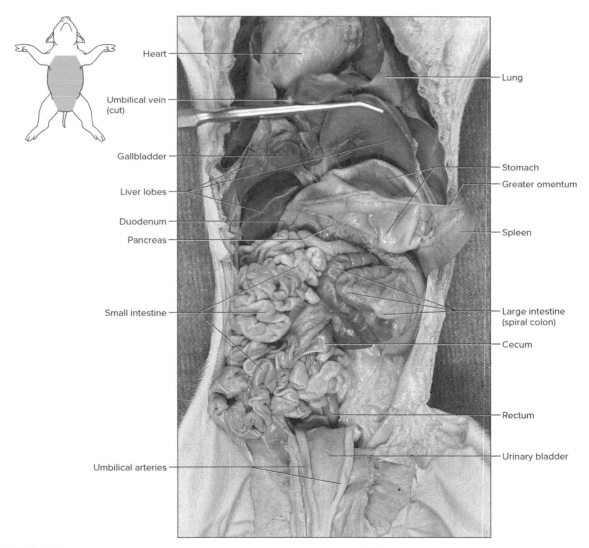

FIGURE 62.3   Abdominal digestive organs of the fetal pig, ventral view.   **APR**   (©J & J Photography)

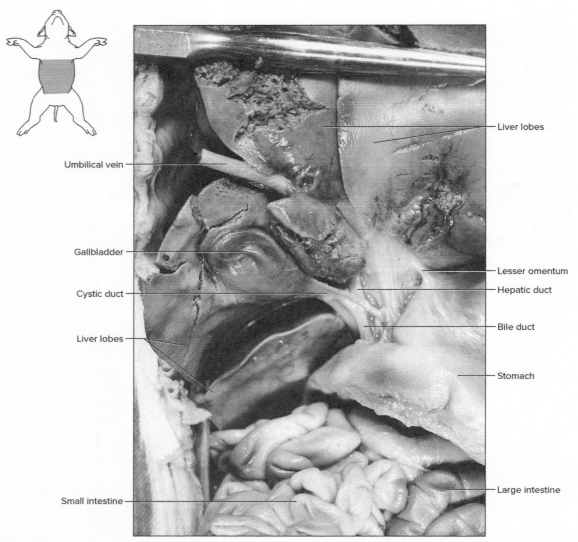

Liver lobes

Umbilical vein

Gallbladder

Lesser omentum

Cystic duct

Hepatic duct

Bile duct

Liver lobes

Stomach

Large intestine

Small intestine

FIGURE 62.4   Digestive organs associated with the gallbladder of a fetal pig, ventral view. The liver is retracted for this view. **APR** (©J & J Photography)

9. Locate the *pancreas* by lifting the stomach and separating some thin peritoneal membrane over the surface of the pancreas. The pancreas extends from the left stomach region into the loop of the duodenum of the small intestine.

10. Trace the *small intestine,* beginning at the pyloric sphincter. The first portion, the *duodenum,* is a short loop that has the bile duct and the pancreatic duct leading into it. The proximal half of the remaining portion of the small intestine is the *jejunum,* and the distal half is the *ileum.* Open the small intestine and note the velvety appearance of the villi. Note how the mesentery supports the small intestine from the dorsal body wall. The small intestine terminates on the left side, where it joins the large intestine.

11. Locate the *large intestine,* and identify the short, blind sac called the *cecum.* Make an incision at the junction between the ileum and cecum, and look for the *ileocecal sphincter.* An appendix is absent in pigs. A characteristic of the large intestine of the pig is the *spiral colon.* This is a tightly coiled mass on the left ventral region of the abdominal cavity. Also locate the *rectum,* which extends through the pelvic cavity to the *anus.*

12. Review the locations of the digestive organs without the aid of the figures. **1**

13. Examine the human torso model along with figures 62.1, 62.3, and 62.4. Identify digestive organs of the pig that correspond with those of the human. **3**

14. Complete Part B of the laboratory report.

63

Fetal Pig Dissection: Respiratory System

MATERIALS NEEDED

Preserved fetal pig
Dissecting tray
Dissecting instruments
Beaker of water
Disposable gloves
Human torso model

SAFETY

- Wear disposable gloves when working on the fetal pig dissection.
- Dispose of tissue remnants and gloves as instructed.
- Wash the dissecting tray and instruments as instructed.
- Wash your laboratory table.
- Wash your hands before leaving the laboratory.

PURPOSE OF THE EXERCISE

To examine and identify the major respiratory organs of the fetal pig and to compare these organs with those of the human.

 ## LEARNING OUTCOMES APR

After completing this exercise, you should be able to

1. Locate and identify the major respiratory organs of the fetal pig.

2. Compare the respiratory system of the human with that of the fetal pig.

3. Identify the corresponding organs in the human torso model.

In this laboratory exercise, you will dissect the major respiratory organs of a preserved fetal pig. As you observe these structures in the fetal pig, compare them with those of the human torso model.

 ## PRACTICE

PROCEDURE—Respiratory System Dissection

1. Place the fetal pig in a dissecting tray with its ventral surface up.
2. Examine the *nostrils (external nares)* located on the flat *rostrum (snout)*.
3. Open the oral cavity wide to expose the *hard palate* and the *soft palate* (fig. 63.1). Cut through the tissues of the soft palate, and observe the *nasopharynx* above it. Locate the small openings of the *auditory tubes* in the lateral walls of the nasopharynx. Insert a probe into an opening of an auditory tube. Examine the *oropharynx* located near the base of the tongue.
4. Pull the tongue posteriorly and locate the *epiglottis* near its base. Also identify the *glottis,* an opening between the *vocal folds (vocal cords)* into the larynx. Locate the *esophagus,* which is dorsal to the larynx.
5. Open the thoracic cavity, and expose its contents. Examine the organs located in the *mediastinum*. The mediastinum separates the right and left lungs and pleural cavities within the thorax.
6. Dissect the *trachea* in the neck, and expose the *larynx* at the anterior end near the base of the tongue. Note the *tracheal rings,* and locate the *thyroid gland* on the trachea, just posterior to the larynx. Locate a small *parathyroid gland* attached to the dorsal surface of the thyroid gland. Also note the *thymus gland,* extending along each side of the trachea into the thorax (fig. 63.2).
7. Examine the larynx by removing any attached muscles, and identify the *thyroid cartilage* and *cricoid cartilage* in its wall. Make a longitudinal incision through the ventral wall of the larynx, and locate the vocal folds (vocal cords) that appear as whitish folds inside.

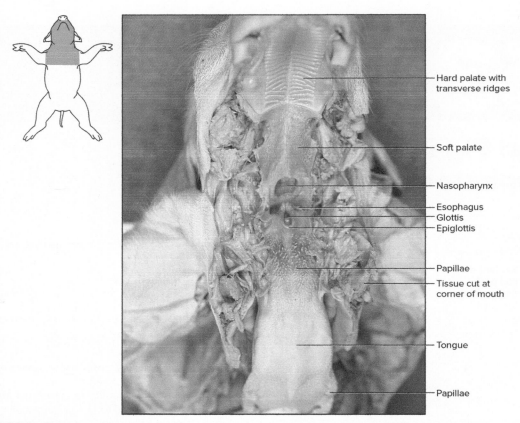

FIGURE 63.1 Oral cavity of the fetal pig with tongue pulled posteriorly, ventral view. **APR** (©J & J Photography)

Labels on figure:
- Hard palate with transverse ridges
- Soft palate
- Nasopharynx
- Esophagus
- Glottis
- Epiglottis
- Papillae
- Tissue cut at corner of mouth
- Tongue
- Papillae

8. Remove the heart and trace the trachea posteriorly to where it divides into *bronchi*, which pass into the lungs (fig. 63.3*a*).

9. Examine the *lungs,* each of which is subdivided into lobes—an *anterior (cranial),* a *middle,* and a *posterior (caudal) lobe.* The right lung has an additional deep *accessory lobe* (fig. 63.3*a* and *b*). Notice the thin membrane, the *visceral pleura,* on the surface of each lung. Also notice the *parietal pleura,* which forms the inner lining of the thoracic wall, and locate the spaces of the *pleural cavities.*

10. Make an incision through a lobe of a lung, and examine its interior. Note the branches of the smaller air passages. You can see the branches of the bronchial tree if you follow a primary bronchus into the lung and gently scrape away lung tissue with a scalpel.

11. Examine the *diaphragm* and locate the *phrenic nerve.* This nerve appears as a white thread passing along the side of the heart to the diaphragm (see fig. 63.2).

12. Remove a small piece of the lung and notice the solid texture of the tissue. Place the piece of lung into a beaker of water. Does the fetal lung tissue float or sink?

_____

ASSESS

CRITICAL THINKING

Under what conditions would a piece of lung tissue sink?

_____
_____
_____
_____

Under what conditions would it float?

_____
_____
_____
_____

13. Review the locations of the respiratory organs without the aid of the figures. **1**

14. Examine the human torso model along with figures 63.2 and 63.3. Identify respiratory organs of the pig that correspond with those of the human. **3**

15. Complete Laboratory Report 63.

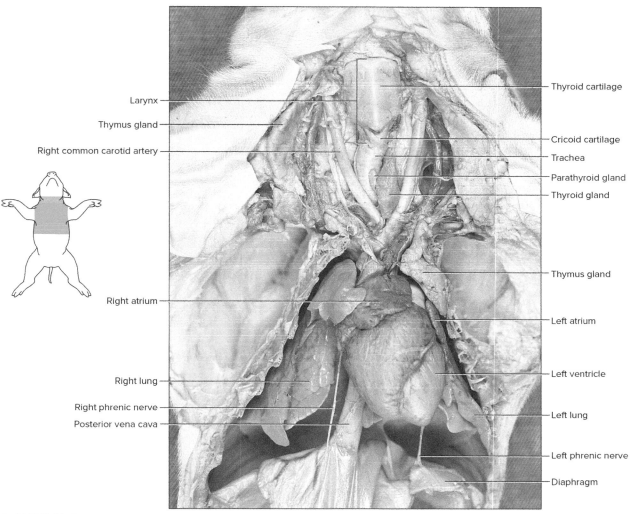

Larynx

Thymus gland

Right common carotid artery

Thyroid cartilage

Cricoid cartilage

Trachea

Parathyroid gland

Thyroid gland

Thymus gland

Right atrium

Left atrium

Right lung

Left ventricle

Right phrenic nerve

Posterior vena cava

Left lung

Left phrenic nerve

Diaphragm

FIGURE 63.2   Respiratory organs and glands of the fetal pig, ventral view. **APR** (©J & J Photography)

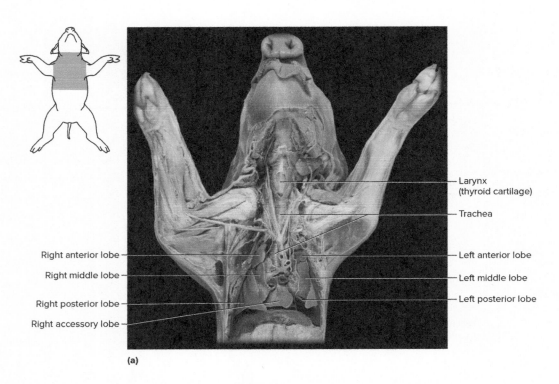

(a)

- Larynx (thyroid cartilage)
- Trachea
- Left anterior lobe
- Left middle lobe
- Left posterior lobe

Right anterior lobe
Right middle lobe
Right posterior lobe
Right accessory lobe

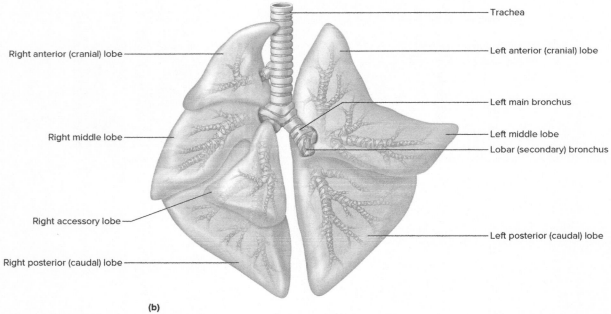

(b)

- Trachea
Right anterior (cranial) lobe
- Left anterior (cranial) lobe
- Left main bronchus
Right middle lobe
- Left middle lobe
- Lobar (secondary) bronchus
Right accessory lobe
- Left posterior (caudal) lobe
Right posterior (caudal) lobe

FIGURE 63.3   Lungs and respiratory tubes of the fetal pig, ventral view: (*a*) dissection; (*b*) illustration.   **APR** ((*a*) McGraw-Hill Education/APR)

Name _____

Date _____

Section _____

The A corresponds to the indicated Learning Outcome(s) found at the beginning of the laboratory exercise.

Fetal Pig Dissection: Respiratory System

Complete the following:

1. What is the purpose of the auditory tubes opening into the nasopharynx? A 1 _____

_____

_____

_____

_____

2. Distinguish between the glottis and the epiglottis. A 1 _____

_____

_____

_____

_____

3. Are the tracheal rings of the fetal pig complete or incomplete circles? A 1 _____ How does this

   feature compare with that of the human? A 2 _____

_____

_____

4. Compare the number of lobes in the human lungs with the number of lobes in the fetal pig lungs. A 2 _____

_____

_____

_____

_____

5. Describe the attachments of the diaphragm. A 1 _____

_____

_____

_____

_____

6. What major structures are located within the mediastinum? A 1 _____

_____ How does this compare to the human mediastinum? A 2 _____

_____

LABORATORY EXERCISE

64

Fetal Pig Dissection: Urinary System

MATERIALS NEEDED

Human torso model
Preserved fetal pig
Dissecting tray
Dissecting instruments
Disposable gloves

⚠ SAFETY

- Wear disposable gloves when working on the fetal pig dissection.
- Dispose of tissue remnants and gloves as instructed.
- Wash the dissecting tray and instruments as instructed.
- Wash your laboratory table.
- Wash your hands before leaving the laboratory.

PURPOSE OF THE EXERCISE

To examine and identify the urinary organs of the fetal pig and to compare them with those of the human.

 ## LEARNING OUTCOMES APR

After completing this exercise, you should be able to

1. Locate and identify the urinary organs of the fetal pig.
2. Compare the urinary organs of the fetal pig with those of the human.
3. Identify the corresponding organs in the human torso model.

In this laboratory exercise, you will dissect the urinary organs of the fetal pig. As you observe these structures, compare them with the corresponding human organs by observing the parts of the human torso model.

 ## PRACTICE

PROCEDURE—Urinary System Dissection

1. Place the fetal pig in a dissecting tray with its ventral surface up.
2. Open the abdominal cavity and remove the liver, stomach, spleen, pancreas, small intestine, and large intestine.
3. Locate the *kidneys* in the dorsal abdominal wall on either side of the vertebral column. The kidneys are located dorsal to the *parietal peritoneum* (retroperitoneal).
4. Carefully remove the parietal peritoneum surrounding the kidneys. Locate the following using figure 64.1 as a guide:

   ureters
   renal arteries
   renal veins

5. Locate the bandlike *adrenal glands,* which lie medially and anteriorly to the kidneys.
6. Expose the ureters by cleaning away the connective tissues along their lengths. They enter the *fetal urinary bladder (allantoic bladder)* on the dorsal surface.
7. Examine the fetal urinary bladder, an elongated, muscular, collapsed sac between two umbilical arteries. Trace the allantoic bladder into the umbilical cord where it becomes the *allantoic stalk.* The fetal urinary bladder becomes a functional urinary bladder after birth when the umbilical cord deteriorates and wastes no longer eliminate through the allantoic stalk.
8. Use a sharp scalpel to open the bladder and examine its interior. Locate the openings of the ureters and the urethra on the inside.
9. Expose the *urethra* at the posterior of the urinary bladder (fig. 64.2). The urethra extends through the *penis* in the male. In the female, the urethra enters the *urogenital sinus* a short distance from the urogenital orifice.

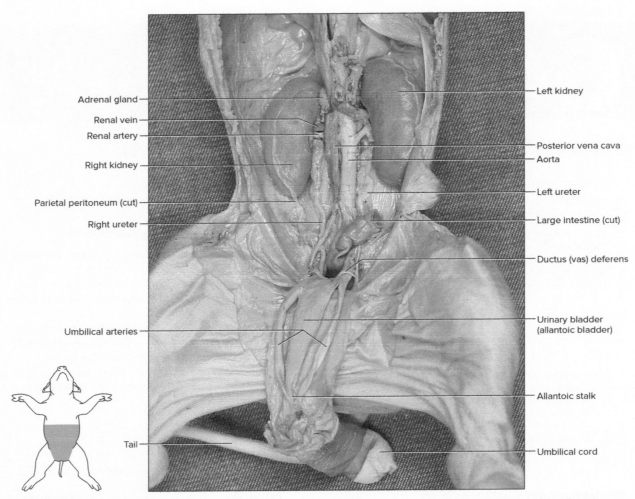

Adrenal gland
Renal vein
Renal artery
Right kidney
Parietal peritoneum (cut)
Right ureter
Umbilical arteries
Tail

Left kidney
Posterior vena cava
Aorta
Left ureter
Large intestine (cut)
Ductus (vas) deferens
Urinary bladder (allantoic bladder)
Allantoic stalk
Umbilical cord

FIGURE 64.1   Urinary system of the male fetal pig with the umbilical cord structures reflected, and most of the digestive organs removed. **APR** (©J & J Photography)

10. Remove one kidney, and section it longitudinally. Use figures 64.2 and 64.3 as a reference for a kidney of a pig. Identify the following features:

fibrous capsule
renal cortex
renal medulla
renal pyramid
minor calyx
major calyx
renal pelvis
renal sinus
hilum of kidney

11. Review the locations of the urinary organs without the aid of the figures. ⚠️

12. Discard the organs and tissues that were removed from the fetal pig, as directed by the laboratory instructor.

13. Examine the human torso model along with figures 64.1–64.3. Identify urinary organs of the pig that correspond with those of the human. 🔺

14. Complete Laboratory Report 64.

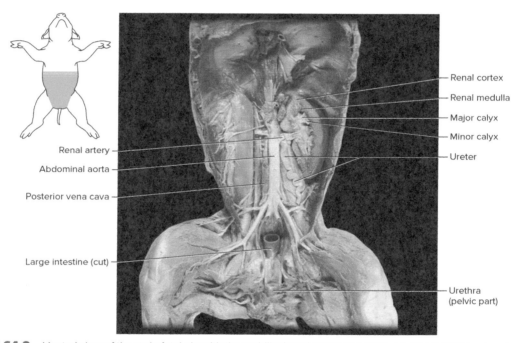

FIGURE 64.2 Ventral view of the male fetal pig with the umbilical cord structures removed. The left kidney has been sectioned to show its internal structures. **APR** (McGraw-Hill Education/APR)

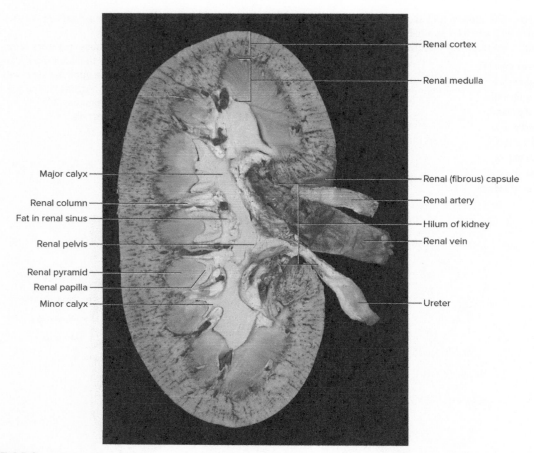

FIGURE 64.3 Longitudinal section of a kidney of a mature pig that has a triple injection of latex (*red* in the renal artery, *blue* in the renal vein, and *yellow* in the ureter and renal pelvis). (*Note:* The kidney of a fetal pig is not as well developed as the kidney of a mature pig.) (©J & J Photography)

Fetal Pig Dissection: Reproductive Systems

MATERIALS NEEDED

Preserved fetal pig
Dissecting tray
Dissecting instruments
Magnifying lens
Disposable gloves
Models of human reproductive systems

 SAFETY

- Wear disposable gloves when working on the fetal pig dissection.
- Dispose of tissue remnants and gloves as instructed.
- Wash the dissecting tray and instruments as instructed.
- Wash your laboratory table.
- Wash your hands before leaving the laboratory.

PURPOSE OF THE EXERCISE

To examine and identify the reproductive organs of the fetal pig and to compare them with the corresponding organs of the human.

LEARNING OUTCOMES APR

After completing this exercise, you should be able to

1. Locate and identify the reproductive organs of a fetal pig.
2. Compare the reproductive organs of the fetal pig with those of the human.
3. Identify the corresponding organs in models of the human reproductive systems.

In this laboratory exercise, you will dissect the reproductive system of the fetal pig. If you have a female fetal pig, begin with Procedure A. If you have a male fetal pig, begin with Procedure B. After completing the dissection, exchange fetal pigs with someone who has dissected one of the opposite sex, and examine its reproductive organs.

As you observe the fetal pig reproductive organs, compare them with the corresponding human organs by examining the models of the human reproductive systems.

 PRACTICE

PROCEDURE A—Female Reproductive System

1. Place the female fetal pig in a dissecting tray with its ventral surface up, and open its abdominal cavity.
2. Locate the small, oval *ovaries* just posterior to the kidneys (fig. 65.1).
3. Near the anterior end of the ovary, locate the funnel-shaped *infundibulum*, which is at the end of the small, coiled *uterine tube (oviduct)*. Note the tiny projections, or *fimbriae*, that create a fringe around the edge of the infundibulum. Trace the uterine tube around the ovary to its nearby connection with the *uterine horn*. These structures can best be located with a magnifying lens in fetal pigs.
4. Examine the uterine horn. It is suspended from the body wall by a mesentery, the *broad ligament*. Also note the fibrous *round ligament*, which extends from the uterine horn laterally and posteriorly to the body wall.
5. Use scissors and cut carefully through the midline of the pelvis to observe the remaining organs of the reproductive system.
6. Trace the relatively long uterine horns posteriorly. They unite to form the *uterine body*, which is located between the urethra and rectum. The Y-shaped uterus in a pig allows the large uterus to contain a litter of pigs. The uterine body is continuous with the *vagina*, which leads to the outside.

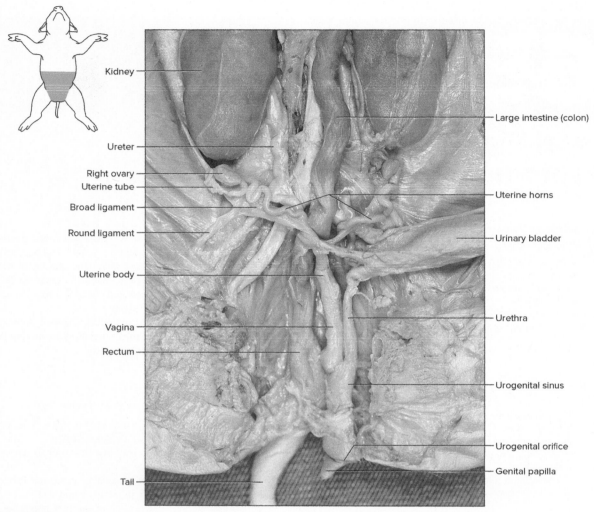

Kidney

Ureter

Right ovary

Uterine tube

Broad ligament

Round ligament

Uterine body

Vagina

Rectum

Tail

Large intestine (colon)

Uterine horns

Urinary bladder

Urethra

Urogenital sinus

Urogenital orifice

Genital papilla

FIGURE 65.1   Reproductive system of the female fetal pig, ventral view.  (©J & J Photography)

7. Trace the *urethra* from the urinary bladder posteriorly. The urethra and the vagina open into a common chamber called the *urogenital sinus.* The opening of this chamber, ventral to the anus, is called the *urogenital orifice.* Locate the *genital papilla,* a small hood over the urogenital orifice.

8. Use scissors to open the vagina along its lateral wall, beginning at the urogenital orifice and continuing to the body of the uterus. Note the *urethral orifice* in the ventral wall of the urogenital sinus, and locate the small, rounded *cervix* of the uterus, which projects into the vagina at its deep end. The *clitoris* is located in the ventral wall of the urogenital sinus near its opening to the outside.

9. Review the locations of the reproductive organs without the aid of the figures. ⚠1

10. Examine the models of the human reproductive systems along with figure 65.1. Identify the reproductive

organs of the pig that correspond with those of the human. ⚠3

11. Complete Part A of Laboratory Report 65.

♻ PRACTICE

PROCEDURE B—Male Reproductive System

1. Place the male fetal pig in a dissecting tray with its ventral surface up.

2. Locate the *scrotum,* which appears as an external pouch just ventral to the anus. During fetal development, the *testes* migrate from posterior to the kidneys through the *inguinal canals* into the scrotum by means of a *gubernaculum.* The gubernaculum, a cord of tissue that extends from the *epididymis* to the scrotum, grows more slowly

than other tissues and "pulls" each testis into the scrotum. Make an incision on the left side of the scrotum and continue the incision anteriorly to open one of the inguinal canals to locate the *spermatic cord* (fig. 65.2).

3. Locate a testis within the inguinal sac. The exact location will vary depending on the degree of development of the fetal pig. Remove the sheath surrounding the testis and locate the convoluted epididymis cupped around the lateral surface of each testis.

4. Locate the spermatic cord on the left side, leading away from the testis. This spermatic cord contains the *ductus deferens (vas deferens)*, which is continuous with the epididymis, as well as with the nerves and blood vessels that supply the testis on that side. Trace the spermatic cord to the body wall, where its contents pass through the inguinal canal and enter the pelvic cavity (fig. 65.2).

5. Locate the *penis* and identify the *prepuce,* which forms a sheath around the entire penis. Make an incision through the skin of the prepuce, and expose the shaft of the penis.

6. Use scissors and cut carefully through the midline of the pelvis to observe the remaining organs of the reproductive system. Trace the ductus deferens from the inguinal canal to the penis. The ductus deferens loops over the ureter within the pelvic cavity and passes downward behind the urinary bladder to join the urethra. Locate the small, paired *seminal vesicles,* which appear as enlargements near the junction of the ductus deferens and urethra. A careful dissection between the seminal vesicles might expose the *prostate gland* near the junction of the urinary bladder and the urethra. (It is difficult to locate in the fetal pig.)

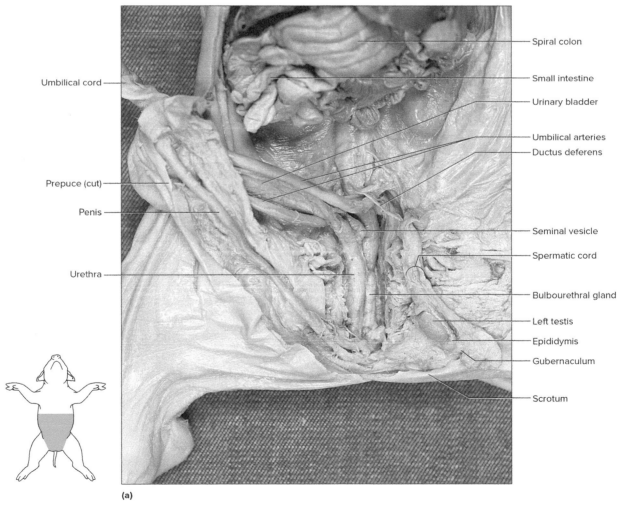

(a)

FIGURE 65.2   Reproductive system of the male fetal pig, ventral view: *(a)* view with many digestive and urinary structures included, and the umbilical cord reflected; *(b)* view with digestive, urinary, and umbilical cord structures removed. **APR**

((a) ©J & J Photography; (b) McGraw-Hill Education/APR)

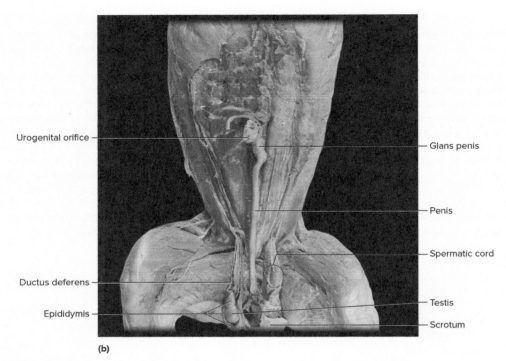

(b)

FIGURE 65.2 *Continued.*

7. Trace the urethra to the penis. Locate the *bulbourethral glands,* which form elongated swellings along the urethra, on either side at the proximal end of the penis.

8. Use a sharp scalpel to cut a transverse section of the penis. Identify the urethra, which serves as a common pathway for urine and semen in male pigs and humans. Locate the *urogenital orifice* at the distal end of the penis.

9. Review the locations of the reproductive organs without the aid of the figures. 🔺1

10. Examine the models of the human reproductive systems along with figure 65.2. Identify the reproductive organs of the pig that correspond with those of the human. 🔺3

11. Complete Part B of the laboratory report.

Supplemental Laboratory Exercises

66

Skeletal Muscle Contraction

MATERIALS NEEDED

Textbook
Recording system (kymograph, Physiograph, etc.)
Stimulator and connecting wires
Live frog
Dissecting tray
Dissecting instruments
Disposable gloves
Probe for pithing
Heavy thread
Frog Ringer's solution

For Demonstration A—The Kymograph:
Kymograph recording system
Electronic stimulator (or inductorium)
Frog muscle (from pithed frog)
Probe for pithing
Dissecting instruments
Frog Ringer's solution

For Demonstration B—The Physiograph:
Physiograph
Myograph and stand
Frog muscle (from pithed frog)
Probe for pithing
Dissecting instruments
Frog Ringer's solution

⚠ SAFETY

- Wear disposable gloves when handling the frogs.
- Dispose of gloves and frogs as instructed.
- Wash your hands before leaving the laboratory.

PURPOSE OF THE EXERCISE

To observe and record skeletal muscle contractions of an isolated frog muscle to electrical stimulation of varying strength and frequency.

LEARNING OUTCOMES

After completing this exercise, you should be able to

1. Match the term with the definition of a muscle response.

2. Determine the threshold level of electrical stimulation in frog muscle.

3. Determine the intensity of stimulation needed for maximal muscle contraction.

4. Record a single muscle twitch and identify its phases.

5. Record the response of a muscle to increasing frequency of stimulation and identify the patterns of tetanic contraction and fatigue.

To study the characteristics of certain physiological events such as muscle contractions, it often is necessary to use a recording device, such as a *kymograph,* a *Physiograph,* or a digital recording device. These devices can provide accurate recordings of various physiological changes.

To observe the phenomenon of skeletal muscle contractions, muscles can be isolated from anesthetized frogs. These muscles can be attached to recording systems and stimulated by electrical stimuli of varying strength, duration, and frequency. Recordings obtained from such procedures can be used to demonstrate the basic characteristics of skeletal muscle contractions.

PRACTICE

DEMONSTRATION A—The Kymograph

1. Observe the kymograph and, at the same time, study figure 66.1 to examine the names of its major parts.

2. Note that the kymograph consists of a cylindrical *drum* around which a sheet of paper is wrapped. The drum is mounted on a motor-driven *shaft,* and the speed of the motor can be varied. Thus, the drum can be rotated rapidly if rapid physiological events are being recorded or rotated slowly for events that occur more slowly.

A *stylus* that can mark on the paper is attached to a *movable lever,* and the lever, in turn, is connected to an isolated muscle. The origin of the muscle is fixed in position by a *clamp,* and its insertion is hooked to the muscle lever. The muscle also is connected by wires to an *electronic stimulator* (or inductorium). The stimulator can deliver single or multiple electrical shocks to the muscle, and it can be adjusted so that the intensity (voltage), duration (milliseconds), and frequency (stimuli per second) can be varied. Another stylus, on the *signal*

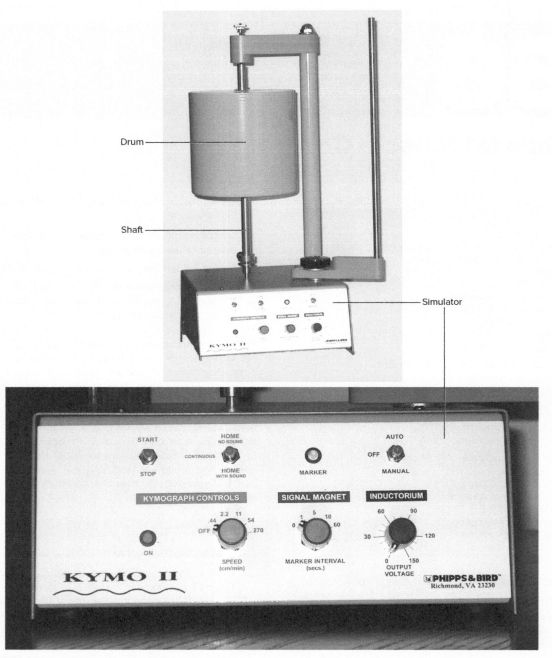

FIGURE 66.1   Kymograph to record frog muscle contractions.   (©Phipps & Bird, Inc., Richmond, VA. Used with permission.)

marker, records the time each stimulus is given to the muscle. As the muscle responds, the duration and relative length of its contraction are recorded by the stylus on the muscle lever.

3. Watch carefully while the laboratory instructor demonstrates the operation of the kymograph to record a frog muscle contraction.

 PRACTICE

DEMONSTRATION B—The Physiograph

1. Observe the Physiograph and, at the same time, study figures 66.2 and 66.3 to examine the names of its major parts.

2. Note that the recording system of the Physiograph includes a transducer, an amplifier, and a recording pen. The *transducer* is a sensing device that can respond

to some kind of physiological change by sending an electrical signal to the amplifier. The *amplifier* increases the strength of the electrical signal and relays it to an electric motor that moves the *recording pen.* As the pen moves, a line is drawn on paper.

To record a frog muscle contraction, a transducer called a *myograph* is used (fig. 66.3). The origin of the muscle is held in a fixed position, and its insertion is attached to a small lever in the myograph by a thread. The myograph, in turn, is connected to the amplifier by a transducer cable. The muscle is also connected by wires to the electronic stimulator, which is part of the Physiograph. This stimulator can be adjusted to deliver single or multiple electrical shocks to the muscle, and the intensity (voltage), duration (milliseconds), and frequency (stimuli per second) can be varied.

FIGURE 66.2 Physiograph to record frog muscle contractions. (©Narco Trace Bio-Systems is a division of International Biomedical Inc.)

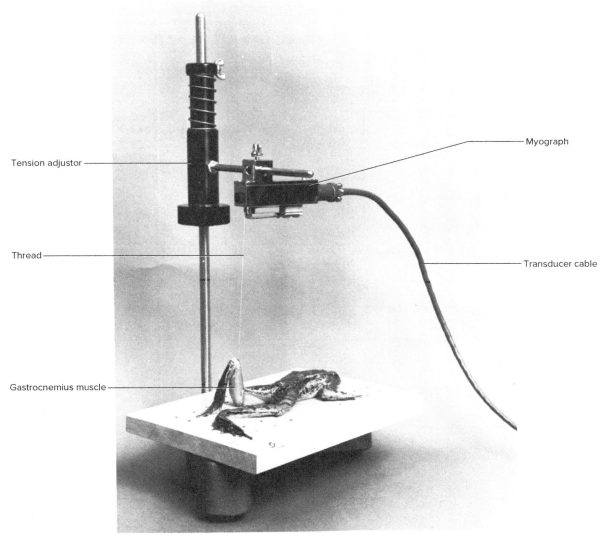

Tension adjustor

Thread

Gastrocnemius muscle

Myograph

Transducer cable

FIGURE 66.3 Myograph attached to frog muscle. (McGraw-Hill Education/Richard Pflanzer, photographer)

The speed at which the paper moves under the recording pen can be controlled. A second pen, driven by a timer, marks time units on the paper and indicates when the stimulator is activated. As the muscle responds to stimuli, the recording pen records the duration and relative length of each muscle contraction.

3. Watch carefully while the laboratory instructor operates the Physiograph to record a frog muscle contraction.

 PRACTICE

PROCEDURE A—Textbook Review

1. Review section 9.4 titled "Muscular Responses" in chapter 9 of the textbook.
2. Complete Part A of Laboratory Report 66.

 PRACTICE

PROCEDURE B—Recording System

1. Set up the recording system and stimulator that you have been provided to record the contractions of a frog muscle according to the directions provided by the laboratory instructor.
2. Obtain a live frog, and prepare one gastrocnemius muscle as described in Procedure C.

 PRACTICE

PROCEDURE C—Muscle Preparation

1. Prepare the live frog by pithing so that it will not have sensations or movements when its muscle is removed. To do this, follow these steps:
   a. Hold the frog securely in one hand so that its limbs are extended downward.
   b. Position the frog's head between your thumb and index finger (fig. 66.4a).
   c. Bend the frog's head forward at an angle of about 90 degrees by pressing on its snout with your index finger.
   d. Use a sharp probe to locate the foramen magnum between the occipital condyles in the midline between the frog's tympanic membranes.
   e. Insert the probe through the skin and into the foramen magnum, and then quickly move the probe from side to side to separate the brain from the spinal cord.
   f. Slide the probe forward into the braincase, and continue to move the probe from side to side to destroy the brain.

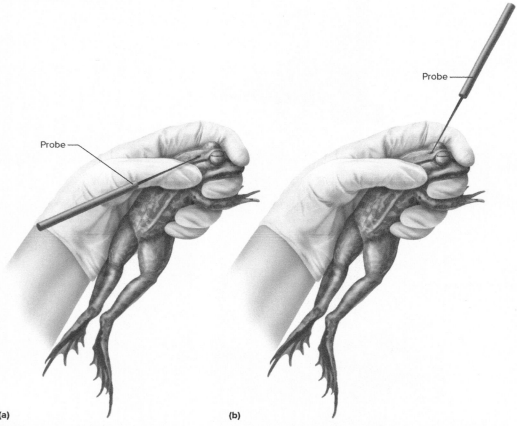

(a)   (b)

FIGURE 66.4 Hold the frog's head between your thumb and index finger to pith (a) its brain and (b) its spinal cord.

g. Remove the probe from the braincase, and insert it into the spinal cord through the same opening in the skin (fig. 66.4*b*).

h. Slide the probe up and down the spinal cord to destroy it. If the frog has been pithed correctly, its legs will be extended and relaxed. Also, the eyes will not respond when touched with a probe.

 ALTERNATIVE PROCEDURE

An anesthetizing agent, tricaine methane sulfonate, can be used to prepare frogs for this lab. This procedure eliminates the need to pith frogs.

2. Isolate the frog's gastrocnemius muscle by proceeding as follows:
  a. Place the pithed frog in a dissecting tray.
  b. Use scissors to cut through the skin completely around the hindlimb in the thigh.
  c. Pull the skin downward and off the hindlimb.
  d. Locate the gastrocnemius muscle in the calf and retain the calcaneal tendon (Achilles tendon) at its distal end (fig. 66.5*a*).
  e. Separate the calcaneal tendon from the underlying tissue, using forceps.
  f. Tie a thread firmly around the tendon (fig. 66.5*b*).

g. When the thread is secure, free the distal end of the tendon by cutting it with scissors.

h. Attach the frog muscle to the recording system in the manner suggested by your laboratory instructor (see figs. 66.1, 66.3, or 66.6).

i. Insert the ends of the stimulator wires into the muscle so that one wire is located on either side of the belly of the muscle.

Keep the frog muscle moist at all times by dripping frog Ringer's solution on it. When the muscle is not being used, cover it with some paper toweling that has been saturated with frog Ringer's solution.

Before you begin operating the recording system and stimulator, have the laboratory instructor inspect your setup.

 PRACTICE

PROCEDURE D—Threshold Stimulation

1. To determine the threshold stimulus or minimal strength of electrical stimulation (voltage) needed to elicit a contraction in the frog muscle, follow these steps:
  a. Set the stimulus duration to a minimum (about 0.1 millisecond).
  b. Set the voltage to a minimum (about 0.1 volt).
  c. Set the stimulator so that it will administer single stimuli.

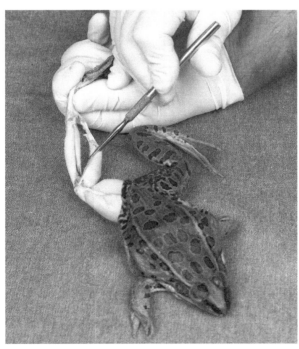

(a)

(b)

FIGURE 66.5   (*a*) Separate the calcaneal (Achilles) tendon from the underlying tissue. (*b*) Tie a thread around the tendon, and cut its distal attachments. (©J & J Photography)

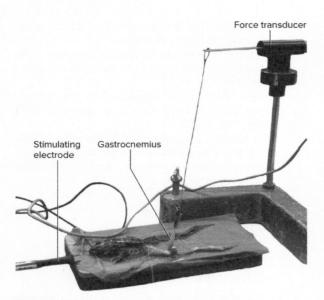

Force transducer

Stimulating electrode   Gastrocnemius

FIGURE 66.6 Typical setup for frog muscle recording. The gastrocnemius is attached to the force transducer that will send signals to the recording system. (©Brian Kipp)

2. Administer a single stimulus to the muscle and watch to see if it responds. If no response is observed, increase the voltage to the next higher setting and repeat the procedure until the muscle responds by contracting.
3. After determining the threshold level of stimulation, continue to increase the voltage in increments of 1 or 2 volts until a maximal muscle contraction is obtained.
4. Complete Part A of the laboratory report.

 PRACTICE

PROCEDURE E—Single Muscle Twitch

1. To record a single muscle twitch, set the voltage at the maximal muscle contraction as determined in Procedure D.
2. Set the recording speed at maximum, and while recording, administer a single electrical stimulus to the frog muscle.
3. Repeat this procedure to obtain several recordings of single muscle twitches.
4. Complete Part C of the laboratory report.

 PRACTICE

PROCEDURE F—Sustained Contraction

1. To record a sustained muscle contraction, follow these steps:
   a. Set the stimulator for continuous stimulation.
   b. Set the voltage for maximal muscle contraction as determined in Procedure D.
   c. Set the frequency of stimulation at a minimum.
   d. Set the recording speed at about 0.05 cm/second.
   e. With the paper moving, administer electrical stimulation and slowly increase the frequency of stimulation until the muscle sustains a contraction (tetanic contraction, or tetanus).
   f. Continue to stimulate the muscle at the frequency that produces sustained contractions until the muscle fatigues and relaxes (this may take several minutes).
2. Every 15 seconds for the next several minutes, stimulate the muscle to see how long it takes to recover from the fatigue.
3. Complete Part D of the laboratory report.

 LEARN: ACTIVITY

To demonstrate the staircase effect (treppe), obtain a fresh frog gastrocnemius muscle and attach it to the recording system as before. Set the paper control for slow speed, and set the stimulator voltage to produce a maximal muscle contraction. Stimulate the muscle once each second for several seconds. How do you explain the differences in the lengths of successive muscle contractions?

_____

_____

_____

_____

_____

Name _____

Date _____

Section _____

The A corresponds to the indicated Learning Outcome(s) found at the beginning of the laboratory exercise.

Skeletal Muscle Contraction

PART A ASSESSMENTS

Match the terms in column A with the definitions in column B. Place the letter of your choice in the space provided. A

Column A	Column B
a. All-or-none	_____ **1.** Minimal intensity of stimulation necessary to trigger a muscle contraction
b. Latent period	_____ **2.** Response of a muscle fiber/motor unit complete contraction if stimulated sufficiently
c. Motor unit	_____ **3.** Consists of a single motor neuron and all of the muscle fibers with which the neuron is associated
d. Muscle tone	_____ **4.** Action of a muscle contraction and immediate relaxation when exposed to a single stimulus
e. Myogram	_____ **5.** Time between stimulation and response
f. Refractory period	_____ **6.** Time following a muscle contraction during which the muscle remains unresponsive to stimulation
g. Tetanic contraction (tetanus)	_____ **7.** Forceful, sustained contraction
h. Threshold stimulus	_____ **8.** Some contraction of muscle fibers when a muscle is at rest
i. Twitch	_____ **9.** Recording of the pattern of a muscle contraction

PART B ASSESSMENTS

Complete the following:

1. What recording system was used for this laboratory exercise? _____

2. What was the threshold voltage for stimulation of the frog gastrocnemius muscle? 2 _____

3. What voltage produced maximal contraction of this muscle? A 3 _____

ASSESS

CRITICAL THINKING

Do you think other frog muscles would respond in an identical way to these voltages of stimulation?

_____

_____

Why or why not?

_____

_____

_____

PART C ASSESSMENTS

Complete the following:

1. Fasten a recording of two single muscle twitches in the following space. **4**

2. On a muscle twitch recording, label the *latent period, period of contraction,* and *period of relaxation,* and indicate the time it took for each of these phases to occur. **4**

3. What differences, if any, do you note in the two myograms of a single muscle twitch? How do you explain these differences? _____

_____

_____

 PART D ASSESSMENTS

Complete the following:

1. Fasten a recording demonstrating wave summation and sustained contraction in the following space. **5**

2. On the recording above, indicate when the muscle twitches began to combine (summate), and label the period of tetanic contraction and the period of fatigue. **5**

3. At what frequency of stimulation did tetanic contraction occur? **5** _____

4. How long did it take for the tetanic muscle to fatigue? **5** _____

5. Is the length of muscle contraction at the beginning of tetanic contraction the same as or different from the length of the single muscle contractions before tetanic contraction occurred? _____ How do you explain this? **5**

_____

_____

6. How long did it take for the fatigued muscle to become responsive again? **5** _____

66-8

LABORATORY EXERCISE

67

Nerve Impulse Stimulation

MATERIALS NEEDED

Textbook
Live frog
Dissecting tray
Dissecting instruments
Disposable gloves
Frog Ringer's solution
Electronic stimulator
Filter paper
Glass rod
Glass plate
Ring stand and ring
Microscope slides
Bunsen burner
Heat-resistant gloves
Ice
1% HCl
1% NaCl
2% Novocain solution (procaine hydrochloride)

⚠ SAFETY

- Wear disposable gloves when handling the frogs and chemicals.
- Keep loose hair and clothes away from the Bunsen burner.
- Wear heat-resistant gloves when heating the glass rod.
- Dispose of gloves, frogs, and chemicals as instructed.
- Wash your hands before leaving the laboratory.

PURPOSE OF THE EXERCISE

To review the characteristics of a nerve impulse and to investigate the effects of certain stimuli on a nerve.

 LEARNING OUTCOMES

After completing this exercise, you should be able to

1. Describe the structures and the events that lead to the stimulation of a nerve impulse.

2. Determine the threshold voltage to stimulate a nerve impulse.

3. Test the effects of various factors on a nerve-muscle preparation.

4. List four types of factors that can stimulate a nerve impulse.

A nerve cell usually is polarized due to an unequal distribution of ions on either side of its membrane. When such a polarized membrane is stimulated at or above its threshold intensity, a wave of action potentials is triggered to move in all directions away from the site of stimulation. This wave constitutes a nerve impulse, and if it reaches a muscle, the muscle may respond by contracting.

 PRACTICE

PROCEDURE—Nerve Impulse Stimulation

1. Review section 10.3 titled "Cell Membrane Potential" in chapter 10 of the textbook.
2. Complete Part A of Laboratory Report 67.
3. Obtain a live frog, and pith its brain and spinal cord as described in Procedure C of Laboratory Exercise 67.

 ALTERNATIVE PROCEDURE

An anesthetizing agent, tricaine methane sulfonate, can be used to prepare frogs for this lab. This procedure eliminates the need to pith frogs.

4. Place the pithed frog in a dissecting tray and remove the skin from its hindlimb, beginning at the waist, as described in Procedure C of Laboratory Exercise 66. (As the skin is removed, keep the exposed tissues moist by flooding them with frog Ringer's solution.)

5. Expose the frog's sciatic nerve. To do this, follow these steps:

    a. Use a glass rod to separate the gastrocnemius muscle from the adjacent muscles.

    b. Locate the calcaneal (Achilles) tendon at the distal end of the gastrocnemius, and cut it with scissors.

    c. Place the frog ventral surface down, and separate the muscles of the thigh to locate the sciatic nerve. The nerve will look like a silvery white thread passing through the thigh, dorsal to the femur (fig. 67.1).

    d. Dissect the nerve to its origin in the spinal cord.

    e. Use scissors to cut the nerve at its origin, and carefully snip off all of the branch nerves in the thigh, leaving only its connection to the gastrocnemius muscle.

    f. Use a scalpel to free the proximal end of the gastrocnemius.

    g. Carefully remove the nerve and attached muscle, and transfer the preparation to a glass plate supported on the ring of a ring stand.

    h. Use a glass rod to position the preparation so that the sciatic nerve is hanging over the edge of the glass plate. (Be sure to keep the preparation moistened with frog Ringer's solution at all times.)

6. Determine the threshold voltage and the voltage needed for maximal muscle contraction by using the electronic stimulator, as described in Procedure D of Laboratory Exercise 66.

7. Expose the cut end of the sciatic nerve to each of the following conditions, and observe the response of the gastrocnemius muscle. Add frog Ringer's solution after each of the experiments.

    a. Firmly pinch the end of the nerve between two glass microscope slides or pinch using forceps.

    b. Touch the cut end with a glass rod that is at room temperature.

    c. Touch the cut end with a glass rod that has been cooled in ice water for 5 minutes.

    d. Touch the cut end with a glass rod that has been heated in the flame of a Bunsen burner. Wear heat-resistant gloves for this procedure.

    e. Dip the cut end in 1% HCl.

    f. Dip the cut end in 1% NaCl.

8. Complete Part B of the laboratory report.

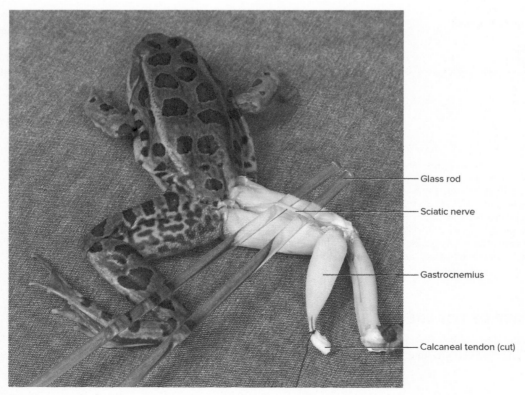

Glass rod

Sciatic nerve

Gastrocnemius

Calcaneal tendon (cut)

FIGURE 67.1   The sciatic nerve appears as a silvery white thread between the muscles of the thigh.   (©J & J Photography)

LEARN: ACTIVITY

Test the effect of Novocain on a frog sciatic nerve. To do this, follow these steps:

1. Place a nerve-muscle preparation on a glass plate supported by the ring of a ring stand, as before.
2. Use the electronic stimulator to determine the voltage needed for maximal muscle contraction.
3. Saturate a small piece of filter paper with 2% Novocain solution, and wrap the paper around the midsection of the sciatic nerve.

4. At 2-minute intervals, stimulate the nerve, using the voltage needed for maximal contraction until the muscle fails to respond.
5. Remove the filter paper, and flood the nerve with frog Ringer's solution.
6. At 2-minute intervals, stimulate the nerve until the muscle responds again. How long did it take for the nerve to recover from the effect of the Novocain?

_____

Notes

Name _____

Date _____

Section _____

The Ⓐ corresponds to the indicated Learning Outcome(s) found at the beginning of the laboratory exercise.

Nerve Impulse Stimulation

PART A ASSESSMENTS

Complete the following statements:

1. _____ ions tend to pass through cell membranes more easily than sodium ions. Ⓐ

2. When a nerve cell is at rest, there is a relatively greater concentration of _____ ions outside its membrane. Ⓐ

3. When sodium ions are actively transported outward through a nerve cell membrane, _____ ions are transported inward. Ⓐ

4. The difference in electrical charge between the inside and the outside of a nerve cell membrane is called the _____ . Ⓐ

5. If the resting potential becomes less negative in response to stimulation, the cell membrane is said to be _____ . Ⓐ

6. As a result of an additive phenomenon called _____ , the threshold potential of a membrane may be reached. Ⓐ

7. Following depolarization, potassium ions diffuse outward and cause the cell membrane to become _____ . Ⓐ

8. An action potential is a rapid sequence of changes involving depolarization and _____ . Ⓐ

9. The moment following the passage of an action potential during which a threshold stimulus will not trigger another impulse is called the _____ . Ⓐ

10. Muscle fiber contraction and nerve impulse conduction are similar in that both are _____ responses. Ⓐ

11. Myelin contains a high proportion of _____ . Ⓐ

12. Nodes of Ranvier occur between adjacent _____ . Ⓐ

13. The type of conduction in which an impulse seems to jump from node to node is called _____ . Ⓐ

14. The greater the diameter of a nerve fiber, the _____ the impulse travels. Ⓐ

PART B ASSESSMENTS

1. What was the threshold voltage for the frog sciatic nerve? **2** _____

2. What was the voltage needed for maximal contraction of the gastrocnemius muscle? **2** _____

3. Complete the following table: **3**

Factor Tested	Muscle Response	Effect on Nerve
Pinching		
Glass rod (room temperature)		
Glass rod (cooled)		
Glass rod (heated)		
1% HCl		
1% NaCl		

4. Summarize the results of these tests. **4**

_____

_____

_____

_____

_____

_____

68

Blood Testing

MATERIALS NEEDED

Mammal blood other than human or contaminant-free human blood is suggested as a substitute for collected blood
Textbook Chapter 14
Sterile disposable blood lancets
Alcohol swabs
Disposable gloves

For Procedure A:
Heparinized microhematocrit capillary tube
Sealing clay (or Critocaps)
Microhematocrit centrifuge
Microhematocrit reader

For Procedure B:
Tallquist test kit
Hemoglobinometer
Lens paper
Hemolysis applicator

For Procedure C:
Capillary tubes (nonheparinized)
Small, triangular file
Timer

⚠ SAFETY

- Review all Laboratory Safety Guidelines in Appendix 1 of your laboratory manual.
- It is important that students learn and practice correct procedures for handling body fluids. Consider using either mammal blood other than human or contaminant-free blood that has been tested and is available from various laboratory supply houses. Some of the procedures might be accomplished as demonstrations only. If student blood is used, it is important that students handle only their own blood.
- Use an appropriate disinfectant to wash the laboratory tables before and after the procedures.
- Wear disposable gloves when handling blood samples.
- Clean the end of a finger with alcohol swabs before the puncture is performed.

- Use the sterile blood lancet only once.
- Dispose of used lancets and blood-contaminated items in an appropriate container (never use the wastebasket).
- Wash your hands before leaving the laboratory.

PURPOSE OF THE EXERCISE

To observe the blood tests used to determine hematocrit, hemoglobin content, and coagulation time.

LEARNING OUTCOMES

After completing this exercise, you should be able to

1. Test and record the hematocrit, hemoglobin, and coagulation in a blood sample.
2. Evaluate the results of the blood tests compared to normal values.
3. Identify the blood tests performed in this laboratory exercise that could indicate anemia.

As an aid in identifying various disease conditions, tests are often performed on blood to determine how its composition compares with normal values. These tests commonly include hematocrit (red blood cell percentage), hemoglobin content, and coagulation. A self-diagnosis should never be made as a result of a test conducted in the biology laboratory. Always obtain proper medical exams and treatments from medical personnel.

PRACTICE

PROCEDURE A—Hematocrit

To determine the hematocrit (percentage of red blood cells) of a whole blood sample, the cells must be separated from the liquid plasma. This separation can be quickly accomplished by placing a tube of blood in a centrifuge. The force created by the spinning motion of the centrifuge causes the cells to be packed into the lower end of the tube. The quantities of cells and plasma can then be measured, and the hematocrit can be calculated. The normal range for men is 42%–52%; the normal range for women is 37%–47%.

1. To determine the hematocrit in a blood sample, follow these steps:

   a. Thoroughly wash hands with soap and water and dry them with paper towels. Don disposable gloves except for on the hand of the person with the finger to be lanced. (If purchased blood is used for this procedure, skip to step 1e.)

   b. Cleanse the end of your middle finger with an alcohol swab and let the finger dry in the air.

   c. Remove a sterile disposable blood lancet from its package without touching the sharp end.

   d. Puncture the skin on the side near the tip of the middle finger with the lancet and properly discard the lancet. Wipe away the first drop of blood with the alcohol swab.

   e. Touch a drop of blood with the colored end of a heparinized capillary tube. Hold the tube tilted slightly upward so that the blood will easily move into it by capillary action (fig. 68.1*a*). To prevent an air bubble, keep the tip of the capillary tube in the blood.

   f. Allow the blood to fill about two-thirds of the length of the tube. Cover the lanced finger location with a bandage.

   g. Hold a finger over the tip of the dry end so that blood will not drain out while you seal the blood end. Plug the blood end of the tube by pushing it with a rotating motion into sealing clay or by adding a plastic Critocap (fig. 68.1*b*).

   h. Place the sealed tube into one of the numbered grooves of a microhematocrit centrifuge. The tube's sealed end should point outward from the center and should touch the rubber lining on the rim of the centrifuge (fig. 68.1*c*).

   i. Balance the centrifuge by placing specimen tubes on opposite sides of the moving head, tighten the inside cover, and securely fasten the outside cover.

   j. Run the centrifuge for 3–5 minutes.

   k. After the centrifuge has stopped, remove the specimen tube. The red blood cells have been packed into the bottom of the tube. The clear liquid on top of the cells is plasma.

   l. Use a microhematocrit reader to determine the percentage of red blood cells in the tube. If a microhematocrit reader is not available, measure the total length of the blood column in millimeters (red cells plus plasma) and the length of the red blood cell column alone in millimeters. Divide the red blood cell length by the total blood column length and multiply the answer by 100 to calculate the percentage of red blood cells.

2. Record the test result in Part A of Laboratory Report 68.

(a)

(b)

(c)

FIGURE 68.1 Steps of the hematocrit (red blood cell percentage) procedure: (*a*) load a heparinized capillary tube with blood; (*b*) plug the blood end of the tube with sealing clay; (*c*) place the tube in a microhematocrit centrifuge.

PRACTICE

PROCEDURE B—Hemoglobin Content

Hemoglobin (Hb) is responsible for binding oxygen in the RBC. It is possible that anemia may be a result of low Hb even if the RBC count is normal. The hemoglobin content of a blood sample can be measured in several ways, including the *Tallquist method* and the *hemoglobinometer method*.

The Tallquist method uses a special test paper and is a simple, less expensive test that can be done more quickly than the hemoglobinometer method. The hemoglobinometer instrument is designed to compare the color of light passing through a hemolyzed blood sample with a standard color. The results of these tests are expressed in grams of hemoglobin per 100 mL (g/dL). The normal range for men is 13–18 g/dL; the normal range for women is 12–16 g/dL.

Tallquist Method

1. To measure the hemoglobin content of a blood sample, follow these steps:
   a. Gather one piece of paper from the Tallquist booklet/kit, a Tallquist hemoglobin scale, and a blood sample.
   b. Obtain a drop of blood from a finger by following the directions in Procedure A.
   c. Place a large drop of blood in the center of the Tallquist paper so that the drop of blood is larger than the holes in the Tallquist hemoglobin scale.
   d. Let the blood dry only long enough to lose its glossy appearance. (It should not dry to a brown color.)
   e. Match the color, using the natural light, of the blood on the Tallquist paper with the most accurate match of the color scale provided with the kit (fig. 68.2).

2. Record the result in Part A of the laboratory report.

Hemoglobinometer Method

1. To measure the hemoglobin content of a blood sample, follow these steps:
   a. Obtain a hemoglobinometer and remove the blood chamber from the slot in its side.
   b. Separate the pieces of glass from the metal clip and clean them with alcohol swabs and lens paper. One of the pieces of glass has two broad, U-shaped areas surrounded by depressions. The other piece is flat on both sides.

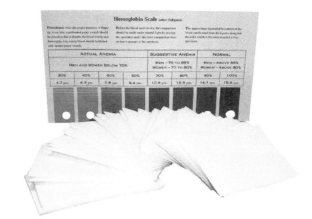

FIGURE 68.2 A simple method of measuring the amount of hemoglobin blood is to compare a piece of Tallquist paper that has been saturated with a sample of blood with a Tallquist color scale. (Kemtec Science)

   c. Obtain a large drop of blood from a finger, by following the directions in Procedure A.
   d. Place the drop of blood on one of the U-shaped areas of the blood chamber glass (fig. 68.3a).
   e. Stir the blood with the tip of a hemolysis applicator until the blood appears clear rather than cloudy. This usually takes about 45 seconds (fig. 68.3b).
   f. Place the flat piece of glass on top of the blood plate and slide both into the metal clip of the blood chamber.
   g. Push the blood chamber into the slot on the side of the hemoglobinometer, making sure that it is in all the way (fig. 68.3c).
   h. Hold the hemoglobinometer in the left hand with the thumb on the light switch on the underside (fig. 68.3d).
   i. Look into the eyepiece and note the green area split in half.
   j. Slowly move the slide on the side of the instrument back and forth with the right hand until the two halves of the green area look the same.
   k. Note the value in the upper scale (grams of hemoglobin per 100 mL of blood), indicated by the mark in the center of the movable slide.

2. Record the test result in Part A of the laboratory report.

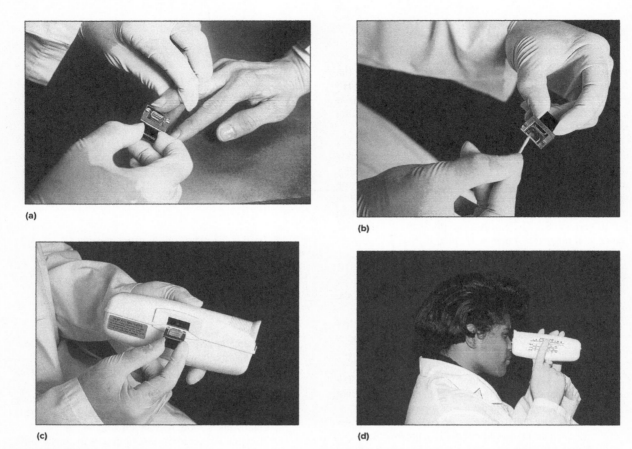

FIGURE 68.3 Steps of the hemoglobin content procedure: (*a*) load the blood chamber with blood; (*b*) stir the blood with a hemolysis applicator; (*c*) place the blood chamber in the slot of the hemoglobinometer; (*d*) match the colors in the green area by moving the slide on the side of the instrument. (James Shaffer/McGraw-Hill Education)

 **PRACTICE**

PROCEDURE C—Coagulation

Coagulation time, often called clotting time, is the time from the onset of bleeding until the clot is formed. Injured tissues and platelets release chemicals that trigger the clotting cascade, which ultimately results in thrombin converting soluble fibrinogen into insoluble fibrin. Fibrin is a mesh that traps RBCs and forms the clot. Clotting time normally ranges from 3 to 8 minutes. This process is prolonged if the person has clotting deficiencies or is being treated with anticoagulants such as heparin (some are produced by basophils and mast cells), warfarin (Coumadin), or aspirin. In this laboratory exercise, we will try to determine the time to the nearest minute.

1. To determine coagulation time, follow these steps:
    a. Prepare the finger to be lanced by following the directions in Procedure A. Lance the end of a finger to obtain a drop of blood. Wipe away the first drop of blood with the alcohol swab.
    b. Touch a drop of blood with one end of a nonheparinized capillary tube. Hold the tube tilted slightly upward so that the blood will easily move into it by capillary action (see fig. 68.1*a*). Keep the tip in the blood until the tube is nearly filled. If the tube is nearly filled, it will allow enough tube length for breaking it several times.
    c. Place the capillary tube on a paper towel. Cover the lanced finger location with a bandage. Record the time: _____
    d. At 1-minute intervals, use the small, triangular file and make a scratch on the capillary tube starting near one end of the tube. Hold the tube with fingers on each side of the scratch, the weakened location of the tube, and break the tube away from you, being careful to keep the two pieces close together after the

break. Gently pull the two ends of the tube apart while observing carefully to see if it breaks cleanly apart. If it breaks cleanly, fibrin has not formed yet (fig. 68.4*a*).

    e. Continue breaking the capillary tube each minute until fibrin is noted spanning the two parts of the capillary tube (fig. 68.4*b*). Note the time for coagulation.

2. Record the test results in Part A of the laboratory report.

3. Complete Part B of the laboratory report.

 ALTERNATIVE PROCEDURE

    Laboratories in modern hospitals and clinics use updated hematology analyzers (fig. 68.5) for evaluations of the blood factors described in Laboratory Exercise 68. The more traditional procedures performed in these laboratory activities will help you to better understand each blood characteristic.

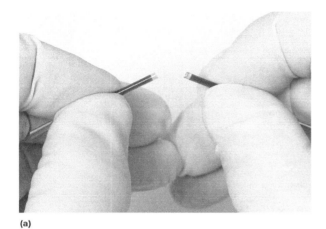

(a)

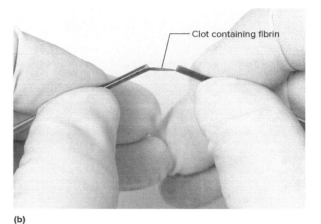

Clot containing fibrin

(b)

FIGURE 68.4 Steps of the blood coagulation procedure: (*a*) clean break of capillary tube before any fibrin formation; (*b*) fibrin (the clot) spans the two ends of the broken capillary tube at coagulation. (©J and J Photography)

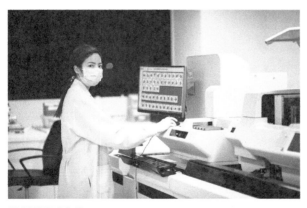

FIGURE 68.5 Modern hematology analyzer being used in the laboratory of a clinic. (*Note*: A tour of a laboratory in a modern hospital or clinic might be arranged.) (Krasaesom/Shutterstock)

Notes

Name _____

Date _____

Section _____

The ⬡ corresponds to the indicated Learning Outcome(s) found at the beginning of the laboratory exercise.

Blood Testing

🔁 PART A ASSESSMENTS

Blood test data: ⬡**1**

Blood Test	Test Results	Normal Values
Hematocrit (mL per 100 mL blood)		Men: 42%–52% Women: 37%–47%
Hemoglobin content (g per 100 mL blood; g/dL)		Men: 13–18 g/dL Women: 12–16 g/dL
Coagulation		3–8 minutes

🔁 PART B ASSESSMENTS

Complete the following:

1. How does the hematocrit from the blood test compare with the normal value? ⬡**2** _____

_____

_____

2. How does the hemoglobin content from the blood test compare with the normal value? ⬡**2** _____

_____

_____

3. How does the coagulation time from the blood test compare with the normal value? ⬡**2** _____

_____

_____

🔁 ASSESS

CRITICAL THINKING

Which blood tests performed in this lab could be used to determine possible anemia? ⬡**3**

_____

_____

_____

_____

LABORATORY EXERCISE

69

Factors Affecting the Cardiac Cycle

MATERIALS NEEDED

Textbook
Physiological recording apparatus such as a kymograph,
    a Physiograph, or a digital recording device
Live frog
Dissecting tray
Dissecting instruments
Dissecting pins
Disposable gloves
Frog Ringer's solution in plastic squeeze bottle
Thread
Small hook
Medicine dropper
Thermometer
Ice
Hot plate
Calcium chloride, 2% solution
Potassium chloride, 5% solution
Epinephrine, 1:10,000 solution
Acetylcholine, 1:10,000 solution
Caffeine, 0.2% solution

⚠ SAFETY

■ Wear disposable gloves when handling the frogs.
■ Dispose of the frogs according to your laboratory
  instructor.
■ Wash your hands before leaving the laboratory.

PURPOSE OF THE EXERCISE

To review the mechanism by which the heartbeat is regulated, to observe the action of a frog heart, and to investigate the effects of various factors on the frog heartbeat.

 LEARNING OUTCOMES

After completing this exercise, you should be able to

1 Describe the mechanism by which the human cardiac
cycle is controlled.

2 Distinguish the atrial and ventricular contractions and
determine the heart rate from a recording of a frog
heartbeat.

3 Assess the effects of various factors on the action of a
frog heart.

Although the cardiac cycle is controlled by the SA node serving as the pacemaker, the rate of heart action can be altered by various other factors. These factors include parasympathetic and sympathetic nerve impulses that originate in the cardiac center of the medulla oblongata, changes in body temperature, and concentrations of certain ions.

 PRACTICE

    **PROCEDURE—Factors Affecting
    the Cardiac Cycle**

1. Review the concept heading titled "Regulation of the Cardiac Cycle" in section 15.3 in chapter 15 of the textbook.
2. Complete Part A of Laboratory Report 69.

 GENERAL SUGGESTION

    Try to become familiar with the content and organization of this laboratory exercise before you pith a frog. If you work quickly, one pithed frog should last for all of the experimental steps.

3. Observe the normal action of a frog heart. To do this, follow these steps:
    a. Obtain a live frog, and pith it according to the directions in Procedure C of Laboratory Exercise 66.
    b. Place the pithed frog in a dissecting tray with its ventral surface up, and pin its jaw and limbs to the tray with dissecting pins (fig. 69.1).

 ALTERNATIVE PROCEDURE

    An anesthetizing agent, tricaine methane sulfonate, can be used to prepare frogs for this lab. This procedure eliminates the need to pith frogs.

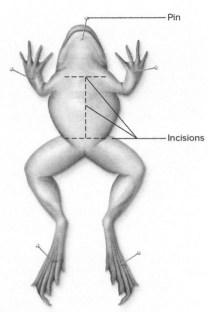

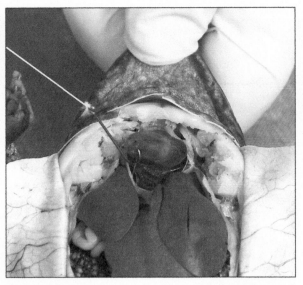

FIGURE 69.2   Attach a hook and thread to the tip of the ventricle.   (©J & J Photography)

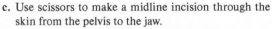

FIGURE 69.1   Pin the frog to the dissecting tray and make incisions through the skin as indicated.

c. Use scissors to make a midline incision through the skin from the pelvis to the jaw.

d. Cut the skin laterally on each side in the pelvic and pectoral regions, and pin the resulting flaps of skin to the tray (fig. 69.1).

e. Remove the exposed pectoral muscles and the sternum, being careful not to injure the underlying organs.

f. Note the beating heart surrounded by the thin-walled pericardium. Use forceps to lift the pericardium upward, and carefully slit it open with scissors, thus exposing the heart.

g. Flood the heart with frog Ringer's solution, and keep it moist throughout this exercise.

h. The frog heart has only three chambers—two atria and a ventricle. Watch the heart carefully as it beats, and note the sequence of chamber movements during a cardiac cycle.

4. Tie a piece of thread about 45 cm long to a small metal hook, and insert the hook into the tip (apex) of the ventricle without penetrating the chamber (fig. 69.2). The laboratory instructor will demonstrate how to connect the thread to a physiological recording apparatus so that you can record the frog heart movements. The thread should be adjusted so that there is no slack in it, but at the same time, it should not be so taut that it pulls the heart out of its normal position (fig. 69.3).

5. Record the movements of the frog heart for 2 to 3 minutes. Identify on the recording the smaller atrial contraction waves and the larger ventricular contraction waves. Also, determine the heart rate (beats per minute) for each minute of recording, and calculate the average rate. Enter the results in Part B of the laboratory report.

6. Test the effect of temperature change on the frog's heart rate. To do this, follow these steps:

a. Remove as much as possible of the Ringer's solution from around the heart, using a medicine dropper.

b. Flood the heart with fresh Ringer's solution that has been cooled in an ice water bath to about 10°C (50°F).

c. Record the heart movements, and determine the heart rate as before.

d. Remove the cool liquid from around the heart, and replace it with room-temperature Ringer's solution.

e. After the heart is beating at its normal rate again, flood it with Ringer's solution that has been heated on a hot plate to about 35°C (95°F).

f. Record the heart movements, and determine the heart rate as before.

g. Enter the results in Part B of the laboratory report.

7. Complete Part B of the laboratory report.

8. Test the effect of an increased concentration of calcium ions on the frog heart. If the frog heart from the previous experiment is still beating, replace the fluid around it with room-temperature Ringer's solution, and wait until its rate is normal. Otherwise, prepare a fresh specimen, and determine its normal rate as before. To perform the test, follow these steps:

a. Flood the frog heart with 2% calcium chloride. (This solution of calcium chloride will allow ionization to occur, providing Ca^{++}.)

b. Record the heartbeat for about 5 minutes, and note any change in rate.

FIGURE 69.3 Attach the thread from the heart to the recording apparatus so that there is no slack in the thread.

c. Flood the heart with fresh Ringer's solution until heart rate returns to normal.

9. Test the effect of an increased concentration of potassium ions on the frog heart. To do this, follow these steps:

   a. Flood the heart with 5% potassium chloride. (This solution of potassium chloride will allow ionization to occur, providing K^+.)

   b. Record the heartbeat for about 5 minutes, and note any change in rate.

10. Complete Part C of the laboratory report.

LEARN ACTIVITY

Plan an experiment to test the effect of an additional factor on the action of a frog heart. For example, you might test the effect of epinephrine, acetylcholine, caffeine, or some other available substance. If the laboratory instructor approves your plan, perform the experiment and record the heart movements. What do you conclude from the results of your experiment?

_____

_____

_____

_____

Notes

Name _____

Date _____

Section _____

The /A\ corresponds to the indicated Learning Outcome(s) found at the beginning of the laboratory exercise.

Factors Affecting
the Cardiac Cycle

PART A ASSESSMENTS

Complete the following statements:

1. The primary function of the heart is to _____ . /1\

2. The _____ normally controls the heart rate. /1\

3. Parasympathetic nerve fibers that supply the heart make up part of the _____ nerve. /1\

4. Endings of parasympathetic nerve fibers secrete _____, which causes the heart rate to decrease. /1\

5. Sympathetic nerve fibers reach the heart by means of _____ nerves. /1\

6. Endings of sympathetic nerve fibers secrete _____, which causes the heart rate to increase. /1\

7. The cardiac control center is located in the _____ of the brainstem. /1\

8. Baroreceptors (pressoreceptors) located in the walls of the aorta and carotid arteries are sensitive to changes in
_____ . /1\

9. If baroreceptors (pressoreceptors) in the walls of the venae cavae are stimulated by stretching, the cardioaccelerator center sends _____ impulses to the heart. /1\

10. Rising body temperature usually causes the heart rate to _____ . /1\

11. Of the ions that affect heart action, the most important ions are calcium and _____ . /1\

PART B ASSESSMENTS

1. Describe the actions of the frog heart chambers during a cardiac cycle. /2\

_____

_____

_____

_____

2. Attach a short segment of the normal frog heart recording in the following space. Label the atrial and ventricular waves of one cardiac cycle. Enter the average heart rate beneath the recording. /2\

3. Temperature effect results: 3

Temperature	Heart Rate
10°C (50°F)	
Room temperature	
35°C (95°F)	

4. Summarize the effect of temperature on the frog's heart action that was demonstrated by this experiment. 3

_____

_____

_____

_____

_____

PART C ASSESSMENTS

Complete the following:

1. Describe the effect of an increased calcium ion (Ca^{++}) concentration on the frog's heart rate. 3 _____

_____

_____

_____

_____

2. Describe the effect of an increased potassium ion (K^{+}) concentration on the frog's heart rate. 3 _____

_____

_____

_____

_____

ASSESS

CRITICAL THINKING

In testing the effects of different ions on heart action, why were chlorides used in each case?

_____

_____

_____

_____

APPENDIX 1

Name_____

Date_____ Class_____

Laboratory Safety Guidelines

Carefully review the following safety guidelines before starting a lab exercise. Safety guideline reminders are also included in the appropriate lab exercise sections. Avoid tardiness to laboratory sessions because specific directions and precautions are often given at the beginning of the period. Your instructor should update you about safety procedures, as your school might have safety regulations or modifications in addition to those listed here.

1. *Become familiar with all room exits and the location and operating procedures of all safety equipment (first aid kit, Material Safety Data Sheets (MSDS), eyewash station, safety shower, fire extinguisher, and fire blanket).*
2. *Read all laboratory exercises prior to starting the procedures. Do not work alone in the laboratory.*
3. *Do not smoke, chew, eat, drink, apply cosmetics or lip balm, handle contact lenses (regular eyeglasses should be worn), or work with open wounds in the laboratory. Consider all materials you use as potentially hazardous.*
4. *Use protective eyewear and laboratory coats during labs when handling dangerous chemicals or when heating dangerous materials. Volatile materials should never be heated using an open flame. When heating materials in a test tube, never point the test tube in the direction of a person or leave heated materials unattended.*
5. *Use a commercially prepared disinfectant or a 10% bleach solution to clean laboratory work surfaces before and after any procedures when using animals, body fluids, and dissection specimens.*
6. *Wear enclosed shoes and disposable gloves when working with dangerous chemicals, body fluids, and dissection specimens.*
7. *If body fluids are being studied, work only with your own. Special precautions to prevent contact with body fluids may include wearing disposable gloves, gowns, aprons, or masks, as directed by your instructor.*
8. *Special precautions during dissections may include wearing laboratory coats and protective eyewear. Always use cutting blades away from yourself and laboratory partners.*
9. *Restrain all loose clothing, long hair, and dangling jewelry during laboratory procedures. Keep all unnecessary materials away from the work area to reduce clutter and the possibility of an accident.*
10. *If you have special needs, are taking medications, experience allergic reactions, are pregnant, or are uncomfortable with the procedures, inform, your instructor.*
11. *Use only a mechanical pipetting device (never your mouth).*
12. *Immediately report any accidents, spills, or damaged equipment to your instructor.*
13. *Use only disposable lancets and needles, and never attempt to bend, cut, or recap them when finished. Place sharp items in a puncture-resistant container that is marked "Biohazard Container."*
14. *Dispose of chemicals, waste material, body fluids, and dissection specimens according to appropriate directions. Any reusable glassware or utensils that have been contaminated with body fluids should be placed in a disinfectant (10% bleach solution) and later autoclaved.*
15. *Thoroughly wash your hands with soap and warm water immediately after removing disposable gloves and before leaving the laboratory.*
16. *Practice any modified or additional safety guidelines required by your instructor.*

APPENDIX 2

Periodic Table of the Elements

Key:

9
F
Fluorine
19.00

— Atomic number (9)
— Atomic mass (19.00)

1 1A																	18 8A
1 **H** Hydrogen 1.008	2 2A											13 3A	14 4A	15 5A	16 6A	17 7A	2 **He** Helium 4.003
3 **Li** Lithium 6.941	4 **Be** Beryllium 9.012											5 **B** Boron 10.81	6 **C** Carbon 12.01	7 **N** Nitrogen 14.01	8 **O** Oxygen 16.00	9 **F** Fluorine 19.00	10 **Ne** Neon 20.18
11 **Na** Sodium 22.99	12 **Mg** Magnesium 24.31	3 3B	4 4B	5 5B	6 6B	7 7B	8	9 8B	10	11 1B	12 2B	13 **Al** Aluminum 26.98	14 **Si** Silicon 28.09	15 **P** Phosphorus 30.97	16 **S** Sulfur 32.07	17 **Cl** Chlorine 35.45	18 **Ar** Argon 39.95
19 **K** Potassium 39.10	20 **Ca** Calcium 40.08	21 **Sc** Scandium 44.96	22 **Ti** Titanium 47.88	23 **V** Vanadium 50.94	24 **Cr** Chromium 52.00	25 **Mn** Manganese 54.94	26 **Fe** Iron 55.85	27 **Co** Cobalt 58.93	28 **Ni** Nickel 58.69	29 **Cu** Copper 63.55	30 **Zn** Zinc 65.39	31 **Ga** Gallium 69.72	32 **Ge** Germanium 72.59	33 **As** Arsenic 74.92	34 **Se** Selenium 78.96	35 **Br** Bromine 79.90	36 **Kr** Krypton 83.80
37 **Rb** Rubidium 85.47	38 **Sr** Strontium 87.62	39 **Y** Yttrium 88.91	40 **Zr** Zirconium 91.22	41 **Nb** Niobium 92.91	42 **Mo** Molybdenum 95.94	43 **Tc** Technetium (98)	44 **Ru** Ruthenium 101.1	45 **Rh** Rhodium 102.9	46 **Pd** Palladium 106.4	47 **Ag** Silver 107.9	48 **Cd** Cadmium 112.4	49 **In** Indium 114.8	50 **Sn** Tin 118.7	51 **Sb** Antimony 121.8	52 **Te** Tellurium 127.6	53 **I** Iodine 126.9	54 **Xe** Xenon 131.3
55 **Cs** Cesium 132.9	56 **Ba** Barium 137.3	57 **La** Lanthanum 138.9	72 **Hf** Hafnium 178.5	73 **Ta** Tantalum 180.9	74 **W** Tungsten 183.9	75 **Re** Rhenium 186.2	76 **Os** Osmium 190.2	77 **Ir** Iridium 192.2	78 **Pt** Platinum 195.1	79 **Au** Gold 197.0	80 **Hg** Mercury 200.6	81 **Tl** Thallium 204.4	82 **Pb** Lead 207.2	83 **Bi** Bismuth 209.0	84 **Po** Polonium (210)	85 **At** Astatine (210)	86 **Rn** Radon (222)
87 **Fr** Francium (223)	88 **Ra** Radium (226)	89 **Ac** Actinium (227)	104 **Rf** Rutherfordium (257)	105 **Db** Dubnium (260)	106 **Sg** Seaborgium (263)	107 **Bh** Bohrium (262)	108 **Hs** Hassium (265)	109 **Mt** Meitnerium (266)	110 **Ds** Darmstadtium (269)	111 **Rg** Roentgenium (272)	112 **Cn** Copernicium (285)	113 **Nh** Nihonium (286)	114 **Fl** Flerovium (289)	115 **Mc** Moscovium (289)	116 **Lv** Livermorium (293)	117 **Ts** Tennessine (294)	118 **Og** Oganesson (294)

Metals
Metalloids
Nonmetals

58 **Ce** Cerium 140.1	59 **Pr** Praseodymium 140.9	60 **Nd** Neodymium 144.2	61 **Pm** Promethium (147)	62 **Sm** Samarium 150.4	63 **Eu** Europium 152.0	64 **Gd** Gadolinium 157.3	65 **Tb** Terbium 158.9	66 **Dy** Dysprosium 162.5	67 **Ho** Holmium 164.9	68 **Er** Erbium 167.3	69 **Tm** Thulium 168.9	70 **Yb** Ytterbium 173.0	71 **Lu** Lutetium 175.0
90 **Th** Thorium 232.0	91 **Pa** Protactinium (231)	92 **U** Uranium 238.0	93 **Np** Neptunium (237)	94 **Pu** Plutonium (242)	95 **Am** Americium (243)	96 **Cm** Curium (247)	97 **Bk** Berkelium (247)	98 **Cf** Californium (249)	99 **Es** Einsteinium (254)	100 **Fm** Fermium (253)	101 **Md** Mendelevium (256)	102 **No** Nobelium (254)	103 **Lr** Lawrencium (257)

The 1–18 group designation has been recommended by the International Union of Pure and Applied Chemistry (IUPAC).

Elements up to atomic number 118 have been confirmed and officially named.

Different sources may report different numbers of elements occurring naturally. According to Los Alamos National Laboratory (New Mexico, US), of elements 92 (Uranium) and below, only Promethium appears to be completely absent on earth outside of the particle physics laboratory. A few elements above and below Uranium are reportedly found naturally only in trace amounts on earth or in spectral analysis of certain stars. Including or excluding such elements would affect the stated total.

APPENDIX 3

Preparation of Solutions

AMYLASE SOLUTION, 0.5%

Place 0.5 g of bacterial amylase in a graduated cylinder or volumetric flask. Add distilled water to the 100 mL level. Stir until dissolved. The amylase should be free of sugar for best results; a low-maltose solution of amylase yields good results. (Store amylase powder in a freezer until mixing this solution.)

BENEDICT'S SOLUTION

Prepared solution is available from various suppliers.

CAFFEINE, 0.2%

Place 0.2 g of caffeine in a graduated cylinder or volumetric flask. Add distilled water to the 100 mL level. Stir until dissolved.

CALCIUM CHLORIDE, 2.0%

Place 2.0 g of calcium chloride in a graduated cylinder or volumetric flask. Add distilled water to the 100 mL level. Stir until dissolved.

CALCIUM HYDROXIDE SOLUTION (LIMEWATER)

Add an excess of calcium hydroxide to 1 L of distilled water. Stopper the bottle and shake thoroughly. Allow the solution to stand for 24 hours. Pour the supernatant fluid through a filter. Store the clear filtrate in a stoppered container.

EPSOM SALT SOLUTION, 0.1%

Place 0.5 g of Epsom salt in a graduated cylinder or volumetric flask. Add distilled water to the 500 mL level. Stir until dissolved.

GLUCOSE SOLUTIONS

1. *1.0% solution.* Place 1 g of glucose in a graduated cylinder or volumetric flask. Add distilled water to the 100 mL level. Stir until dissolved.
2. *10% solution.* Place 10 g of glucose in a graduated cylinder or volumetric flask. Add distilled water to the 100 mL level. Stir until dissolved.

IODINE-POTASSIUM-IODIDE (IKI SOLUTION)

Add 20 g of potassium iodide to 1 L of distilled water, and stir until dissolved. Then add 4.0 g of iodine, and stir again until dissolved. Solution should be stored in a dark, stoppered bottle.

METHYLENE BLUE

Dissolve 0.3 g of methylene blue powder in 30 mL of 95% ethyl alcohol. In a separate container, dissolve 0.01 g of potassium hydroxide in 100 mL of distilled water. Mix the two solutions. (Prepared solution is available from various suppliers.)

MONOSODIUM GLUTAMATE (MSG) SOLUTION, 1%

Place 1 g of monosodium glutamate in a graduated cylinder or volumetric flask. Add distilled water to the 100 mL level. Stir until dissolved.

PHYSIOLOGICAL SALINE SOLUTION

Place 0.9 g of sodium chloride in a graduated cylinder or volumetric flask. Add distilled water to the 100 mL level. Stir until dissolved.

POTASSIUM CHLORIDE, 5%

Place 5.0 g of potassium chloride in a graduated cylinder or volumetric flask. Add distilled water to the 100 mL level. Stir until dissolved.

QUININE SULFATE, 0.5%

Place 0.5 g of quinine sulfate in a graduated cylinder or volumetric flask. Add distilled water to the 100 mL level. Stir until dissolved.

RINGER'S SOLUTION (FROG)

Dissolve the following salts in 1 L of distilled water:

    6.50 g sodium chloride
    0.20 g sodium bicarbonate
    0.14 g potassium chloride
    0.12 g calcium chloride

SODIUM CHLORIDE SOLUTIONS

1. *0.9% solution.* Place 0.9 g of sodium chloride in a graduated cylinder or volumetric flask. Add distilled water to the 100 mL level. Stir until dissolved.
2. *1.0% solution.* Place 1.0 g of sodium chloride in a graduated cylinder or volumetric flask. Add distilled water to the 100 mL level. Stir until dissolved.
3. *3.0% solution.* Place 3.0 g of sodium chloride in a graduated cylinder or volumetric flask. Add distilled water to the 100 mL level. Stir until dissolved.
4. *5.0% solution.* Place 5.0 g of sodium chloride in a graduated cylinder or volumetric flask. Add distilled water to the 100 mL level. Stir until dissolved.

STARCH SOLUTIONS

1. *0.5% solution.* Add 5 g of cornstarch to 1 L of distilled water. Heat until the mixture boils. Cool the liquid, and pour it through a filter. Store the filtrate in a refrigerator.
2. *1.0% solution.* Add 10 g of cornstarch to 1 L of distilled water. Heat until the mixture boils. Cool the liquid, and pour it through a filter. Store the filtrate in a refrigerator.
3. *10% solution.* Add 100 g of cornstarch to 1 L of distilled water. Heat until the mixture boils. Cool the liquid, and pour it through a filter. Store the filtrate in a refrigerator.

SUCROSE, 5% SOLUTION

Place 5.0 g of sucrose in a graduated cylinder or volumetric flask. Add distilled water to the 100 mL level. Stir until dissolved.

WRIGHT'S STAIN

Prepared solution is available from various suppliers.

APPENDIX 4

Assessments of Laboratory Reports

Many assessment models can be used for laboratory reports. A rubric, which can be used for performance assessments, contains a description of the elements (requirements or criteria) of success to various degrees. The term *rubric* originated from *rubrica terra,* which is Latin for the application of red earth to indicate anything of importance. A rubric used for assessment contains elements for judging student performance, with points awarded for varying degrees of success in meeting the learning outcomes. The content and the quality level necessary to attain certain points are indicated in the rubric. It is effective if the assessment tool is shared with the students before the laboratory exercise is performed.

Following are two sample rubrics that can easily be modified to meet the needs of a specific course. Some of the elements for these sample rubrics may not be necessary for every laboratory exercise. The generalized rubric needs to contain the possible assessment points that correspond to learning outcomes for a specific course. The point value for each element may vary. The specific rubric example contains performance levels for laboratory reports. The elements and the point values can easily be altered to meet the value placed on laboratory reports for a specific course.

Assessment: Generalized Laboratory Report Rubric

Element	Assessment Points Possible	Assessment Points Earned
1. Figures are completely and accurately labeled.		
2. Sketches are accurate, contain proper labels, and are of sufficient detail.		
3. Colored pencils were used extensively to differentiate structures on illustrations.		
4. Matching and fill-in-the-blank answers are completed and accurate.		
5. Short-answer/discussion questions contain complete, thorough, and accurate answers. Some elaboration is evident for some answers.		
6. Data collected are complete, are accurately displayed, and contain a valid explanation.		

TOTAL POINTS: POSSIBLE _____ EARNED _____

Assessment: Specific Laboratory Report Rubric

Element	Excellent Performance (4 points)	Proficient Performance (3 points)	Marginal Performance (2 points)	Novice Performance (1 point)	Points Earned
Figure labels	Labels completed with ≥ 90% accuracy.	Labels completed with 80%–89% accuracy.	Labels completed with 70%–79% accuracy.	Labels <70% accurate.	
Sketches	Accurate use of scale, details illustrated, and all structures labeled accurately.	Minor errors in sketches. Missing or inaccurate labels on one or more structures.	Sketch not realistic. Missing or inaccurate labels on two or more structures.	Several missing or inaccurate labels.	
Matching and fill-in-the-blanks	All completed and accurate.	One or two errors or omissions.	Three or four errors or omissions.	Five or more errors or omissions.	
Short-answer and discussion questions	Answers complete, valid, and contain some elaboration. No misinterpretations noted.	Answers generally complete and valid. Only minor inaccuracies noted. Minimal elaboration.	Marginal answers to the questions and contains inaccurate information.	Many answers incorrect or fail to address the topic. May be misinterpretations.	
Data collection and analysis	Data complete and displayed with a valid interpretation.	Only minor data missing or a slight misinterpretation.	Some omissions. Not displayed or interpreted accurately.	Data incomplete or show serious misinterpretations.	

TOTAL POINTS EARNED _____

INDEX

superior border of scapula, 117, 119f
superior colliculus, 218
superior lobes of lungs, 366, 367f
superior mesenteric artery, 322
superior mesenteric vein, 328
superior nasal concha, 97, 242f
superior oblique muscle, 265, 268t
superior orbital fissure, 102f, 105f
superior pubic ramus, 406f
superior rectus muscle, 265, 268t
superior sagittal sinus, 224f
superior vena cava, 304, 307, 308f, 321, 328
supinator muscle, 161
supraclavicular fossa, 189f
supraorbital foramen, 97
supraspinatus muscle, 511
  human, 161
  in cat, 448, 449f, 450f, 450t, 451f, 452f
supraspinous fossa, 117
suprasternal notch, 189f, 312f
surface anatomy, 187, 188f–193f, 194
surgical neck of humerus, 118, 120f
suspensory ligaments, 267, 270, 276, 413
sustentacular cells, 408
sutural bone, 106f
sutures, 137
swallowing, 348
sweat glands, 81f, 82f
sweet sensation, 241
sympathetic nervous system, 311
symphyses, 137
synarthroses, 137
synchondroses, 137
syndesmoses, 137
synergistic actions, 150
synovial joints, 137, 138, 139f
syringe, 376, 377f
systemic circuit, 319, 321, 323f
systole, 311
systolic pressure, 336

T

tactile corpuscles, 235, 236f
taeniae coli, 351
tail
  fetal pig, 506, 506f
talus, 132
tapetum fibrosum, 270, 271f
tarsal bones, 91, 132
taste buds, 241, 243f, 244f
taste hairs, 244

taste pores, 244
taste sense, 241, 243–244, 243f, 244f
tattoos, 80
teeth
  as major organ, 12
  of cat, 476
  structure, 345, 346f, 347f
telophase, 56f
temperature
  blood vessel response to changes, 321
  effects on digestion, 359, 361
  metric units, 2t
  sensory receptors, 237
temporal bones, 97
temporal lobe, 207, 213, 214f, 215, 216f
temporal process, 101
temporalis muscle, 153
temporomandibular joint, 188f
tendons
  as major part of muscular system, 12
  of lower limbs, 176f, 181f, 191f, 193f
  origin and insertion, 148
tensor fasciae latae muscle, 175, 449f, 454, 455f, 456f, 457t
tensor tympani muscle, 249
tentorium cerebelli, 224f
tenuissimus muscle, 456, 457f, 458f, 459t
teres major muscle, 511
  human, 161
  in cat, 443f, 446, 447t, 448, 448t, 449f, 449t, 450f, 450t, 451f, 452f
teres minor muscle, 161
tertiary bronchi, 366
testes, 546, 547f
  as major organs, 12
  in cat, 498, 499f, 500f, 501f
  location in human, 282
  structure, 405, 406f, 407
testosterone, 408
thalamus, 208, 218
thenar eminence, 191f
theories, 1
thigh muscles
  fetal pig, 506, 506f
  human, 175, 177f–178f
  in cat, 456, 457f, 458f
third ventricle, 208
thoracic aorta, 322, 466f
thoracic cage, 91, 109, 113, 114f
thoracic cavity, 9, 11f, 519
thoracic curvature, 109, 110f
thoracic duct, 339, 340f

thoracic muscles in cat, 443f, 446, 446f, 447f, 448f, 449t
thoracic vertebrae, 109, 112f
thorax muscles, 511t
thymus, 520
  as major organ, 12
  location, 282
  of cat, 464f, 465f, 483, 485f
  structure, 341, 342f
thyrocervical artery, 325f, 464f, 465
thyrohyoid muscle, 444, 444t
thyroid cartilage, 188f, 365, 483, 485f
thyroid gland
  as major organ, 12
  examining, 284
  location in human, 11f, 282
  of cat, 483, 485f
thyroid hormone, 284
tibia, 91, 131f, 132, 179f
tibial arteries, 322
tibial collateral ligament, 138f
tibial tuberosity, 132, 192f
tibial veins, 330f
tibialis anterior muscle, 515
  human, 175, 186f, 192f
  in cat, 455f, 456, 457f, 459t
tibialis posterior muscle, 179
tidal volume, 376–377, 378f, 379t
time, metric units, 2t
tissues
  blood, 293–297
  epithelial, 63, 64f
  joint classification by, 137
  major types, 63
  muscle, 75–76, 147–148
  nervous, 75–76, 199, 203
  respiratory, 365–369, 370f
tongue
  digestive system role, 12
  fetal pig, 506, 506f
  of cat, 476, 477f, 484f
  rollers and nonrollers, 430, 431f
  structure, 345
  taste buds on, 241, 243f
  testing gustation on, 243–244
tonsils, 345, 476
total lung capacity, 378
total magnification, 32
touch receptors, 80f, 235–237
trabeculae, 87f
trachea
  as major organ, 12
  of cat, 465f, 483, 484f, 485f, 486f
  structure, 348f, 366
tracheal rings, 483

traits, 429, 430–433
transitional epithelium, 394
transmembrane proteins, 41f
transparency of urine, 400
transverse arch, 192f
transverse colon, 351, 478
transverse fissure, 207
transverse foramina, 111f, 112f, 113
transverse plane, 13f
transverse processes, 113, 114f
transverse tubules, 150
transversus abdominis muscle, 511
  in cat, 445f, 447, 448f
transversus costarum muscle, 445, 446f, 447t
trapezium, 119, 142f
trapezius muscles
  basic functions, 161
  location, 162f
  surface anatomy, 188f, 189f
trapezoid, 119
triceps brachii muscle
  in cat, 443f, 449f
  location in human, 161, 164f
  reflexes, 230
  surface anatomy, 190f
triceps reflex, 230, 232f
Trichomonas, 402
tricuspid valve, 304, 307–308, 312
trigeminal nerves, 210, 217f, 218
triquetrum, 119, 122f
trochanters, 94, 132
trochlea, 118, 120f, 265, 266f
trochlear nerves, 210, 217f, 218
trochlear notch, 118
trophoblasts, 425
true ribs, 113
true vocal cords, 365
trypsin, 360
TSH, 282
tubercles
  defined, 94
  of humerus, 118, 120f, 190f
  of ribs, 113, 114f
  of sacrum, 113
tuberosities, 94, 118, 132
tunica albuginea, 405, 407, 408
tunica externa, 321, 321f
tunica interna, 320, 321f
tunica media, 320, 321f
two-point threshold, 236–237
tympanic cavity, 249, 250f
tympanic membrane, 249
type 1 diabetes mellitus, 282
type 2 diabetes mellitus, 282
tyrosine crystals, 402f

Credits